COMMON FOREST TREES of HAWAII

Native and Introduced

ELBERT L. LITTLE, Jr.
Chief Dendrologist (retired)
Timber Management Research
USDA Forest Service, Washington, DC

ROGER G. SKOLMEN
Principal Silviculturist (retired)
Institute of Pacific Islands Forestry
Pacific Southwest Forest and Range Experiment Station
USDA Forest Service, Honolulu, Hawaii

Originally published in May 1989 as Agriculture Handbook No. 679

Library of Congress Cataloging-in-Publication Data

Little, Elbert L., 1907-
    Common forest trees of Hawaii / Elbert L. Little, Jr., Roger G. Skolmen.
       p.   cm. -- (Agriculture handbook / United States. Dept. of Agriculture ; no. 679)
    Bibliography: p.
    Includes index.
    1. Trees--Hawaii--Identification.   2. Forest flora--Hawaii--Identification.   I. Skolmen, Roger G. II. Title. III. Series: Agriculture handbook (United States. Dept. of Agriculture) ; no. 679.

QK473.H4L58 1989
582.1609969--dc19
                                                                                                               89-7592
                                                                                                                     CIP

A PATHFINDER BOOK REPRINT EDITION

Printed in the United States of America

ISBN-13: 978-1951682453

Publisher's Note: In this reprint edition the color plates of the original publication (pages 31-42) are reproduced in black and white.

CONTENTS

List of tree species   *1*
List of maps   *6*
List of color plates   *6*
Introduction   *7*
    Geography, climate, and vegetation   *7*
    Other publications   *9*
    Preparation of this handbook   *9*
    Plan   *10*
    Hawaiian names of places and trees   *11*
    Illustrations   *11*
    How to use this handbook   *12*
    Summary of Common Forest Trees of Hawaii   *13*
    Origin of Hawaiian trees   *13*
Special lists   *15*
    Special areas   *15*
    Champion trees of Hawaii   *15*
    Weed trees   *16*
    Poisonous trees   *16*
Forests and forestry in Hawaii   *18*
    Forest types   *18*
    Forestry   *25*
Color plates   *31*
Acknowledgments   *43*
Tree species, descriptions and illustrations   *44*
Selected references   *308*
Index of common and scientific names   *315*

ABSTRACT

Little, Elbert L., Jr., and Roger G. Skolmen.
  1989. Common forest trees of Hawaii (native and introduced). U.S. Department of Agriculture, Agric. Handb. 679, 321 p.

This handbook provides an illustrated reference for identifying the common trees in the forests of Hawaii. Useful information about each species is also compiled, including Hawaiian, English, and scientific names; description; distribution within the islands and beyond; uses of wood and other products; and additional notes.

The 152 species described and illustrated by line drawings comprise 60 native species (including 53 that are endemic), 85 species introduced after the arrival of Europeans, and 7 species introduced apparently by the early Hawaiians. The native tree species of Hawaii are mostly scattered in distribution and of small size. Only two native tree species presently are commercially important for wood, because of their abundance and large size: 'ōhi'a lehua, *Metrosideros polymorpha*, and koa, *Acacia koa*. Of the two, only koa is considered to be of high value.

The introduced species described include 13 species of eucalypts (*Eucalyptus*), 5 species of pines (*Pinus*), and 11 other conifers. Two plant families are well represented, the myrtle family (Myrtaceae), with 25 species, and the legume family (Leguminosae), with 15.

One chapter is devoted to forests and forestry in Hawaii. Maps of the Hawaiian Islands show the physical features and place names, major forest types, and forest reserves and conservation districts.

KEYWORDS: Hawaii, trees, forests.

LIST OF TREE SPECIES

Treefern family (Dicksoniaceae)   44
    1. Hāpu'u i'i, Hawaiian treefern, *Cibotium chamissoi* Kaulf.   44
    2. Hāpu'u pulu, Hawaiian treefern, *Cibotium glaucum* (Sm.) Hook. & Arn.   46
    3. Meu, Hawaiian treefern, *Cibotium hawaiiense* Nakai & Ogura   46

Blechnum fern family (Blechnaceae)   48
    4. 'Ama'u, sadleria, *Sadleria cyatheoides* Kaulf.   48

Araucaria family (Araucariaceae*)   50
    5. Parana-pine, *Araucaria angustifolia* (Bert.) Kuntze*   50
    6. Columnar araucaria, *Araucaria columnaris* (G. Forst.) Hook.*   50
    7. Hoop-pine, *Araucaria cunninghamii* D. Don*   54

Pine family (Pinaceae)   56
    8. Slash pine, *Pinus elliottii* Engelm.*   56
    9. Jelecote pine, *Pinus patula* Schiede & Deppe*   58
    10. Cluster pine, *Pinus pinaster* Ait.*   58
    11. Monterey pine, *Pinus radiata* D. Don*   60
    12. Loblolly pine, *Pinus taeda* L.*   60

Redwood family (Taxodiaceae*)   64
    13. Sugi, cryptomeria, *Cryptomeria japonica* (L. f.) D. Don*   64
    14. Redwood, *Sequoia sempervirens* (D. Don) Endl.*   66

Cypress family (Cupressaceae*)   66
    15. Arizona cypress, *Cupressus arizonica* Greene*   66
    16. Mexican cypress, *Cupressus lusitanica* Mill.*   68
    17. Monterey cypress, *Cupressus macrocarpa* Hartw.*   70
    18. Italian cypress, *Cupressus sempervirens* L.*   70
    19. Bermuda juniper, *Juniperus bermudiana* L.*   74

Screwpine family (Pandanaceae)   74
    20. Hala, screwpine, *Pandanus tectorius* Parkins.   74

Grass family (Gramineae)   76
    21. 'Ohe, common bamboo, *Bambusa vulgaris* Schrad. ex Wendl.*   76

Palm family (Palmae)   78
    22. Coconut, niu, *Cocos nucifera* L.**   78
    23. Loulu, pritchardia, *Pritchardia* spp.   82

Agave family (Agavaceae)   84
    24. Ti, common dracaena, *Cordyline fruticosa* (L.) Chev.**   84
    25. Halapepe, golden dracaena, *Pleomele aurea* (Mann) N. E. Br.*   86

Casuarina family (Casuarinaceae*)   86
    26. River-oak casuarina, *Casuarina cunninghamiana* Miq.*   86
    27. Horsetail casuarina, *Casuarina equisetifolia* L. ex J. R. & G. Forst.*   90
    28. Longleaf casuarina, *Casuarina glauca* Sieber ex Spreng.*   90

Bayberry (waxmyrtle) family (Myricaceae*)   92
    29. Firetree, *Myrica faya* Ait.*   92

Birch family (Betulaceae*)   96
    30. Nepal alder, *Alnus nepalensis* D. Don*   96

Species (or families) with scientific names followed by an asterisk (*) are introduced (or exotic) and are not native to Hawaii. Those with two asterisks (**) were introduced, apparently by the early Hawaiians.

Elm family (Ulmaceae)   96
    31. Gunpowder-tree, *Trema orientalis* (L.) Blume*   96
Mulberry family (Moraceae)   98
    32. 'Ulu, breadfruit, *Artocarpus altilis* (Parkins.) Fosberg**   98
    33. Trumpet-tree, *Cecropia obtusifolia* Bertol.*   100
    34. Chinese banyan, *Ficus microcarpa* L. f.*   102
    35. A'ia'i, Hawaiian false-mulberry, *Streblus pendulinus* (Endl.) F. Muell.   104
Nettle family (Urticaceae)   104
    36. Mamaki, *Pipturus albidus* (Hook. & Arn.) Gray   104
    37. Ōpuhe, *Urera glabra* (Hook. & Arn.) Wedd.   106
Protea family (Proteaceae*)   106
    38. Kahili-flower, *Grevillea banksii* R. Br.*   106
    39. Silk-oak, *Grevillea robusta* A. Cunn.*   110
Sandalwood family (Santalaceae)   112
    40. 'Iliahi-a-lo'e, coast sandalwood, *Santalum ellipticum* Gaud.   112
    41. 'Iliahi, Freycinet sandalwood, *Santalum freycinetianum* Gaud.   114
Buckwheat family (Polygonaceae)   114
    42. Seagrape, *Coccoloba uvifera* (L.) L.*   114
Amaranth family (Amaranthaceae)   116
    43. Pāpala, *Charpentiera obovata* Gaud.   116
Four-o'clock family (Nyctaginaceae)   118
    44. Pāpala kēpau, *Pisonia brunoniana* Endl.   118
    45. Āulu, *Pisonia sandwicensis* Hillebr.   118
Laurel family (Lauraceae)   122
    46. Camphor-tree, *Cinnamomum camphora* (L.) J. S. Presl*   122
    47. Avocado, *Persea americana* Mill.*   124
Saxifrage family (Saxifragaceae)   124
    48. Kanawao, *Broussaisia arguta* Gaud.   124
Pittosporum family (Pittosporaceae)   126
    49. Hō'awa, *Pittosporum confertiflorum* Gray   126
Legume family (Leguminosae)   128
    50. Formosa koa, *Acacia confusa* Merr.*   128
    51. Koa, *Acacia koa* Gray   128
    52. Black-wattle acacia, *Acacia mearnsii* De Wild.*   132
    53. Blackwood acacia, *Acacia melanoxylon* Br.*   134
    54. Molucca albizia, *Albizia falcataria* (L.) Fosberg*   134
    55. Siamese cassia, *Cassia siamea* Lam.*   136
    56. Kolomona, scrambled-eggs, *Cassia surattensis* Burm. f.*   138
    57. Wiliwili, *Erythrina sandwicensis* Deg.   138
    58. India coralbean, *Erythrina variegata* L.*   142
    59. Koa haole, leucaena, *Leucaena leucocephala* (Lam.) de Wit*   144
    60. 'Opiuma, *Pithecellobium dulce* (Roxb.) Benth.*   146
    61. Monkey-pod, 'ōhai, *Pithecellobium saman* (Jacq.) Benth.*   146
    62. Kiawe, algaroba, *Prosopis pallida* (Humb. & Bonpl. ex Willd.) H.B.K.*   150
    63. Mamane, *Sophora chrysophylla* (Salisb.) Seem.   152
    64. Tamarind, *Tamarindus indica* L.*   154
Rue or citrus family (Rutaceae)   156
    65. Queensland-maple, *Flindersia brayleyana* F. Muell.*   156
    66. Mokihana, *Pelea anisata* Mann   156
    67. Alani, Clusia-leaf pelea, *Pelea clusiifolia* Gray   158
    68. Pilo kea, spatula-leaf platydesma, *Platydesma spathulatum* (Gray) Stone   158
    69. A'e, *Zanthoxylum oahuense* Hillebr.   162

Mahogany family (Meliaceae*)   162
    70. Chinaberry, pride-of-India, *Melia azedarach* L.*   162
    71. Australian toon, *Toona ciliata* M. Roem.*   164
Spurge family (Euphorbiaceae)   166
    72. Kukui, candlenut-tree, *Aleurites moluccana* (L.) Willd.**   166
    73. Hame, *Antidesma platyphyllum* Mann   168
Cashew family (Anacardiaceae)   170
    74. Mango, manako, *Mangifera indica* L.*   170
    75. Neneleau, Hawaiian sumac, *Rhus sandwicensis* Gray   172
    76. Christmas-berry, *Schinus terebinthifolia* Raddi*   172
Karaka family (Corynocarpaceae*)   176
    77. Karaka, *Corynocarpus laevigatus* J. R. & G. Forst.*   176
Holly family (Aquifoliaceae)   178
    78. Kāwa'u, Hawaiian holly, *Ilex anomala* Hook. & Arn.   178
Bittersweet family (Celastraceae)   180
    79. Olomea, *Perrottetia sandwicensis* Gray   180
Soapberry family (Sapindaceae)   180
    80. 'A'ali'i, *Dodonaea viscosa* Jacq.   180
    81. Āulu, *Sapindus oahuensis* Hillebr.   182
    82. Wingleaf soapberry, mānele, *Sapindus saponaria* L.   182
Buckthorn family (Rhamnaceae)   186
    83. Kauila, *Alphitonia ponderosa* Hillebr.   186
Elaeocarpus family (Elaeocarpaceae)   188
    84. Kalia, *Elaeocarpus bifidus* Hook. & Arn.   188
Mallow family (Malvaceae)   188
    85. Koki'o ke'oke'o, native white hibiscus, *Hibiscus arnottianus* Gray   188
    86. Mahoe, Cuban-bast, *Hibiscus elatus* Sw.*   190
    87. Hau, sea hibiscus, *Hibiscus tiliaceus* L.   192
    88. Milo, portiatree, *Thespesia populnea* (L.) Soland. ex Correa**   194
Chocolate family (Sterculiaceae)   196
    89. Melochia, *Melochia umbellata* (Houtt.) Stapf*   196
Mangosteen family (Guttiferae*)   196
    90. Kamani, *Calophyllum inophyllum* L.**   196
Flacourtia family (Flacourtiaceae)   200
    91. Maua, xylosma, *Xylosma hawaiiense* Seem.   200
Mezereum family (Thymelaeceae)   202
    92. 'Ākia, *Wikstroemia oahuensis* (Gray) Rock   202
Mangrove family (Rhizophoraceae*)   204
    93. Mangrove, *Rhizophora mangle* L.*   204
Combretum family (Combretaceae*)   206
    94. Tropical-almond, false kamani, *Terminalia catappa* L.*   206
    95. Jhalna, *Terminalia myriocarpa* Heurck & Muell.-Arg.*   208
Myrtle family (Myrtaceae)   208
    96. Lanceleaf gum-myrtle, *Angophora costata* Domin*   208
    97-109. Eucalyptus, eucalypt, *Eucalyptus**   210
    97. Bangalay eucalyptus, *Eucalyptus botryoides* Sm.*   214
    98. River-redgum eucalyptus, *Eucalyptus camaldulensis* Dehn.*   214
    99. Lemon-gum eucalyptus, *Eucalyptus citriodora* Hook.*   216
    100. Bagras eucalyptus, *Eucalyptus deglupta* Blume*   218
    101. Bluegum eucalyptus, *Eucalyptus globulus* Labill.*   220
    102. Rosegum eucalyptus, *Eucalyptus grandis* W. Hill ex Maid.*   222
    103. Tallowwood eucalyptus, *Eucalyptus microcorys* F. Muell.*   222

104. Gray-ironbark eucalyptus, *Eucalyptus paniculata* Sm.*   *224*
105. Blackbutt eucalyptus, *Eucalyptus pilularis* Sm.*   *226*
106. Kinogum eucalyptus, *Eucalyptus resinifera* Sm.*   *226*
107. Robusta eucalyptus, *Eucalyptus robusta* Sm.*   *230*
108. Saligna eucalyptus, *Eucalyptus saligna* Sm.*   *232*
109. Red-ironbark eucalyptus, *Eucalyptus sideroxylon* A. Cunn. ex Woolls*   *234*
110. Java-plum, *Eugenia cumini* (L.) Druce*   *234*
111. Rose-apple, 'ōhi'a loke, *Eugenia jambos* L.*   *236*
112. 'Ōhi'a ai, mountain-apple, *Eugenia malaccensis* L.**   *238*
113. 'Ōhi'a ha, *Eugenia sandwicensis* Gray   *240*
114. Manuka, *Leptospermum scoparium* J. R. & G. Forst.*   *240*
115. Paperbark, cajeput-tree, *Melaleuca quinquenervia* (Cav.) S. T. Blake*   *240*
116. 'Ōhi'a lehua, *Metrosideros polymorpha* Gaud.   *244*
117. Strawberry guava, *Psidium cattleianum* Sabine*   *248*
118. Guava, kuawa, *Psidium guajava* L.*   *248*
119. Turpentine-tree, *Syncarpia glomulifera* (Sm.) Niedz.*   *250*
120. Brushbox, *Tristania conferta* R. Br.*   *250*

Melastome family (Melastomataceae*)   *254*
121. Glorybush, *Tibouchina urvilleana* (DC.) Cogn.*   *254*

Ginseng family (Araliaceae)   *256*
122. Lapalapa, *Cheirodendron platyphyllum* (Hook. & Arn.) Seem.   *256*
123. 'Ōlapa, common cheirodendron, *Cheirodendron trigynum* (Gaud.) Heller   *258*
124. 'Ohe makai, Hawaiian reynoldsia, *Reynoldsia sandwicensis* Gray   *258*
125. Octopus-tree, *Schefflera actinophylla* (Endl.) Harms*   *260*
126. 'Ohe'ohe, *Tetraplasandra hawaiiensis* Gray   *262*

Epacris family (Epacridaceae)   *266*
127. Pūkiawe, *Styphelia tameiameiae* (Cham. & Schlecht.) F. Muell.   *266*

Myrsine family (Myrsinaceae)   *266*
128. Shoebutton ardisia, *Ardisia elliptica* Thunb.*   *266*
129. Kōlea, *Myrsine lessertiana* A. DC.   *268*
130. Kōlea lau-li'i, *Myrsine sandwicensis* A. DC.   *268*

Sapodilla family (Sapotaceae)   *270*
131. Keahi, Hawaiian nesoluma, *Nesoluma polynesicum* (Hillebr.) Baill.   *270*
132. 'Āla'a, aulu, *Pouteria sandwicensis* (Gray) Baehni & Deg.   *270*

Ebony family (Ebenaceae)   *274*
133. Lama, Hillebrand persimmon, *Diospyros hillebrandii* (Seem.) Fosberg   *274*
134. Lama, *Diospyros sandwicensis* (A. DC.) Fosberg   *276*

Olive family (Oleaceae)   *276*
135. Tropical ash, *Fraxinus uhdei* (Wenzig) Lingelsh.*   *276*
136. Olopua, pua, *Osmanthus sandwicensis* (Gray) Knobl.   *278*

Dogbane family (Apocynaceae)   *280*
137. Hao, Hawaiian rauvolfia, *Rauvolfia sandwicensis* A. DC.   *280*

Borage family (Boraginaceae)   *280*
138. Kou, *Cordia subcordata* Lam.**   *280*
139. Tree-heliotrope, *Tournefortia argentea* L. f.*   *284*

Verbena family (Verbenaceae*)   *286*
140. Fiddlewood, *Citharexylum caudatum* L.*   *286*

Bignonia family (Bignoniaceae*)   *286*
141. Primavera, goldtree, *Roseodendron donnell-smithii* (Rose) Miranda*   *286*
142. African tuliptree, *Spathodea campanulata* Beauv.*   *288*

Myoporum family (Myoporaceae)   *290*
143. Naio, false-sandalwood, *Myoporum sandwicense* Gray   *290*

Madder or coffee family (Rubiaceae)   292
    144. 'Ahakea, *Bobea sandwicensis* (Gray) Hillebr.   292
    145. Alahe'e, *Canthium odoratum* (G. Forst.) Seem.   292
    146. Coffee, *Coffea arabica* L.*   296
    147. Pilo, *Coprosma montana* Hillebr.   298
    148. Manono, *Gouldia affinis* (DC.) Wilbur   298
    149. Noni, Indian-mulberry, *Morinda citrifolia* L.**   300
    150. Kōpiko, *Psychotria hawaiiensis* (Gray) Fosberg   300
Bellflower family (Campanulaceae)   304
    151. 'Ōhawai, haha, tree clermontia, *Clermontia arborescens* (Mann) Hillebr.   304
Goodenia or naupaka family (Goodeniaceae)   306
    152. Naupaka kuahiwi, mountain naupaka, *Scaevola gaudichaudiana* Cham.   306

LIST OF MAPS

Figure 1--State of Hawaii, showing the eight principal islands (Nelson and Wheeler 1963).   8

Figure 2--Forest types on Kauai and Oahu.   20

Figure 3--Forest types on Molokai and Maui.   21

Figure 4--Forest types on Niihau, Lanai, and Kahoolawe.   22

Figure 5--Forest types on Hawaii.   23

Figure 6--Forest Reserves on Kauai and Oahu.   26

Figure 7--Forest Reserves on Molokai and Maui.   28

Figure 8--Forest Reserves on Hawaii.   29

LIST OF COLOR PLATES
Twelve paintings by Isabella Sinclair (1885)

41.   'Iliahi, Freycinet sandalwood, *Santalum freycinetianum* Gaud.   31

43.   Pāpala, *Charpentiera obovata* Gaud.   32

48.   Kanawao, *Broussaisia arguta* Gaud.   33

57.   Wiliwili, *Erythrina sandwicensis* Deg.   34

80.   'A'ali'i, *Dodonaea viscosa* Jacq.   35

83.   Kauila, *Alphitonia ponderosa* Hillebr.   36

87.   Hau, sea hibiscus, *Hibiscus tiliaceus* L.   37

88.   Milo, portiatree, *Thespesia populnea* (L.) Soland. ex Correa**   38

112.   Ōhi'a ai, mountain-apple, *Eugenia malaccensis* L.**   39

116.   'Ōhi'a lehua, *Metrosideros polymorpha* Gaud.   40

138.   Kou, *Cordia subcordata* Lam.**   41

149.   Noni, Indian-mulberry, *Morinda citrifolia* L.**   42

INTRODUCTION

The main objective of this handbook is to provide an illustrated reference for identifying the common trees in the forests through the various islands within the State of Hawaii. Another is to compile useful information about each species, including current Hawaiian, English, and scientific names; description; distribution within the islands and beyond; uses of wood and other products; and additional notes. Many of these introduced trees of forest plantations are not described in available references. In addition to serving as a source book of names and identification of the trees, this handbook also provides some information on their management and use in Hawaii.

The term *forest* is used here in the broad sense to include wild lands with growing trees. Common examples are natural forests, cutover areas, second growth, and forest plantations. Also covered are trees bordering roadsides and coasts or shorelines, as well as weed trees spreading in waste places and old fields or abandoned lands.

Common species of both native and introduced trees are described in this handbook. The tree species native to the Hawaiian Islands are not numerous, and some attain only small size or are shrubby. Many native tree species are uncommon or rare, and a few are proposed or listed as threatened or endangered and thus outside the scope of this handbook.

The introduced trees, or exotics, described here are mostly those successful in forest plantations out of more than a thousand species tested or planted. Also included are hardy ornamental, shade, and fruit trees that have escaped from cultivation and have become naturalized, spreading from seeds as though wild. Numerous other trees cultivated for flowers, shade, and fruit are outside the scope of this handbook and may be found in special publications.

Measurements (trees and their parts, horizontal distances, and altitudes) are given in the English system (feet and inches) first, then in the metric system in parentheses. These approximate equivalents may be noted: 1 foot (ft) is 0.3048 meter (m), 1 inch (in) is 2.54 centimeters (cm) or 25.4 millimeters (mm), and 1/8 inch (in) is 3 millimeters (mm). Also, 1 meter (m) equals 3.28 feet or 39.37 inches. One mile (5,280 ft) is 1.6 kilometers (km) or 1,609 meters (m). One acre (43,560 square feet) is 0.45 hectares (94,047 square meters).

Trees are defined as woody plants having one erect perennial stem or trunk at least 3 in (7.5 cm) in diameter at breast height (4 1/2 ft or 1.4 m), a more or less definitely formed crown of foliage, and a height of at least 13 ft (4 m).

GEOGRAPHY, CLIMATE, AND VEGETATION

The Hawaiian Islands, which form Hawaii, the Fiftieth State of the United States of America, are unique. A review of their geography will serve as a background. These volcanic islands are among the most isolated in the world, located near the center of the largest ocean, the Pacific, and among the farthest from the continents. North America is the nearest continent. California is 2,400 miles (3,862 km) to the northeast and southern Alaska slightly more northward. Japan is 3,400 miles (5,472 km) away to the west. Scattered oceanic atolls and volcanoes dot the southwest Pacific, the closest within 500 miles (805 km).

The precise location of the Hawaiian Islands in the North Pacific Ocean is within latitude 18 degrees 54 minutes to 22 deg. 14 min. N. and longitude 154 deg. 48 min. to 160 deg. 15 min. W. The islands lie south of the Tropic of Cancer, near the northern border of the tropics and are subject to the Northeast Trade Winds and the North Equatorial Current, which flows westward.

The Hawaiian archipelago comprises eight principal volcanic islands (*fig. 1*), totaling 6,450 sq mi (2,490 sq km) and scattered over a distance of nearly 400 mi (644 km) from southeast to northwest (Hawaii University 1983) as well as a chain of isles and reefs extending 1,200 mi (1,931 km) northwest to Kure Atoll (Midway is administered separately and not considered part of the State). [Names and dates in parentheses refer to Selected References.]

At the southeast end is the Island of Hawaii, nicknamed the "Big Island," with an area of 4,038 sq mi (1,559 sq km), about the size of Connecticut and slightly larger than Puerto Rico. It is somewhat larger than all the other Hawaiian Islands combined. The other seven major islands, in approximate order toward the northwest, are Kahoolawe, Maui, Lanai, Molokai, Oahu, Kauai, and Niihau.

Few places on earth have as great diversity of topography and climate within such a small area. The geology and climate of the Hawaiian Islands have been summarized by Carlquist (1965, 1970, 1974) in his references on their natural history and on island life. Briefly, these relatively young islands formed as volcanoes from the ocean floor at a "hot spot" that lies beneath the Pacific Plate near the present location of the Island of Hawaii. The islands have each moved off towards the northwest with the movement of the plate, forming a

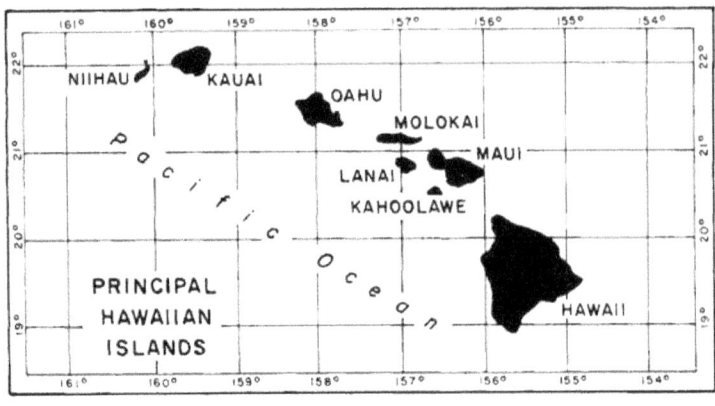

Figure 1--*State of Hawaii, showing the principal islands (Nelson and Wheeler 1963).*

string of islands and coral reefs that extends 1,500 mi (2,414 km) to Kure. Beyond Kure, the Emperor Seamounts, also formed at the hot spot but eroded to below sea level, continue beneath the sea for another 1,500 mi (2,414 km) toward Kamchatka Peninsula (Ballard 1983).

The Island of Hawaii, about a million years old, has the State's two loftiest peaks. Mauna Kea, altitude 13,796 ft (4,205 m), rises nearly 30,000 ft (8,144 m) above the ocean floor. From its base, it is the earth's highest and largest mountain. Nearby Mauna Loa (altitude 13,677 ft or 4,169 m) is still active, with craters, molten lava, and occasional lava flows. Haleakala, altitude 10,023 ft (3,055 m), on the island of Maui is the third highest Hawaiian peak.

Topography of the Hawaiian Islands varies from sandy beaches and long sea cliffs to deep canyons and gorges separated by knife-edge ridges and divides. Mountain summits range from cinder cones to jagged peaks and flattened domes. Barren lava flows of historically dated age to the present also descend several slopes of Mauna Loa.

Many different climates from over the globe are compressed and scattered along the mountainsides. Generally, the windward slopes, which face the trade winds from the northeast, create updrafts that cool the moist air, causing ample rainfall. Waialeale Peak, altitude 5,148 ft (1,569 m) at the summit of Kauai may be the world's rainiest spot. It has a mean annual rainfall of 460 in (38.3 ft, 11.7 m, or 11,684 mm), and a daily average of 1.26 in (32 mm)! Leeward slopes away from the wind and the lowlands nearly everywhere are dry, nearly deserts, with yearly rainfalls as low as 20 in (508 mm).

Temperatures in the tropical lowlands are hot but decrease gradually with altitude, perhaps more than the usual 1 degree Fahrenheit (F) to 300 ft (1 degree centigrade [C] to 51 m). The upland mountain slopes become progressively cooler and have some freezing weather in winter. For short periods each year, the rounded treeless summits of Mauna Kea and Mauna Loa have snow caps. The climate at higher elevations is not temperate, however. It is cool all year without the pronounced seasonal changes many temperate plants require to stimulate normal growth and flowering responses. Also, day length variation is much smaller throughout the year.

Geographic isolation is provided not only by the eight islands but by the higher peaks. Each volcanic peak rising more than about 2,500 ft (762 m) above sea level creates a separate, climatic "island" with a cooler oceanic montane climate above. Thus, the Island of Hawaii has the equivalent of four geographically isolated, mountainous islands, Maui has two, and Oahu two ranges.

The complex vegetation that has developed because of Hawaii's assorted topography and climate is equally varied. The number of habitats or ecological niches for plant and animal species is very high. Vegetation zones and types have been described and mapped by Ripperton and Hosaka (1942), Carlquist (1970), Knapp (1975), and Küchler in the National Atlas (U.S. Geological Survey 1969).

A forest survey and forest type maps were prepared by the U.S. Department of Agriculture Forest Service (Nelson and Wheeler 1963; Nelson 1967). The forest types are further described here in the chapter entitled Forests and Forestry. Hawaii's three highest peaks have a timberline at about 8,000 ft (2,348 m) and a zone of shrubs beyond. The alpine zone above 10,000 ft (3,048 m) is composed of barren rocky outcrops, mosses, and lichens.

OTHER PUBLICATIONS

Various other publications describing the trees of Hawaii are cited here and under Selected References. These references have been helpful in the preparation of this handbook and may be consulted for identification of Hawaiian trees outside its scope.

"Flora of the Hawaiian Islands" was prepared more than a century ago by William Hillebrand (1821-86), a German physician who resided in Honolulu for 20 years (1851-71). Published posthumously by his son in 1888 and reprinted in 1965, it still is the only descriptive flora of Hawaii. However, the Bishop Museum is preparing a new flora: "Manual of the Flowering Plants of Hawaii" by Warren L. Wagner, Derral R. Herbst, and Seymour H. Sohmer (1989). The latest compilation is the "List and Summary of the Flowering Plants in the Hawaiian Islands" by Harold St. John (1973). It includes nomenclature, common names, and summary of distribution, as well as whether the species is native, adventive, or cultivated.

Otto Degener (1899--1987) has prepared a valuable "Flora Hawaiiensis" (1933--73) in looseleaf form with a drawing and description of each species. Published in parts, it contains many trees but is incomplete in 7 volumes. A list of the looseleaf sheets has been compiled (Mill, Wagner, and Herbst 1985). "Plants of Hawaii National Parks" (Degener 1973) provided illustrations, descriptions, and abundant information on the uses of many Hawaiian plants, including trees. "In Gardens of Hawaii" by Marie C. Neal (1965) stressed the cultivated plants of gardens, including ornamental trees. A revision by Bishop Museum and contributors is in progress.

The most comprehensive reference on the native trees of Hawaii is the classic, scholarly work by Joseph F. Rock (1884--1962), entitled "The Indigenous Trees of the Hawaiian Islands" (1913). It contained descriptions of about 230 native species, including 1 new genus, 22 new species, and 31 new varieties and was illustrated by 215 photographs, nearly all by the author. That large volume was based upon 5 years of thorough field exploration and was published when its author was only 29 years old. Fortunately, a reprint has the nomenclature brought up-to-date by Derral Herbst (Rock 1974).

Rock prepared that publication and several others while he was a consulting botanist with the Hawaii Division of Forestry (1911--21) and botanist with the University of Hawaii. "The Ornamental Trees of Hawaii" (Rock 1917a) was a companion volume. He wrote several taxonomic monographs including the sandalwoods, *Santalum* (Rock 1916), the 'ōhi'a lehua trees, *Metrosideros* (Rock 1917b), the lobelia tribe Lobelioideae of the bluebell family, Campanulaceae (the largest native family of seed plants in these islands) (Rock 1919a), and the native and introduced leguminous plants, Leguminosae (Rock 1919a, 1920). He collaborated with Beccari on a monograph on the native palm genus loulu, *Pritchardia* (Beccari and Rock 1921).

A list of about 230 species of native forest trees of Hawaii, based mainly upon Rock's book, was compiled by MacCaughey (1917; Rock 1917c). A checklist of some common native and introduced forest plants in Hawaii was issued by the Forest Service (Bryan and Walker 1962, 1966).

The compilation of the flora of Hawaii by St. John (1973) contains about 370 species of native trees, as defined here. Most later additions apparently have resulted from splitting or dividing species and naming smaller and smaller variations. According to a conservative estimate, the number of native species is about 300.

"Native Trees and Shrubs of the Hawaiian Islands" by Samuel H. Lamb (1981) is fully illustrated by photographs and contains original notes on woods. Earlier, he prepared a list of the trees of Hawaii Volcanoes National Park (Lamb 1936). Dorothy and Bob Hargreaves have two publications filled with color photos, "Tropical Trees of Hawaii" (1964) and "Tropical Trees of the Pacific" (1970). Two illustrated guides, "Hawaiian Forest Plants" (Merlin 1976) and "Hawaiian Coastal Plants" (Merlin 1978) describe numerous plants, including trees. There is also "Trailside Plants of Hawaii's National Parks" (Lamoureux 1976) with photos and descriptions of many native trees and other plants. Another guide with illustrations in color is "Hawaiian Flowers and Flowering Trees" by Loraine E. Kuck and Richard C. Tongg (1958). The latest book with color photos is "Plants and Flowers of Hawaii" by Shomer and Gustafson (1987).

PREPARATION OF THIS HANDBOOK

This handbook follows the similar two-volume Forest Service handbook on the trees of Puerto Rico and the Virgin Islands (Little and Wadsworth 1964; Little, Wadsworth, and Marrero 1967; Little, Woodbury, and Wadsworth 1974). It is based upon many published sources as well as the experiences of the authors. For about 40 introduced species, the text and illustrations have been adapted from the Puerto Rico books.

The first author made trips to Hawaii in 1968, 1976, 1978, and 1987. In August--September 1976, he did field work on the six large islands with local foresters, collecting more than 300 numbers of tree specimens.

Duplicates have been deposited in the herbaria of the United States National Museum of Natural History (US) and Bernice P. Bishop Museum (BISH).

The second author has been a resident of Hawaii since 1960. From 1960 until his retirement in 1986, he conducted research in silviculture and forest products at the Institute of Pacific Islands Forestry, Pacific Southwest Forest and Range Experiment Station, Forest Service, U.S. Department of Agriculture, headquartered in Honolulu. His investigations, mainly on Hawaiian woods and forest management, have been reported in more than 50 publications. Pertinent ones are listed under Selected References (Skolmen 1963a, 1963b, 1964, 1967a, 1967b, 1968, 1973, 1974a, 1974b, 1976, 1978, 1980a, 1980b, 1983, 1986a, 1986b).

PLAN

The plan of this handbook follows that of the two-volume Forest Service handbook for Puerto Rico and the Virgin Islands previously cited. In this handbook, the species are grouped by plant families botanically and within each family alphabetically by scientific name. Species with scientific names followed by an asterisk (*) were introduced to Hawaii after the arrival of Captain James Cook in 1778, either intentionally or accidentally, and thus are not native. Those with two asterisks (**) were apparently introduced by the early Hawaiians at various times during several hundred years, beginning about 300 to 500 A.D.

Names--For each tree species, the heading at left consists of plant family name in English and Latin and the preferred common name, most often in Hawaiian or English. The accepted scientific name at right consists of the generic name, which is capitalized, and the specific epithet. In references and technical publications, the name of the author--the botanist who first named the species--is added, usually in abbreviated form if common or long. Where the scientific name has been transferred from another genus or combination, the original author is placed in parentheses and followed by the second author, who made the change. Any other common names and botanical synonyms are listed at the end of the write-up. Spanish names from Puerto Rico have been added for introduced trees. Common names used in American Samoa and in many islands of western Pacific that are associated with the United States are provided from two compilations (Falanruw et al. 1989; Amerson et al. 1982). These names are included in the Index of Common and Scientific Names.

Tree descriptions--The size and appearance of mature trees are noted first. Trees are classed as evergreen if in full leaf through the year or deciduous if leafless or nearly so for a brief period, such as in the dry season. Average size at maturity is given as small (less than 30 ft or 9 m tall), medium (30--70 ft or 9--21 m), or large (more than 70 ft or 21 m tall). Trunk measurements are diameters at breast height (d.b.h.) or 4 1/2 ft (1.4 m) above ground level.

Notes on bark include color of surface and the texture, whether smooth or rough, and if fissured (with many narrow thin cracks) or furrowed (with broad deep grooves). Other details are added if helpful in identification; these include color, taste, and odor of inner bark (the living tissue exposed by cutting), and presence of colored sap or latex.

Leaves, flowers, fruits--Because of their importance for identification, the leaves, flowers, and fruits (including seeds) are described in detail, mostly in nontechnical terms, with measurements for reference. Any unfamiliar technical terms may be found in dictionaries or botanical references.

Wood and uses--Information on wood and its uses is given for many species, mostly compiled from publications by the second author and others of the Institute of Pacific Islands Forestry (Brown 1922; Gerhards 1964, 1965, 1966a, 1966b, 1967; Skolmen op. cit.; Youngs 1960, 1964). Some notes on introduced trees are from studies made elsewhere. The uses listed are primarily those in Hawaii, but special uses elsewhere are added. Most native woods are available only in limited quantities and not for export. Nevertheless, many of these woods could be utilized in additional ways.

Economic notes--Other uses and economic notes have been compiled from various sources. Many kinds of trees are planted for more than one purpose, such as wood, shade, ornament, fruit, or flowers. Others have bark that yields fibers, dyes or tannin, or have parts employed in home remedies. Some are honey plants, because of their flowers secrete nectar in quantities or produce pollen and attract bees. In special topics are Weed Trees and Poisonous Trees.

Other notes--For some species, particularly those tested in forest plantations, brief notes on performance, propagation, and growth rate have been added. Special uses by native Hawaiians have been compiled, and Hawaiian legends about some native trees have been repeated. Other notes explain origins of common names and derivations of scientific names. In some genera, related or similar species are mentioned or compared.

Distribution--For each tree species the natural distribution, or range, is stated, both in Hawaii and beyond. Distribution in Hawaii includes abundance, forest types, altitudinal limits, and names of islands (if a species is not found throughout). For introduced trees the native home is cited. Introduced species are noted as *cultivated*, as in forestry tests or plantations; *escaped*, if spreading; or *naturalized*, if common and established as though wild by reproducing naturally or becoming a weed. Also mentioned are areas where these introduced species also occur in continental United States, mainly subtropical parts of Florida, Texas, and California.

Champions--The largest individual tree in Hawaii is cited for many species, as explained under the topic Champion Trees.

Special areas--Distribution of most species is cited also by parks, botanical gardens, and arboreta where trees can be seen, as listed under Special Areas.

HAWAIIAN NAMES OF PLACES AND TREES

Because most names of places and trees in the Hawaiian Islands are in the Hawaiian language, a summary of the pronunciation is included here, adapted largely from "Hawaiian Dictionary" by Pukui and Elbert (1986). That reference contains English meanings and scientific names for many trees. "Place Names of Hawaii" (Pukui, Elbert, and Mookini 1974) has additional information.

Early missionaries provided the first orthography for the Hawaiian language. The language is phonetically simple, with relatively few sounds and no closed syllables. In 1829, the missionaries by vote adopted an alphabet with vowels corresponding to the continental pronunciation and with consonants pronounced about as in English. This short alphabet with regular spelling and no silent letters helped the islanders to advance readily to books.

The Hawaiian alphabet has only 12 letters. The 5 vowels, **a**, **e**, **i**, **o**, and **u**, are pronounced about as follows: **a** as in above and afar; **e** as in bet; **i** as in it and police; **o** as in sole; and **u** as in blue. There are only 7 consonants: **h**, **k**, **l**, **m**, **n**, **p**, **w**. All are pronounced about as in English, except that **w** may be like **v** or **w**, according to the preceding vowel. There are 8 diphthongs, always stressed on the first letter: **ei**, **eu**, **oi**, **ou**, **ai**, **ae**, **ao**, **au**.

Consonants are single and separated by vowels, never double or two together. Vowels may be single or as many as four together. Words always end in a vowel.

Two diacritical marks or signs are inserted to aid pronunciation. The **hamza** ' (printed as a beginning single quotation mark or reversed apostrophe), or **glottal stop**, is similar to the sound between the oh's in English oh-oh. The **macron** (a line or bar over a vowel like that for a long sound in English) as in ā, indicates stress or accent and often a longer sound. Also, individual words combined to make a name are separated here by a space but may also be joined by a hyphen.

Accents are on the next-to-last syllable and alternating syllables unless otherwise indicated by vowels marked with macrons.

Hawaiian place names adopted into the English language commonly drop the diacritical marks for practical reasons, as do anglicized words from other languages. Publications and maps of the United States Government omit the Hawaiian hamza and macron because of legal requirements--see for example, the National Zip Code Directory (U.S. Postal Service 1988). Uniform name usage is maintained by the U.S. Board on Geographical Names.

Accordingly, in this handbook Hawaiian place names omit any diacritical marks. For example, Hawaii, not Hawai'i; Oahu, not O'ahu; Lanai, not Lāna'i, Haleakala, not Haleakalā. However, the Hawaiian common names of trees in this handbook will follow local usage with hamzas (glottal stops) and macrons. The standard reference for these diacritical marks is that cited above.

The distinctive names in the Hawaiian language for most native trees show that the first settlers were good naturalists. The Hawaiians mentioned trees in songs and legends and did not destroy their forest resources.

ILLUSTRATIONS

The illustrations in this handbook follow the plan of the similar two-volume reference on the trees of Puerto Rico and the Virgin Islands previously cited. The 14 maps are noted under List of Maps. The first shows the 8 principal Hawaiian Islands. Others are maps of forest types by islands, slightly revised for those published in forest surveys and maps of Forest Reserves. All except the first were drafted by Barbara H. Honkala.

Large line drawings showing foliage and flowers and usually fruits face their respective descriptions of the 152 tree species cited in List of Species. Most are natural size (1 X) for easy comparison with specimens. Some have been reduced to 2/3 X or other scales as indicated in legends. Twelve color paintings by Isabella Sinclair (1885), cited below, are reprinted here natural size (1 X).

More than one-third of the drawings are originals made from fresh specimens collected by the first author in Hawaii. Ronald L. Walker made 40 drawings and Ruby Rice Little 15, while Barbara H. Honkala added 5. The Puerto Rican tree handbooks were the source for illustrations of many introduced species, 25 in the first volume and 9 in the second (Little and Wadsworth 1964; Little, Woodbury, and Wadsworth 1976).

Many additional drawings are from publications by Otto Degener, as credited in the legends and gratefully acknowledged here. These include 5 from Degener (1930) and 34 from Degener (1933--73), who was assisted financially by the National Science Foundation.

Other drawings are from published sources acknowledged here and in the legends. Most drawings of *Eucalyptus* and several other Australian trees are from two classic Australian references. These 7 are from the 10-volume monograph by Mueller (1879-84): *Eucalyptus botryoides*, *E. camaldulensis* (*rostrata*), *E. globulus*, *E. microcorys*, *E. paniculata*, *E. pilularis*, and *E. robusta*. Illustrations of the 10 following Australian trees are from Maiden (1902-24): *Acacia melanoxylon*, *Angophora costata*, *Casuarina cunninghamiana*, *C. glauca*, *Eucalyptus resinifera*, *E. saligna*, *E. sideroxylon*, *Pisonia brunoniana* (*inermis*), *Syncarpia glomulifera* (*laurifolia*) and *Tristania conferta*. The drawing of *Roseodendron donnell-smithii* is the original by C. E. Faxon (Rose 1892). Cones of *Araucaria angustifolia* and *A. cunninghamii* are from Barrett (1958).

Twelve color paintings by Isabella Sinclair (Mrs. Francis Sinclair, Jr. 1885) are reprinted to aid identification and to portray the beauty of the tree flowers. Her folio volume of 44 plates, the first book with color illustrations of Hawaiian plants, was reviewed many years later by St. John (1954). The color plates, with Hawaiian names, were accompanied by the author's interesting descriptive text with notes on uses including wood. The Bishop Museum in Honolulu and the National Agricultural Library have copies of this rare work, apparently printed privately in a small edition.

The artist lived on Niihau and Kauai and earlier in New Zealand. According to St. John, in 1864 King Kamehameha V sold the island of Niihau to Eliza McHutcheson Sinclair (Mrs. Francis Sinclair, Sr.) for her two sons, J. and F. Sinclair. It has remained partly in the family and the estate of Aubrey Robinson. Isabella was the daughter-in-law of the original white settlers. Obviously she had artistic ability, a love for flowers, and talent for writing. Her voucher specimens were identified by the British authority Joseph D. Hooker and were deposited in Kew Herbarium, where St. John reexamined them many years later.

The 12 color plates of trees are reproduced natural size (1 X) here together on pages 31--42. The number at left is the species number in this book. As noted by St. John, her list includes some economic species introduced by the Hawaiians (Nos. 88, 112, 138, 149); other lowland species native on both Niihau and Kauai (Nos. 57, 80, 87), and species of higher altitudes native on Kauai but not Niihau (Nos. 41, 43, 48, 83, 116). Additional line drawings of the same species appear facing their text descriptions.

HOW TO USE THIS HANDBOOK

This handbook can be used for tree identification in several ways. In many situations, if you can obtain a local common name you can consult the Index of Common and Scientific Names directly to find the illustration, description, and scientific name.

To avoid errors, a clue from a common name should always be verified by inspecting the drawing and comparing the specimen with the detailed description of leaves, flowers, and fruits. Otherwise, the use of the same common name for unrelated tree species or a misapplication may lead to confusion.

For positive identification, it is desirable to compare a specimen with flowers and fruits as well as foliage, because many kinds of trees have leaves of similar shape. However, collection of specimens is not permitted in National Parks, botanic gardens, arboreta, and some other areas.

If you cannot obtain a name, you may be able to identify many of the trees by reference to the drawings, descriptions, and distribution notes. However, the illustrations alone may not emphasize differences among closely related species. Distribution notes may aid indirectly, as a species may be known only from a certain island. A ruler (last page) and a hand lens are helpful.

The List of Tree Species may help, because related trees are grouped together in the same plant family. If the family is recognized, others can be compared. Likewise, an unknown tree resembling one that is known should be sought under the same family. A key to the 13 species of eucalypts (*Eucalyptus*) has been added.

You may find trees not included in this handbook in other technical and popular references (Selected References). You can also obtain assistance from park naturalists and rangers, botanists, foresters, and researchers and teachers at museums and universities.

SUMMARY OF COMMON FOREST TREES OF HAWAII

Hawaii's trees are unique in several ways. The 152 tree species described and illustrated by line drawings may be grouped as to origin, as follows: native to Hawaii, 60 species (53 endemic); introduced after the arrival of Europeans in 1778 (indicated by an asterisk *), 85 species; and introduced apparently by early Hawaiians (designated by two asterisks **), 7 species. These 152 tree species are classified botanically into 106 genera (including 53 introduced) and 58 plant families (including 16 introduced).

The flora of the Hawaiian islands is very impoverished for a tropical area. The small number of native tree species apparently is related to the great isolation of the islands and their young geologic age. Many widespread tropical plant families and genera are absent from Hawaii. For instance, there are no native conifers, mangroves, oaks, or figs. The mahogany, bombax, and bignonia families are absent.

Many named native tree species in Hawaii differ only slightly from others and might be united as varieties. Some of these variations may have evolved from a single ancestor, the hypothetical original immigrant to Hawaii.

The total number of species of native trees in Hawaii has been estimated here as about 300. Thus, Hawaii, with its mostly tropical climate and great altitudinal range, ranks first among the fifty states in number of native tree species. Florida, with a partly subtropical climate, ranks second with about 258 native tree species. In contrast, Puerto Rico and the Virgin Islands, tropical islands of comparable size, have about 545 native tree species, and Puerto Rico alone, about 535. Though much nearer to a continent than Hawaii and perhaps ten times as old, Puerto Rico has a slightly impoverished flora, lacking a few widespread tree families and common genera. Most native tree species of Hawaii are not widespread within the islands but are limited to one or few of the six large islands. The Island of Hawaii, with about the same area as Puerto Rico, has probably fewer than 150 native tree species, or slightly more than one-fourth as many as Puerto Rico. In the comparison of endemic species, however, Hawaii leads the earth with more than 95 percent of its species of seed plants endemic, or native nowhere else. Puerto Rico has about 25 percent of its tree species endemic.

Most of the native tree species of Hawaii are scattered or uncommon and of small size. Few would be rated as common. This handbook includes 60 of these, about a fifth of the total of about 300. Relatively few of the native tree species of Hawaii would be classed as large trees. Of the 60 native tree species in this handbook, only about 35 have dimensions reaching 30 ft (9 m) in height and 1 ft (0.3 m) d.b.h., or large enough to produce a sawlog. The average size usually is smaller. One major exception, however, is 'ōhi'a lehua, which is ubiquitous in most native forests on all islands and so common that it is often left out of type descriptions.

At present, only two of Hawaii's native tree species are commercially important for wood, because of their abundance and large size: 'ōhi'a lehua, *Metrosideros polymorpha*, and koa, *Acacia koa*. However, in the early 1800's, sandalwood, *Santalum* spp., was harvested nearly to extinction as a specialty wood. 'Ōhi'a and koa form luxuriant rain forests on the island of Hawaii, with an attractive understory of treeferns (*Cibotium* spp.). However, the very small number of tree species in the forest canopy contrasts with the tropical rain forests of numerous species in other parts of the world.

According to the forest surveys, the Island of Hawaii is the only one of the six islands that has commercially important stands of native tree species. Kauai, which has one of the world's rainiest areas, has no commercial native forests. Neither has Oahu, the island with the city of Honolulu and the greatest population.

Thus, this handbook on common forest trees of Hawaii describes somewhat more introduced species (92) than native (60). There is some justification for testing in forest plantations many exotics, including pines (*Pinus*; 5 species described, also 11 other conifers) and eucalypts (*Eucalyptus*; 13 species described). Incidentally, some introduced species may excel in performance in forest plantations away from the diseases and insect pests of their native home.

Two plant families of trees are well represented both in native and introduced tree stands. The myrtle family (Myrtaceae), with 25 species in this handbook, has the native 'ōhi'a lehua plus the introduced eucalypts, Java-plum, rose-apple, mountain-apple, guava, and strawberry guava. The legume family (Leguminosae), with 15 species in this handbook, has the native koa, wiliwili, and mamane, plus koa haole (or leucaena) and kiawe (or algarroba) both naturalized in dry lowland areas.

ORIGIN OF HAWAIIAN TREES

The origin of Hawaii's trees and other plants is a fascinating subject. New Hawaiian volcanoes of barren lava arising from the sea were plantless. Seeds had to be transported over long expanses of ocean. Some could float or ride on driftwood. Others could hitchhike on birds as seeds of water plants on muddy feet, sticky or hooked seeds clinging to feathers, or hard seeds eaten in fleshy fruits. Tiny seeds could be carried long distances by the wind, especially hurricanes.

The origin of the Hawaiian flora can be traced by comparison with the closest relatives in other lands. Several similar species of one Hawaiian tree genus apparently descended from the same extinct ancestor, or original immigrant. For example, Hawaii's native palms, about 18 species, are all in the genus *Pritchardia*. This genus is known elsewhere only from a few species in the South Pacific islands and is of Indo-Pacific affinities.

Hawaii's native tree species can be arranged in groups, each apparently descended from one original immigrant, as did Fosberg (1948) for the Hawaiian flora. Thus, the total number of about 300 native tree species evolved from about 78 species of original immigrants, an average of about 5 from each (Little 1969). Based upon the archipelago's estimated age of 5 to 10 million years, the present number of tree species could have been derived from only one successful immigrant landing at intervals of roughly 60,000 to 120,000 years!

These 78 original immigrants and groups of relative Hawaiian tree species can be arranged according to their geographic affinities, or regions with closest relatives. These regions and approximate numbers of original immigrants are a) the Indo-Pacific region, including Indonesia, or southeast Asia and Pacific, west and southwest of Hawaii, 45 (nearly three-fifths); b) the Austral region, or South Pacific, from Australia to Patagonia, 16 species (one-fifth); c) the American region, 7 (almost one-tenth); d) the Pantropic region, or wide tropical or cosmopolitan, 2; and e) obscure regions, 8 (one-tenth). Thus, the relationships are strongly Indo-Pacific, more than half the original immigrants having come from the west and southwest and others from the south. However, the small American element is noteworthy.

SPECIAL LISTS

SPECIAL AREAS

The heading Special Areas under the Hawaiian distribution of each tree species (with 10 exceptions) contains names from a selected list of 18 parks, gardens, trails, and roads where that species is recorded. These trees can be seen for identification and study, and many are protected in their natural habitats. However, a special search may be needed for minor species. Some of these National, State, and City parks, botanical gardens, and arboreta have trails with labeled specimens and guidebooks.

The assistance of directors of these Special Areas in providing tree species lists is gratefully acknowledged. The data are subject to revision, and detailed species lists have not been compiled for some localities. Certain areas require permission to enter. Trees may or may not be labeled. Collecting specimens is not permitted in the National Parks and most other places listed.

Additional areas are being developed or are not open to the public at present. The Harold L. Lyon Arboretum, University of Hawaii, is open to specialists and students by appointment. It has a published checklist of plants in cultivation (Hirano and Nagata 1972).

A list of the 18 Special Areas cited in this handbook follows. They are located on four islands: Kauai, Oahu, Maui, and Hawaii. Two others, Kula Botanical Garden on Maui and Lawai, Pacific Tropical Botanical Garden on Kauai, have plants of many tree species mentioned in this handbook.

Eighteen Special Areas are listed below by islands. The single word used in the text is followed by full name and notes.

Kauai
Keahua--Keahua Forestry Arboretum.
Kokee--Kokee State Park. Primarily along Awaawapuhi Trail.
Oahu
Aiea--Aiea Loop Trail.
Foster--Foster Botanic Garden, Honolulu.
Iolani--Iolani Palace grounds, Honolulu.
Koko--Koko Crater Botanic Garden.
Tantalus--Tantalus Round Top Drive.
Wahiawa--Wahiawa Botanic Garden.
Waimea Arboretum--Waimea Arboretum, Waimea Falls Park. Admission charge. (Not to be confused with Waimea Canyon State Park on Kauai or Waimea Village on the island of Hawaii.)
Maui
Haleakala--Haleakala National Park.
Kula--Kula Forest Reserve (Poli Poli Spring).
Waihou--Waihou Spring Forest Reserve (Olinda).
Hawaii
City--Puu Honua o Honaunau (City of Refuge) National Historic Park.
Kalopa--Kalopa State Park.
Kipuka Puaula--Kipuka Puaula (Bird Park) Trail, Hawaii Volcanoes National Park.
Pepeekeo--Pepeekeo Arboretum. Obtain directions and permission from Division of Forestry and Wildlife.
Volcanoes--Hawaii Volcanoes National Park.
Waiakea--Waiakea Arboretum. Obtain permission from Division of Forestry and Wildlife.

CHAMPION TREES OF HAWAII

The champions, or largest individuals in Hawaii, have been located and measured for about 200 tree species, both native and introduced, including about 90 listed in this handbook. Information on the size of each champion tree has been taken from the National Register of Big Trees compiled by the American Forestry Association and published in "American Forests" (Littlecott 1969; American Forestry Association 1974; Hunt 1986).

Lester W. "Bill" Bryan, late retired Deputy State Forester, located and measured these big trees of Hawaii in special searches in 1968 and 1974. For these records, credit is due him, several persons who assisted, and the American Forestry Association.

Hawaiians are encouraged to continue the search for champions of other species, for larger trees to replace those listed, and for successors to any that may have died. Also, existing champions merit remeasuring after a period of years. Some relatively young champions of introduced species may not have attained maximum size. Information may be sent to the Hawaii Division of Forestry and Wildlife, 1151 Punchbowl St., Honolulu, HI 96813.

Measurements recorded are height, circumference (girth) at breast height (c.b.h.) or 4 1/2 ft (1.4 m), and crown spread (average diameter). For comparison, figures of circumferences can be converted into average diameters at breast height (d.b.h.) through division by *pi* (3.1416 or 22/7).

A few giants merit special mention here as well as under their species descriptions. The largest native tree in Hawaii measured in 1969 was a koa (*Acacia koa*) 140 ft (42.7 m) high, 37.3 ft (11.4 m) in circumference, and 148 ft (45.1 m) in spread. Unfortunately, this tree split in two because of heart rot, and the remnant tree is much smaller. A new champion koa has not been identified. Another giant is the largest monkeypod (*Pithecellobium saman*), an introduced species. Its dimensions reported in 1969 were height 104 ft (31.7 m), circumference 30.8 ft (9.4 m), and spread 140 ft (42.7 m).

The species of *Eucalyptus* are the tallest trees here as well as in their native home in Australia. Hawaii's tallest tree, also the tallest hardwood in the United States, is a planted saligna eucalyptus (*Eucalyptus saligna*) 276 ft (84 m) high in 1980 and still growing at about 50 years of age (Skolmen 1983).

Banyans, introduced relatives of figs, have the greatest trunk circumference as well as broadest crown spread. Their massive trunks and air roots grow together to form even wider columns. The spreading crown of nearly horizontal large branches is supported by additional trunks from more air roots. Hawaii's champion Chinese banyan (*Ficus microcarpa*) in 1969 measured 104 ft (31.7 m) in height, 90.1 ft (27.5 m) in trunk circumference, and 195 ft (59.4 m) in spread.

Hawaii's naive tree champions could also be regarded as national champions, as these tropical species (with two exceptions) are not native in continental United States. Among introduced trees, Hawaii's champions compete with those of the same species in southern Florida. For example, Hawaii's largest coconut (*Cocos nucifera*) outranks Florida's.

WEED TREES

In this handbook, tree species classed as weeds are so designated under distribution notes. These weed trees grow outside their normal habitat in other places where not wanted and are included in references on Hawaiian weeds. "Common Weeds of Hawaii" (Hawaii Weed Conference 1957) listed about 25 tree species also in this handbook, nearly all introduced, but a few native. Additional information about weed trees is contained in "Handbook of Hawaiian Weeds" (Hasselwood and Motter 1966).

Weed trees may be undesirable because they replace useful plants or are poisonous to persons or animals. Generally, they reproduce well by seeds or sprouts, are hardy, and grow rapidly. Thus, they invade disturbed areas, including waste places, roadsides, pastures and rangelands, and cultivated or abandoned lands. Then they become thoroughly established and naturalized. Aggressive exotics may even invade native vegetation.

Fortunately, several species of naturalized trees are beneficial. Kiawe or algarroba (*Prosopis pallida*) has become perhaps the most useful as well as most common tree of the dry lowlands of Hawaii. It is Hawaii's principal honey tree and has a durable wood prized for fenceposts. The pods are feed for livestock in pastures or rangelands, and the foliage is browsed too. Other naturalized trees bear edible fruits abundantly, for example, guava (*Psidium guajava*) and strawberry guava (*P. cattleianum*). Both are obnoxious weeds.

Unfortunately, other tree species originally introduced for useful purposes have become weeds. For example, firetree (*Myrica faya*), was brought into Hawaii as an ornamental about 1900 and has become naturalized in moist areas. It is classed as a weed in pastures, rangelands, and waste places and has been declared noxious. Another noxious tree, silk-oak (*Grevillea robusta*), has taken over large areas of pasture in Ka'u, Kona, and near Waimea on Kauai. Christmas-berry (*Schinus terebinthifolia*) was imported as an ornamental before 1917. It has escaped widely in dry lowlands and now is rated as a weed and is poisonous because of its sap. Eradication of these undesirable immigrants would be difficult and expensive.

POISONOUS TREES

Several native and introduced tree species of Hawaii are poisonous or injurious to persons or animals. These trees have been included with herbs and shrubs in lists of poisonous plants, for example, by Arnold (1944). Though not a serious threat to residents and tourists, poisonous trees deserve mention.

A list of the 10 species of this handbook known to be poisonous, or otherwise injurious, follows. Additional notes are summarized in the text.

38. kahili-flower, *Grevillea banksii**
57. wiliwili, *Erythrina sandwicensis*
59. koa haole, leucaena, *Leucaena leucocephala**

66. mokihana, *Pelea anisata*
70. chinaberry, pride-of-India, *Melia azedarach**
72. kukui, candlenut-tree, *Aleurites moluccana***
74. mango, manako, *Mangifera indica**
76. Christmas-berry, *Schinus terebinthifolia**
82. wingleaf soapberry, manele, *Sapindus saponaria*
92. 'ākia, *Wikstroemia oahuensis*

Other introduced trees not in this handbook, such as ornamentals, are poisonous according to Arnold (1944). Several others with caustic sap or stinging hairs are contact irritants. The most frequent cause of fatal or dangerous poisoning on Oahu is yellow-oleander, *Cascabela thevetia* (L.) Lippold (*Thevetia peruviana* (Pers.) K. Schum.), known also as nohomālie, be-still-tree, and campanilla, and elsewhere as lucky-nut. This introduced small tree has showy bell-shaped yellow flowers, very narrow shiny leaves, triangular-shaped poisonous nuts, and very poisonous milky juice.

Two general rules regarding poisonous plants, stated by Arnold (1944), merit repetition here, especially for newcomers and tourists: (1) NEVER EAT OR TASTE ANY STRANGE FRUIT, LEAF, OR ROOT. (2) Usually, plants that are closely related botanically have similar properties.

FORESTS AND FORESTRY IN HAWAII

When Captain James Cook arrived in Hawaii in 1778, the islands had forests that were more extensive than those that exist today. Nelson (1967) reported that 200 years ago large areas of land cover had been altered by nearly 1,000 years of use by the large Hawaiian population. However, it was subsequent to the arrival of Europeans that the great inroads into the native forest vegetation were made--by browsing animals, clearing for pasture, invasion of exotic plants, fuelwood cutting, and logging. These changes have been abundantly and repeatedly documented and need not be reviewed further.

Although the Hawaiians had introduced a number of plants, including trees such as breadfruit (*Artocarpus altilis*), coconut (*Cocos nucifera*), hau (*Hibiscus tiliaceus*), kukui (*Aleurites moluccana*), and 'ōhi'a ai (*Eugenia malaccensis*), the coming of Europeans really accelerated the introduction of plants. As early as 1850, a whole new vegetation type had been created by one tree species, kiawe (*Prosopis pallida*), which happened to be placed in a niche perfectly suited to its Peruvian-derived survival characteristics. Other species such as guava (*Psidium guajava*), mango (*Mangifera indica*), and false kamani (*Terminalia catappa*) very early became prominent components of the lowland forests. At present, in many lowland areas, it is difficult to find a native plant left among the host of introduced plants occupying the roadsides. But in the upland forests, particularly the rain forests that are presently very lightly visited by people, there are still large areas of essentially pristine native vegetation.

FOREST TYPES

The forests of Hawaii occur in diverse mixtures of species, often called communities or ecosystems. The makeup of these communities varies with the abilities of different plants to cope with site variables such as moisture availability, temperature, light, soil nutrients, disturbance by humans and animals, salt spray, volcanic fumes, and the like. In Hawaii, many indigenous plants have evolved into various forms capable of occupying different site conditions than those used by the plant that first arrived by bird, wind, or ocean flotation. But during the last 1,000 years, especially the last 180, people have purposely introduced thousands of new plants. These new plants were usually those with particularly desirable qualities--fast growth, abundant flowering and seeding, ability to occupy difficult sites, and the like. Consequently, the recently introduced plants, today often called *aliens* (a pleasingly derogatory term to differentiate them from indigenous plants) are rapidly taking over almost all the lowland forests, and are encroaching rapidly in those of the uplands.

There are 4.1 million acres (1.66 million ha) of land in the eight principal high islands of Hawaii. Of the total, 2 million acres (0.8 million ha) are forested. The forests of indigenous plants today are largely confined to the uplands, mostly on the Island of Hawaii, and difficult of access accept by off-highway vehicle or foot trail. It is possible today to drive entirely around the island of Oahu and never see 'ōhi'a lehua, the most common native tree, at a close enough distance to be recognizable.

In this handbook the forest types of Hawaii are limited to only nine very generalized named communities for clarity of mapping (*figs. 2-5*). Types are given simple names generally descriptive of the principal tree or trees found in them. The native forest types are **'ōhi'a--hāpu'u, koa--'ōhi'a, mamane--naio,** and **native dry forest**. The forest types of introduced trees are **eucalyptus, mixed exotic hardwoods, guava, kiawe--leucaena,** and **conifers**. Each name is a brief identifier for a variable mixture of plants associated with a certain site situation. The guava type, for example, might be better described as mixed lowland wet primarily nonindigenous forest. The classifications used are generally adapted from those of Ripperton and Hosaka (1942) and Küchler (1970).

The forests of Kahoolawe and Niihau (fig. 4) are given only one type, kiawe--leucaena. The other islands all have three or more types. The kiawe--leucaena, eucalyptus, and conifer types are generally similar on all islands and will be described first. The other types, which vary from island to island, will be described separately for each island.

Kiawe--leucaena is common to dry leeward coasts of all islands and to parts of the windward shores. Kiawe (*Prosopis pallida*) is not present in the type on windward shores except on Lanai and Kahoolawe. Generally, on the leeward shores *Prosopis* begins just beyond the salt spray zone of the shoreline and may extend inland to an elevation of 600 ft (182 m) at some locations. *Leucaena leucocephala* may occur in mixture with *Prosopis*, or may occupy a site exclusively. Generally, *Prosopis* grows near the coast with occasional *Acacia farnesiana* and *Pithecellobium dulce*. *Leucaena* replaces *Prosopis* on hillsides up to elevations where increased moisture causes soils to become too acidic for it to compete with other plants. In coastal flats where ground water is available, as on Molokai, kiawe grows to impressive size. Also on Molokai, the type contains prominent shoreline stands of mangrove (*Rhizophora*) taking over old fish ponds. Except on Molokai, this type is probably the most threatened of any by current shoreline development, by frequent fire (which kills kiawe), and by a recently introduced insect, a psyllid that defoliates *Leucaena*.

Eucalyptus plantations are common on the five principal islands. Most frequently, they are single-species plantings of *Eucalyptus robusta*, a tree with thick reddish brown bark. Since 1960, *E. saligna*, a tree with smooth bluish gray bark has been planted extensively and now rivals *E. robusta* in quantity. Along highways, at low elevation the most common other species seen are *E. citriodora*, with gray dimpled bark, and *E. deglupta*, with pink and green scaly bark. At higher elevations, the common species are *E. sideroxylon*, with black bark, and *E. camaldulensis*, with bark mottled gray and brown. In wet areas, the older *Eucalyptus* stands often have a dense understory of strawberry guava (*Psidium cattleianum*) while in drier locations they have little or no ground cover. Hawaii has more than 90 species of *Eucalyptus* as well as many closely related Australian Myrtaceae, and they are often found in mixed plantings.

The conifer type is mapped on Maui, Kauai, and Molokai. At the Kula Forest Reserve on Maui there is a mixture of conifers planted in contiguous stands, principally *Cryptomeria*, *Sequoia*, and *Pinus*. Also on Maui in the upper Waikamoi watershed is a large planting of *Pinus elliottii*, *P. taeda*, *P. radiata*, and *P. pinaster* growing in mixture with *Styphelia* and *Vaccinium* shrubs. On Molokai and Kauai, the conifer plantations are mostly *Pinus elliottii* and *P. taeda*, although some trees of *P. radiata* and *P. pinaster* are also present. Prominent on all the islands, but not shown in the maps, are stands and individual trees of *Araucaria*. The island of Hawaii also has several large stands of *Cryptomeria*, not shown on the maps.

Kauai--On the island of Kauai (fig. 2), the 'ōhi'a--hāpu'u type occupies the wet central part of the island above elevations of about 3,000 ft (914 m). Principal genera throughout most of the type are *Metrosideros*, *Cibotium*, *Cheirodendron*, *Pelea*, *Gouldia*, *Coprosma*, and the fern *Dicranopteris*. The type on this island includes bog communities of the herbs *Dubautia*, *Rynchospora*, and *Oreobolus* with and without *Metrosideros*. 'Ōhi'a lehua in these forests varies from dwarf to about 30 ft (9 m) in height.

The koa--'ōhi'a type on Kauai is confined to a small area in and about Kokee State Park. At this location, much of the type is second-growth on land heavily grazed until the early part of this century and contains many introduced plants, particularly *Grevillea robusta*, and considerable koa, planted as well as natural. Much of the *Acacia* in the Park is *A. kauaiensis*, distinguished when in flower by its panicles. The portion of the type on the drier lee side of Puu Ka Pele ridge contains numerous large specimens of *Alphitonia ponderosa*, almost absent elsewhere in Hawaii's forests. Also present are *Antidesma*, *Pelea*, *Santalum*, and *Tetraplasandra*. There is less *Metrosideros* and *Cibotium* than present in the wetter forests of this type on other islands.

The guava type on Kauai includes a mixture of *Hibiscus*, *Eugenia*, *Psidium*, *Schinus*, *Melastoma*, and on the Na Pali Coast, a preponderance of *Pandanus* with *Aleurites* in the valley bottoms. Occasional clumps of *Mangifera* occur in the vicinity of old home sites.

The term *mixed exotic forest* is used to describe small contiguous plantations of most of the introduced species in this book. In the northern area *Albizia falcataria* is most common. In the eastern patches, *Melaleuca quinquenervia* and *Casuarina* spp. are frequent, and in the southwestern area, *Grevillea*.

Oahu--On the island of Oahu (fig. 2), the 'ōhi'a--hāpu'u type is mapped along the uplands of the Koolau Range and near the top of Mt. Kaala. Neither of these areas fits well into the category named, although there is always some treefern present and 'ōhi'a lehua is the dominant tree. Mt. Kaala has a unique bog forest that is included in the type and in which associated trees are *Cheirodendron*, *Coprosma*, *Gouldia*, and *Myrsine*. In the Koolau range, *Acacia* is associated with *Metrosideros* and *Cibotium* as well as *Antidesma*, *Perottetia*, *Ilex*, *Cheirodendron*, *Elaeocarpus*, and *Pittosporum*.

The area in the central Koolau Range designated koa-'ōhi'a contains *Acacia* as the dominant tree associating with the species listed above for the surrounding 'ōhi'a--hāpu'u type. Throughout both types are large areas of the fern *Dicranopteris*.

The native dry forest in the northwest of the Waianae Range is dominated by *Diospyros*, which associates with *Canthium*, *Caesalpinia*, *Nothocestrum*, *Pleomele*, and *Reynoldsia*.

The guava type along the windward Koolau and the north end of the Waianae Range is a mixture of *Psidium* with *Eugenia*, *Schinus*, occasional small plantings of *Casuarina*, *Araucaria*, patches of *Mangifera*, and remnants of native forest.

The mixed exotic forest type of Oahu includes extensive plantings of native *Acacia koa* in the Honolulu and Honouliuli Forest Reserves, as well as *Aleurites*, *Albizia*, *Angophora*, *Syncarpia*, *Tristania*, *Toona*, and innumerable other introduced trees. Throughout the type, patches of native forests are still present because the aim of the planting was always to replace ground cover where missing, not to replace existing forest. A good example of this is along the Aiea Loop Trail, where introduced trees occupy the formerly badly eroded ridges and native plants, mixed with introduced escapes, the gullies.

Molokai--On Molokai (fig. 3), the 'ōhi'a--hāpu'u type designation applies well for the forest of the upper slopes, which on this island is mostly pristine. There are small bogs containing dwarf 'ōhi'a on some of the ridges. The type where mapped extends to the south

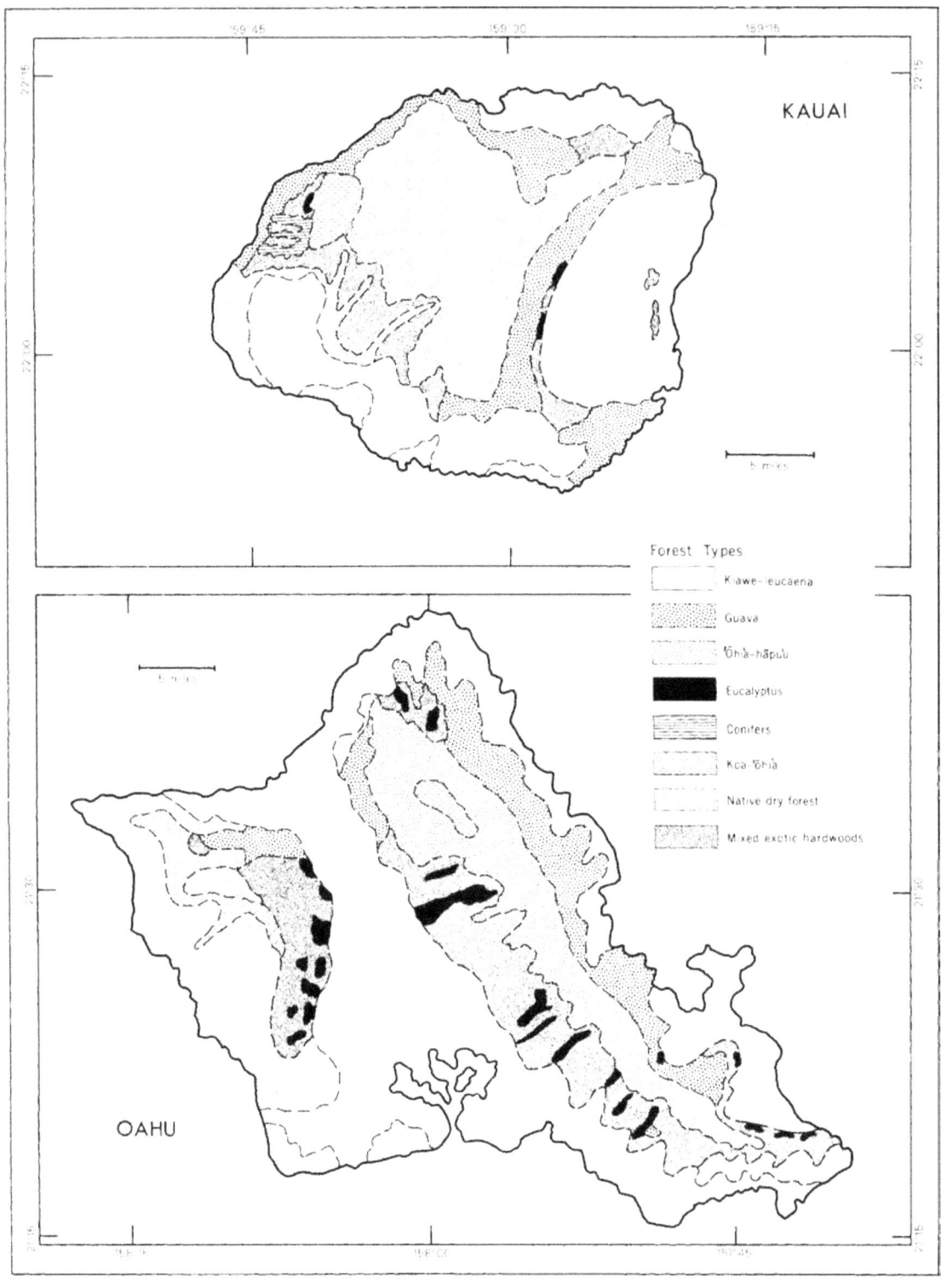

Figure 2--*Forest types on Kauai and Oahu.*

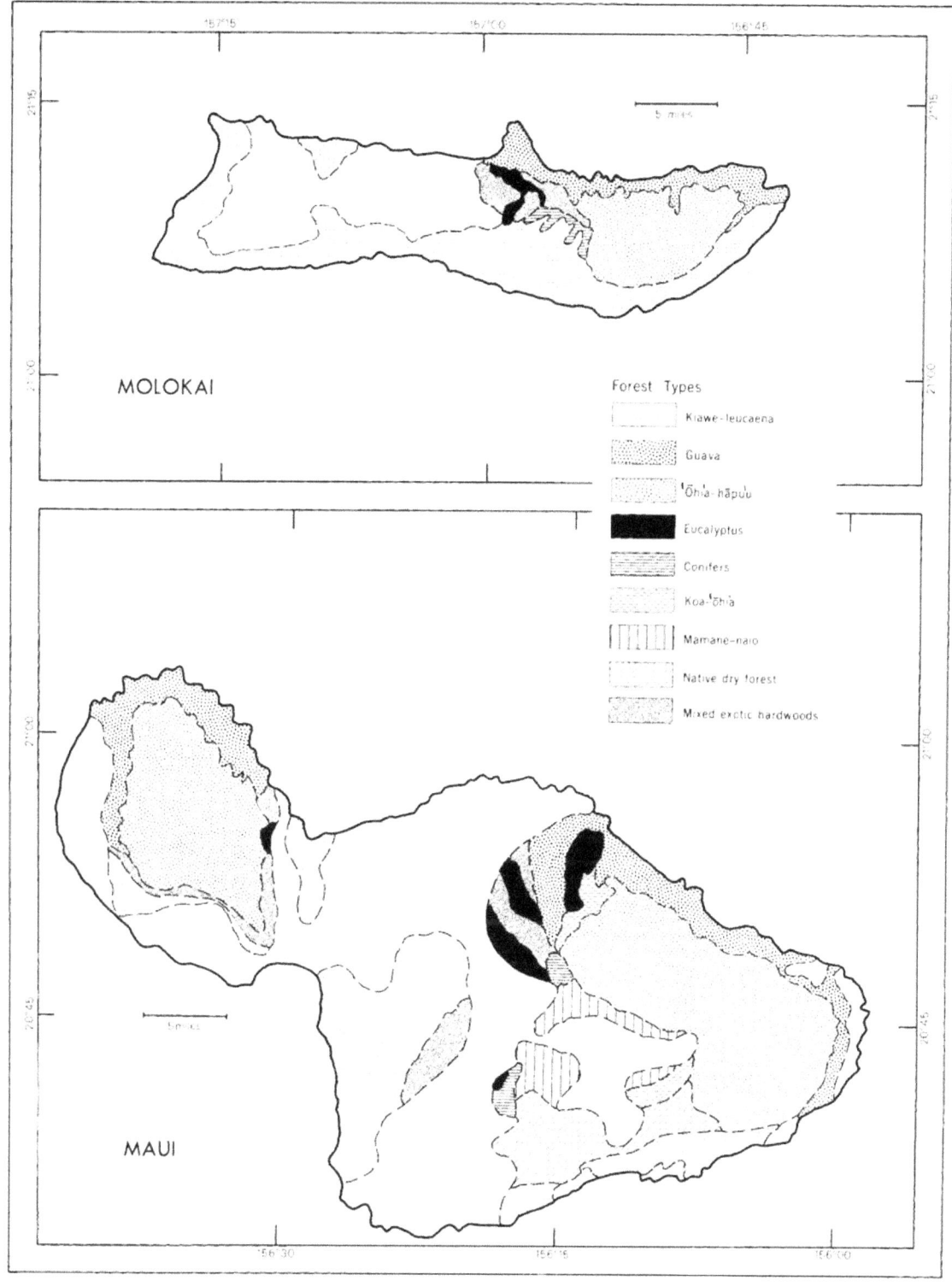

Figure 3--*Forest types on Molokai and Maui.*

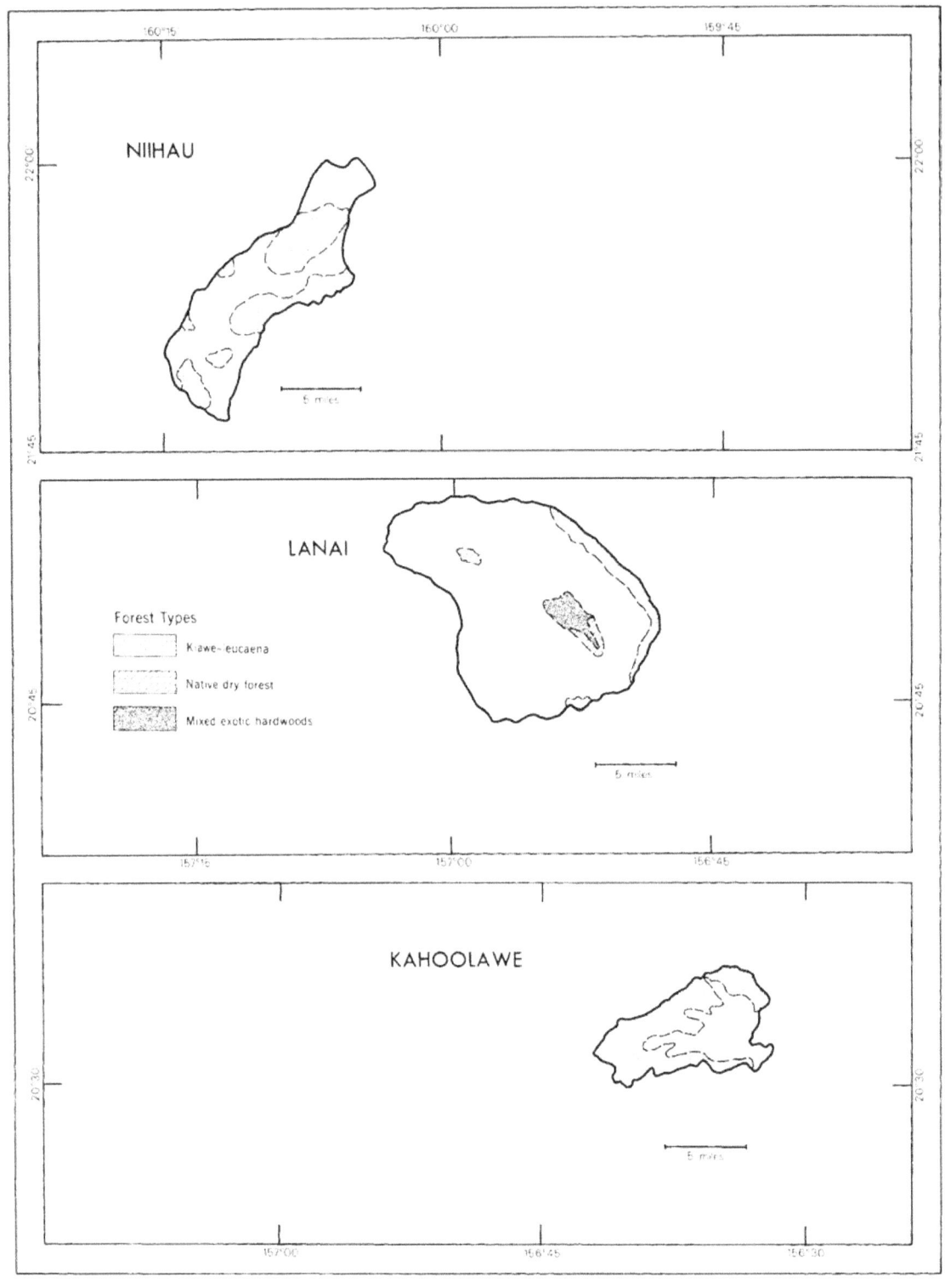

Figure 4--*Forest types on Niihau, Lanai, and Kahoolawe.*

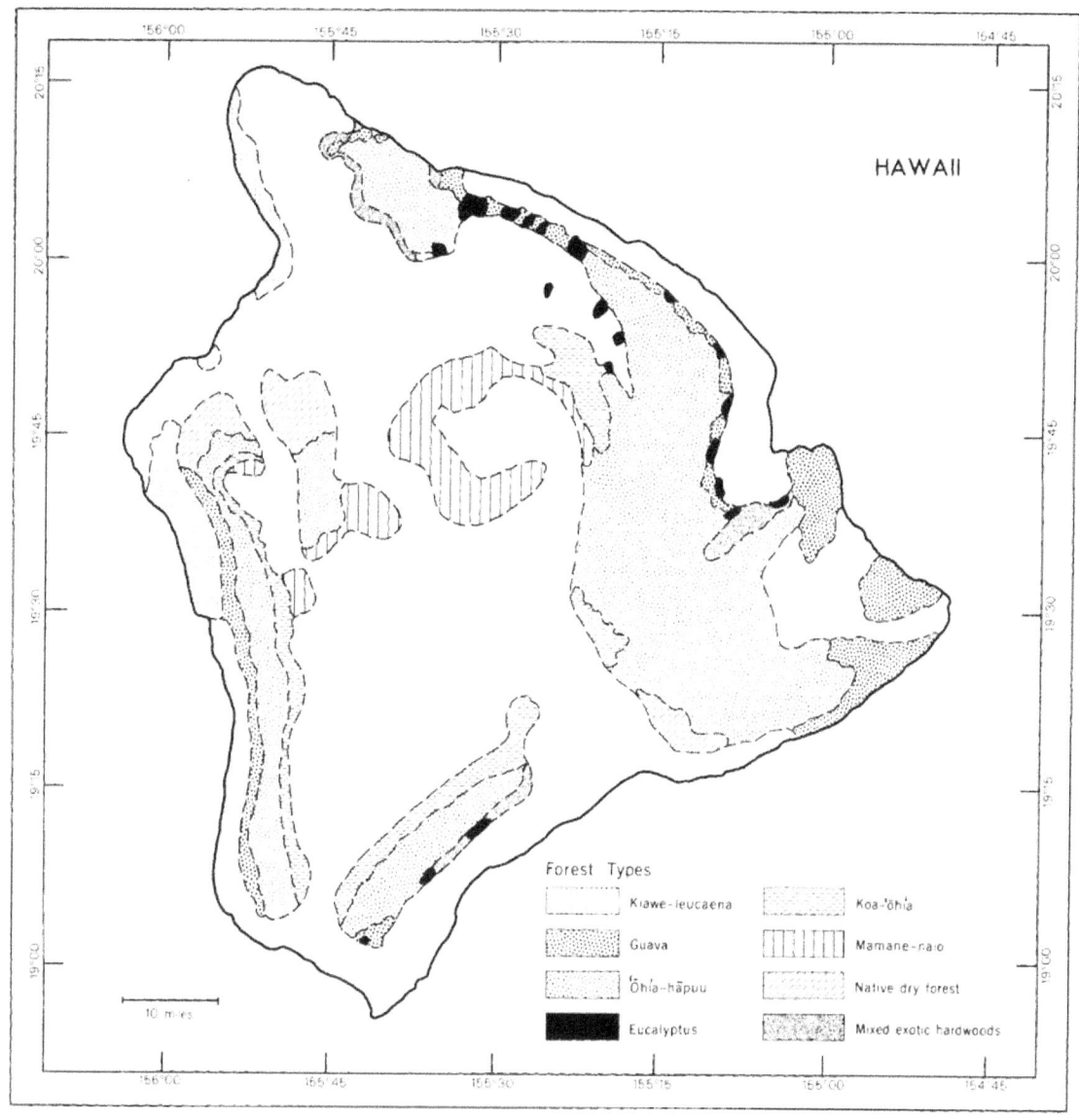

Figure 5--*Forest types on Hawaii.*

into an area, the upper part of which was Styphelia-Dodonaea brushland that is now almost denuded of tree cover by the large goat and deer population. In draws in the lower part of this area, there are a few remaining badly chewed trees of Sesbania arborea.

The guava type on Molokai is simply a catchall term, including the cover mostly of Lantana at Kalalau and that mostly of Eugenia and Mangifera in Halawa Valley. In between Kalalau and Halawa, the valleys contain many trees, the most obvious being Pandanus, Aleurites, and Mangifera.

Mixed exotic forest is largely Acacia confusa and Casuarina spp. along the highway near Kalalau lookout and mixtures of Grevillea and several Myrtaceae in the Forest Reserve.

Lanai--On Lanai (fig. 4), the mixed exotic hardwoods type is confined to the slopes and top of Lanaihale. It includes Casuarina, Toona, Fraxinus, and many trees of Eucalyptus. A single row of Araucaria traverses the ridge top, which is occupied by a native wetland shrub type in which Scaevola is the most notable genus. Leptospermum scoparium, a very large shrub or tree, is rapidly taking over the top of the mountain.

At the southeast edge of the mixed exotic hardwoods type is a strip shown as kiawe--leucaena. This is actually a shrub type of Dodonaea.

The tiny area of native dry forest shown in the north center of the island is a remnant Diospyros type unique to the island and it is fenced to protect it from deer.

Maui--On Maui (fig. 3), the 'ōhi'a--hāpu'u type is mapped in the central highland peaks and valleys of West Maui and the windward and eastern slopes of Haleakala. The top of West Maui, where it is not sharp ridges, is bog. Puu Kukui has the largest of these bogs, and its vegetation is well described and illustrated by Carlquist (1970). The slopes of West Maui are clothed generally in Metrosideros and the ferns Cibotium and Dicranopteris. This type includes also open Metrosideros forest with an understory of native trees, Charpentiera, Psychotria, Pelea, Eleaocarpus, and others, without treefern.

On the slopes of Haleakala, the 'ōhi'a--hāpu'u type is mostly extremely wet, with Cheirodendron, Coprosma, Ilex, Pelea, and Myrsine as understory trees. In its upper portion, Rubus hawaiiensis becomes a principal understory component, and above this the forest rapidly changes to a Styphelia--Vaccinium shrub type. In the upper Kipahulu Valley area some Acacia also occurs and this portion could be segregated as the koa--'ōhi'a type, were it larger in size.

The mamane--naio type shown near the summit of Haleakala is largely Styphelia, Vaccinium, Coprosma, and other shrubs, with occasional Sophora and Myoporum, also as shrubs. There are few Sophora of tree size in the area above the Kula Forest Reserve.

The small area of koa-'ōhi'a is an Acacia--grass--native shrub type heavily affected by goats. It is an open forest of short, crooked Acacia with few Metrosideros.

The large area mapped as native dry forest on the south side of Haleakala is unfortunately extremely degraded by grazing animals. The principal tree genera are Diospyros, Erythrina, Pleomele, and Reynoldsia. A few Metrosideros are mixed with the diverse dryland tree species, but the entire area has largely been taken over by grasses and shrubs.

The guava type on Maui is again a diverse mixture. In the drier portions of this lowland and coastal type, Psidium and Schinus are most common. In wetter areas, especially along the Hana Road, Pandanus is very common along the sides of gulches and Aleurites in the bottoms. The floor of the inland portion of Keanae Valley is covered by a dense stand of Hibiscus tiliaceus. Numerous small plantings of introduced trees and bamboo are common in the East Maui portion of the type.

The West Maui mixed exotic hardwood type is mostly Casuarina. The area shown as this type in the Kula vicinity represents the now largely urban thickets of Acacia decurrens and small ornamental plantings.

Hawaii--The forest types of the Island of Hawaii (fig. 5) are generally less complex because of the younger geologic age of the island. The 'ōhi'a-hāpu'u type on the windward slopes of Mauna Loa grows entirely on relatively young to very young soils developed on aa and pahoehoe lava flows. In this area, the height of the Metrosideros overstory and vigor of the Cibotium understory seems to be related to soil age and depth. On the shallow soils developed on pahoehoe are small 'ōhi'a with extensive Dicranopteris and, at higher elevations, Sadleria, openings. Deeper, older aa soils have tall 'ōhi'a and treefern. Common understory tree genera are Ilex, Coprosma, and Myrsine.

On the windward slopes of Mauna Kea, the soils are developed on volcanic cinder throughout the 'ōhi'a-hāpu'u type. In the wettest portions of the type are numerous bogs largely covered by the fern Dicranopteris but with occasional Metrosideros and Cibotium. In the well-drained parts of these very wet forests, Acacia koa may be found as a co-dominant. Towards the northern part of this extensive type, Cheirodendron becomes the most common understory tree (after Cibotium), with Ilex perhaps the next most common.

The Kohala 'ōhi'a--hāpu'u type is similar to that of West Maui, with almost treeless bogs at and near the summits that are dominated by Lobelioideae and (on this island only) the peatmoss, Sphagnum.

The Kona and Ka'u 'ōhi'a--hāpu'u types are somewhat similar, although the soils of Kona are much younger. The stilt-rooted form of 'ōhi'a is less common in these forests, as is the makeup of the understory. The Cibotium layer is still present, but Psychotria becomes the dominant understory tree. Acacia koa is occasional throughout the type in these two locations.

Above the 'ōhi'a--hāpu'u type, generally starting at 4,000 ft (1,219 m) on the windward side and at 3,500 ft (1,067 m) in Ka'u and Kona, *Acacia koa* becomes the dominant species. These are the areas typed as koa--'ōhi'a. With increasing elevation, *Metrosideros* and *Cibotium* drop out in the type, leaving, at the uppermost sites, only *Acacia* with an occasional *Sophora* or scrub *Myoporum*, intermixed with *Rubus* and *Vaccinium* ground cover. Very little of the koa--'ōhi'a type is untouched by cattle, and in much of it pigs regularly turn over the soil as they root for worms and insects.

The mamane--naio type has been heavily impacted by feral animals, causing *Myoporum* to be more common than the more palatable *Sophora*. The best remaining forest in this type is on the northwest slope of Mauna Kea, the habitat of the endangered bird, palila.

The native dry forest in the vicinity of Puuwaawaa and the north slope of Hualalai is the only location of *Colubrina oppositifolia*. This dry forest contains *Metrosideros* as a common component in addition to *Diospyros*, *Erythrina*, *Caesalpinia*, *Nothocestrum*, *Pleomele*, and *Canthium*. The native dry forest type in Hawaii Volcanoes National Park in the southeast of the island does not contain *Metrosideros*, but rather *Diospyros* and *Canthium* as the dominants.

The guava type near Hilo and Honolulu Landing is primarily mixed *Melastoma*, *Metrosideros*, and *Pandanus*. The type in South Puna is complex: *Pandanus* and *Metrosideros* are probably the most common species in a region much disturbed by ancient as well as current occupation and farming.

The mixed exotic hardwood type on this island frequently includes rather large single-species block plantings of trees other than *Eucalyptus*. In the Waiakea Forest Reserve, the trees most often planted are *Eucalyptus* and *Toona*. In the area shown in Kohala, *Casuarina* is most common, and in Ka'u, *Grevillea*.

FORESTRY

In Hawaii, the word *forestry* had always meant environmental protection and improvement until the 1960's, when an effort to develop a better timber resource and to encourage development of a timber industry began. Prior to the 1960's, essentially all the work undertaken by the Division of Forestry, Department of Land and Natural Resources, was concentrated on reforestation of ravaged forest land, fencing and patrolling forest reserves, eradicating feral animals, and developing Forest Reserves.

During the 1960's, the Division of Forestry carried out a program of converting existing healthy forest to exotic timber species on all islands, but centered primarily in the Waiakea Forest Reserve on Hawaii. In the late 1960's, environmental concern started a shift away from this focus on economic development towards a focus on forest protection during the 1970's. In 1979, the Division of Forestry became the Division of Forestry and Wildlife.

Forestry has a long and well-documented history in Hawaii. It began in the 1880's with the creation and fencing of the privately owned Kau Forest Reserve by Hutchinson Sugar Co. and with the planting of eucalypts by the Kingdom on the former Coney Estate on Tantalus--the area now called Eucalyptus Ridge. It continued through the 1890's with plantings of *Casuarina* and *Eucalyptus* in Nuuanu Valley, which was at that time bare of the trees, and along Mud Lane above Kukuihaele on Hawaii.

In 1903, the agency now called Division of Forestry and Wildlife was created, and the work of establishing Forest Reserves, fencing them, and planting trees began in earnest. In 1904, the first two Forest Reserves were set aside, the Kaipapau on Oahu and the Hamakua Pali on Hawaii. Both public and private lands were given over to Reserve status and eventually 1.2 million acres (485,830 ha), one-fourth of the land in Hawaii, were included in the system. "Reserve" status did not confer a "hands off" type of land management to the forests. Rather, the Division of Forestry took on the task of putting trees back on those parts of the reserves where they had been destroyed by cattle, fire, and other causes.

The Forest Reserves as they were in 1961, before various changes in Hawaii's laws altered their size and boundaries, serve as a guide for locating many of the species and plantations described here (*figs. 6-8*). Since 1961, almost all the privately owned land formerly dedicated to Forest Reserve status has been withdrawn. Also, about 25,000 acres (10,000 ha) of "State land," which was actually "Hawaiian Homes Land," has been withdrawn. The title "Forest Reserve" has become an official designation only for the State-owned land. The land in other ownerships (*figs. 6-8*) is almost all within "Conservation Districts" and subject to various use restrictions intended to maintain its cover type, but it is no longer managed and maintained by the Division of Forestry and Wildlife as it was before withdrawal. Many of the forest plantings mentioned here were made by the government on the land in other ownerships when it was Forest Reserve. So, for simplicity, and because most people still regard the entire area as Forest Reserve regardless of ownership, the entire land areas are called Forest Reserves in this handbook.

In 1904, the Division of Forestry began planting in Makiki Valley. At that time, it was almost bare of trees,

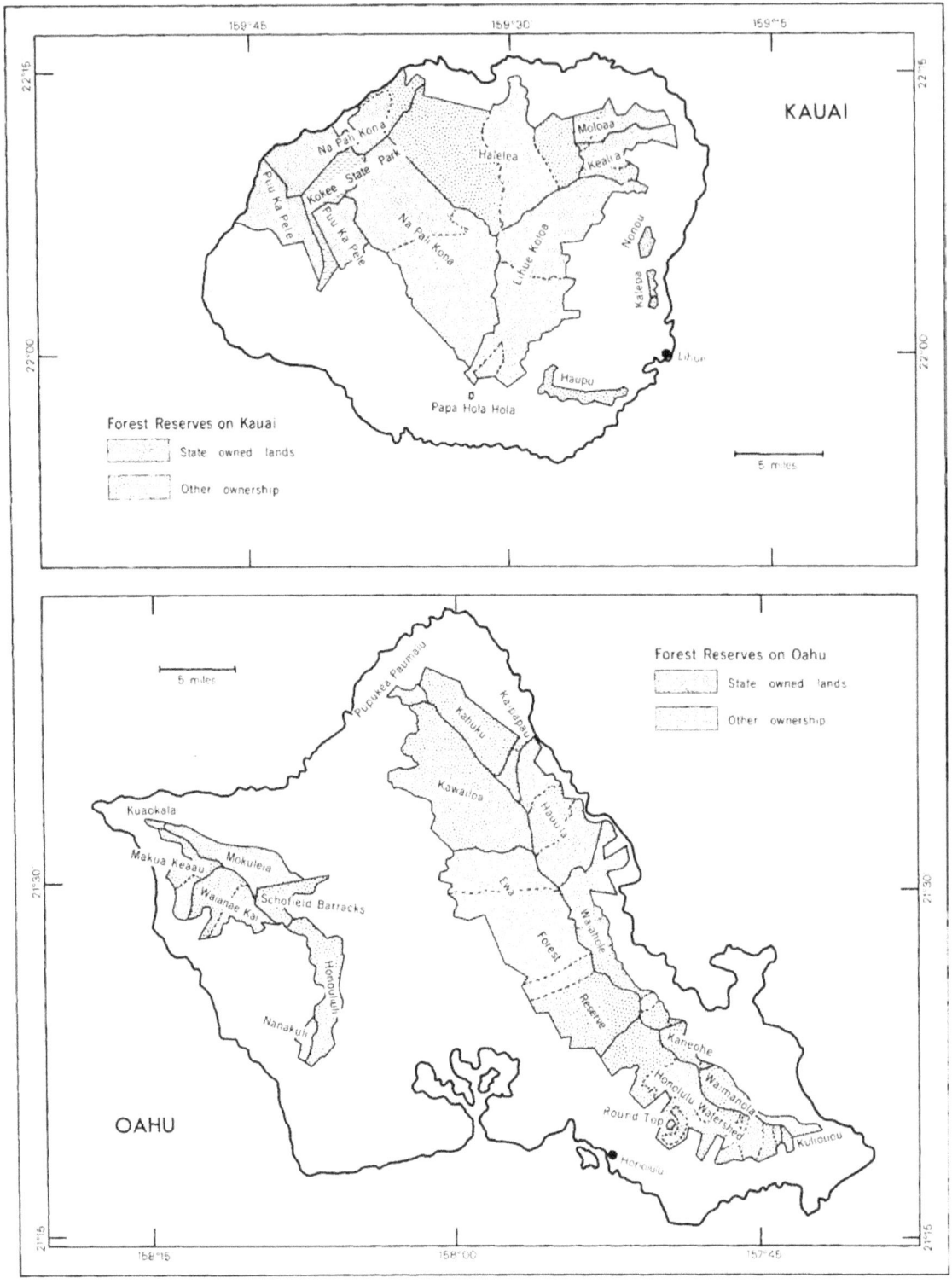

Figure 6--*Forest Reserves on Kauai and Oahu.*

because it had supplied the City of Honolulu with firewood for the previous century. They planted kukui (*Aleurites moluccana*) and koa (*Acacia koa*) at first but soon found that these species did not do well in the eroded soil and greatly modified environment. It was necessary to use exotic species to establish new trees quickly. Thus began the pattern that exists at present. Although the native species *Acacia koa* is numerically the fourth most commonly planted tree in Hawaii, most of the successful plantations are *Eucalyptus*, *Grevillea*, *Albizia*, and other fast-growing exotics.

Today, it is hard to visualize the impact these plantings have had. Imagine the mountains behind Honolulu or Nuuanu and Kalihi Valleys without any trees, as all were bare only 80 years ago. In 1924, there were 5,000 head of cattle on the 5,000-acre (2,024-ha) Honouliuli Forest Reserve above Kunia, Oahu, with no trees at all and hardly a blade of grass. By 1938, planting of this Reserve was completed and today many of the trees of *Eucalyptus*, *Grevillea*, and *Acacia* are large enough to supply sawlogs if desired. Similarly on other islands, it is hard to imagine Kokee and Kauai without koa and silk-oak, the Makawao area of Maui without eucalypts, or the Volcano Road without its forest parks, but all are partly or entirely planted forests of relatively recent origin.

The bulk of the planting in the Forest Reserves came about in the 1930's as part of the Civilian Conservation Corps program. Most of the forest plantations in the State, particularly the eucalyptus plantings, date from this period. These plantings were made simply to get cover on the land. There was no intention at that time of producing a timber resource, although many earlier plantings had been made by the ranches and sugar companies to produce firewood and fenceposts.

In 1957, the Territory of Hawaii requested the assistance of the USDA Forest Service in conducting a timber survey and determining the research needs to enhance both the watershed and timber values of the forest lands. The Forest Service assigned Robert E. Nelson to Hawaii to plan and lead the survey, evaluate other research needs, and provide on-the-ground support for various federally funded forestry support programs.

At the numerous meetings and discussions that were held, the principal forestry concern of land managers in Hawaii was watershed protection. The possibility of timber harvesting sparked interest, but not if it meant damaging watersheds that had taken half a century of effort to get back in order. Thus the first researchers to join Nelson worked on watershed problems--soil trafficability and evapotranspiration.

A comprehensive research plan was developed jointly by local land managers, primarily from private industry, working with the State and Federal agencies involved. This plan, which began to be implemented in earnest in 1960, called for extensive research to solve watershed management problems, develop a thriving timber industry, and determine how best to manage the native and exotic forests. The plan was followed throughout the 1960's, with the bulk of the needed research being carried out by the Forest Service research unit, which became known as the Institute of Pacific Islands Forestry.

By 1961, the forest survey had made it apparent that Hawaii had a sizable timber resource available for harvest. The 20,000 acres (8,097 ha) of planted forest contained almost one-fourth of the total sawtimber volume in the State. It became the policy of the Division of Forestry to increase this level further. This was done by clearing and planting native forest on Hawaii and shrub exotic forest on the other islands at a rate of about 2,000 acres (810 ha) a year from 1961 to 1968. The principal trees planted were *Toona*, *Eucalyptus*, *Flindersia*, *Pinus*, and *Fraxinus*. At the same time, the Forest Service researchers began to evaluate these and alternative species to determine their management characteristics.

The exotic species used for timber production in Hawaii are unusual as compared with elsewhere in the world. The primary reason for this is that most of the forest land available for growing tree crops is at relatively high elevations above 2,000 ft (610 m), where the temperature is too cool for the more common tropical timber species such as teak and mahogany to grow well. The climate at these higher elevations is not a temperate climate, however, because there is very little seasonal variation in temperature. It is an oceanic montane environment at a latitude having few other similar high island groups. Consequently, trees that have grown well after introduction to Hawaii's forests are often not widely known or used elsewhere. Secondly, because for many years the sole objective of the Division of Forestry was to reforest denuded watersheds with any trees that would grow, and preferably with trees that would not later be cut for industrial use, a great many unusual species were tried in forest plantations. Some of these, such as *Toona* and *Flindersia* spp., happened to be good timber trees.

Hawaii in 1987 has about 46,000 acres (18,623 ha) of planted forest on commercial forest land (Metcalf et al. 1978). Half is in eucalypts and the rest in other exotic species. These forests contain about 69 million cu ft (1.95 million cu m) in trees of 5.0 in (13 cm) diameter and larger. The eucalyptus plantations contain 88 percent of this volume.

Only two native tree species are considered commercial timber. These are 'ōhi'a and koa. There are about 115 million cu ft (3.25 million cu m) of 'ōhi'a and 25 million cu ft (0.71 million cu m) of koa in Hawaii (Metcalf et al. 1978). The 'ōhi'a commercial forest occupies about 174,000 acres (70,445 ha) and the koa and 'ōhi'a--koa commercial types, 62,000 acres (25,101 ha).

The Institute of Pacific Islands Forestry grew during the 1970's to an organization of about 30 people, with 12 to 15 scientists engaged in research in several subjects

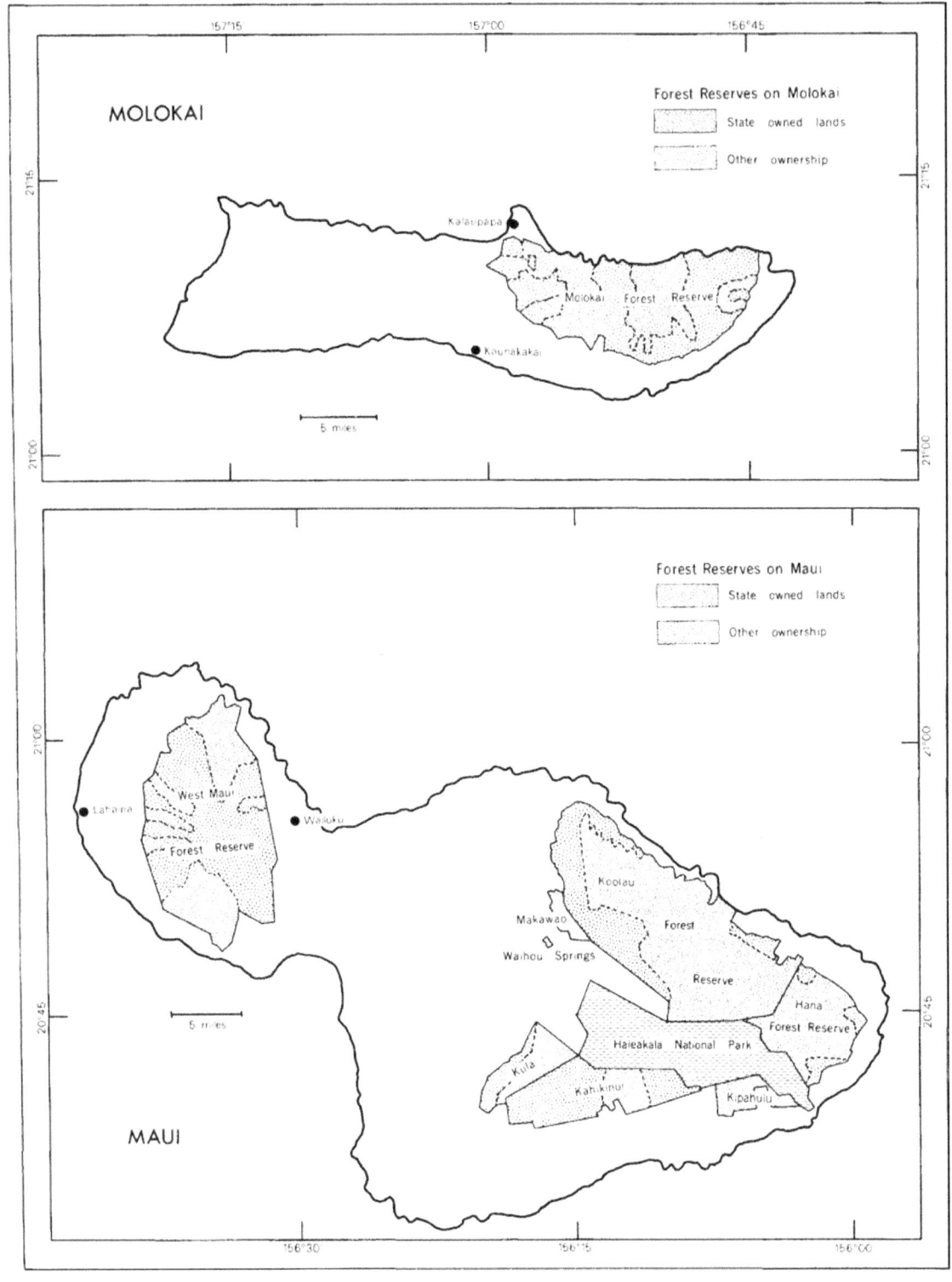

Figure 7--*Forest Reserves on Molokai and Maui.*

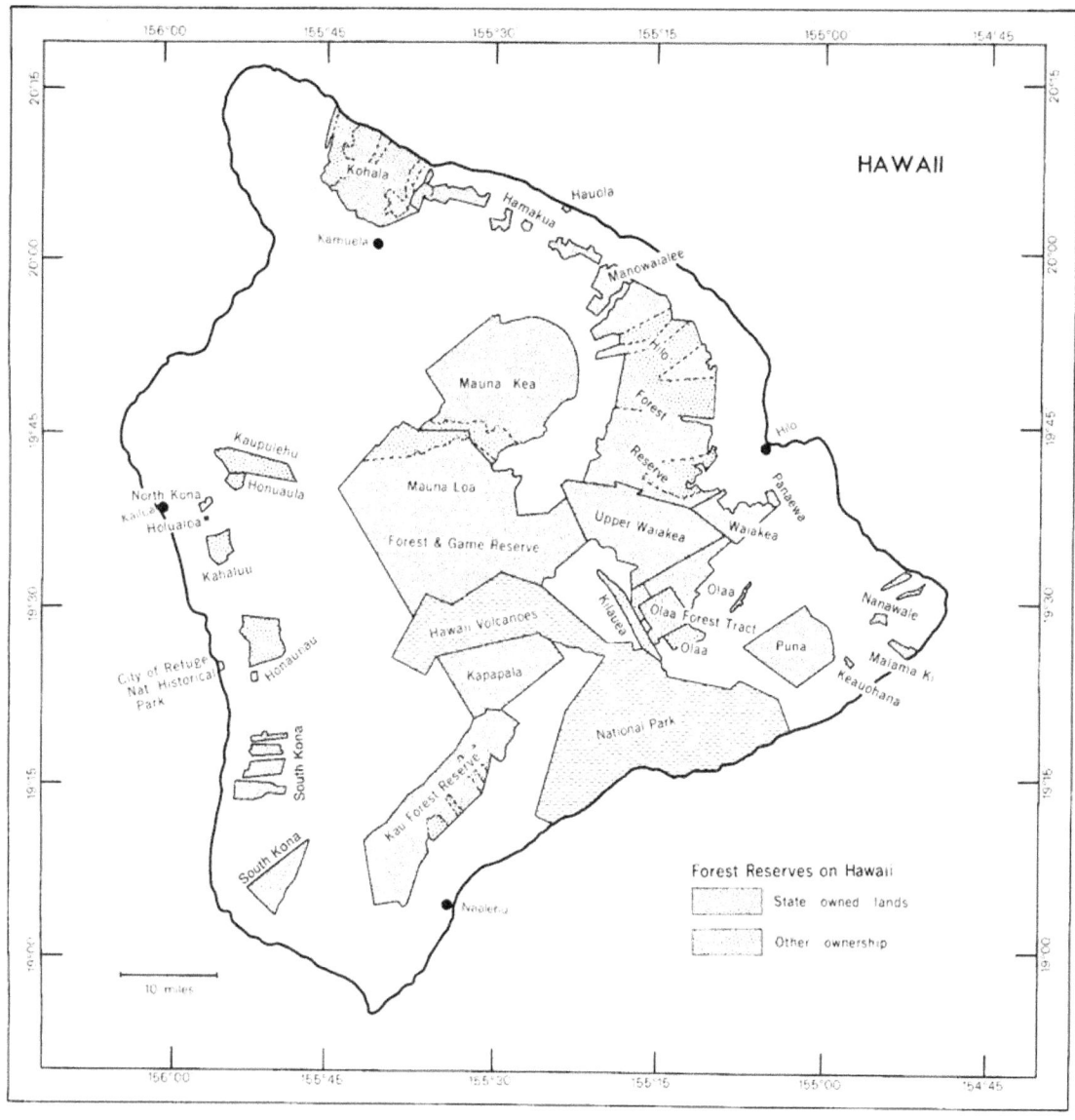

Figure 8--*Forest Reserves on Hawaii.*

including forest pathology (particularly as related to a serious decline of 'ōhi'a), modern container tree nursery development, tissue culture propagation of koa and eucalyptus, biomass production by intensive cultivation of eucalypts, streamflow sediment loading resultings for the 'ōhi'a decline, the habitat of endangered birds, and a host of other subjects, including an entire project devoted to the forestry problems of Micronesia.

This research was in part always funded by the State of Hawaii and in Hawaii carried out primarily on State-owned land, frequently with the help of State employees and equipment. The Institute of Pacific Islands Forestry has currently passed the 30-year mark of Forest Service presence in Hawaii and can point to more than 400 publications that have resulted from its research over the years. Many of these publications are listed in the Selected References.

One long-term task of the Institute that aided materially in providing numerical counts and location information for this publication was the compilation of the planting records of the Division of Forestry for 1910--1960. This staff effort was summarized by Nelson (1965) and completed by Skolmen (1980).

A small forest industry has been present in Hawaii since about 1850. The mainstay of this industry has been the native koa, an excellent cabinetwood. 'Ōhi'a, the principal species available, has a very heavy, hard wood of limited usefulness. Over the years, the industry has changed considerably. At one time, from 1909 to 1917, there was a large mill in Puna cutting railroad ties of 'ōhi'a. More recently, in 1968, there was a veneer and plywood plant in Kawaihae. In the late 1970's, there were three sawmills and one chipping plant in Hawaii. Two sawmills produced lumber from koa and the other cut *Eucalyptus*. The chipping company chipped *Eucalyptus* for shipment to Japan for use as paper pulp. In the 1980's, the industry reverted to 4 or 5 very small mills cutting koa and on-again, off-again attempts to log and utilize eucalyptus chips for fuel in electrical generation.

There is continuing interest in growing *Eucalyptus* on short rotations using intensive agricultural techniques to supply either the pulpwood market or fuelwood for electrical energy production in place of oil. The U.S. Department of Energy has for several years funded a large research project in eucalyptus biomass production.

Hawaii's forests are important in other ways than as a source of timber and will continue to be managed for other purposes. The forests maintain the water supply for irrigation and domestic water. They reduce erosion, preventing stream siltation and killing of coral surrounding some of the islands. They supply the habitat for feral pigs, an important game resource. Although they are rapidly changing because of the invasion of introduced plants, the native forests still supply the habitat for native birds, land snails, and plants found nowhere else in the world. Managing these forests to preserve these unique organisms, some of which are threatened and endangered species, must remain a major goal of the forestry effort.

The forests also supply the scenic backdrop to beaches, cities, roads, and residential areas that, as well as providing recreation, cause Hawaii to be the uniquely beautiful place that it is. Tourists in increasing numbers continue to enjoy Hawaii's tropical trees, both native and exotic. These forests must continue to receive careful management.

41. ʻIliahi, Freycinet sandalwood   *Santalum freycinetianum* Gaud.

43. Pāpala *Charpentiera obovata* Gaud.

48. Kanawao   *Broussaisia arguta* Gaud.

57. Wiliwili   *Erythrina sandwicensis* Deg.

80. ʻAʻaliʻi   *Dodonaea viscosa* Jacq.

83. Kauila   *Alphitonia ponderosa* Hillebr.

87. Hau, sea hibiscus *Hibiscus tiliaceus* L.

88. Milo, portiatree *Thespesia populnea* (L.) Soland. ex Correa**

112. 'Ōhi'a ai, mountain-apple *Eugenia malaccensis* L.**

116. ʻŌhiʻa lehua   *Metrosideros polymorpha* Gaud.

138. Kou *Cordia subcordata* Lam.**

149. Noni, Indian-mulberry Morinda citrifolia L. **

ACKNOWLEDGMENTS

Many persons have assisted the authors in various ways during the preparation of this handbook. Valuable help has been provided by many foresters, past and present, on the staffs of the two main government agencies listed here. The work was done mostly at the Institute of Pacific Islands Forestry, U.S. Department of Agriculture Forest Service, headquartered at 1151 Punchbowl St., Honolulu, HI 96813. Robert E. Nelson, retired director, provided support during early stages. Particularly in field work, assistance was contributed by the staffs of the State forestry agency, the Division of Forestry and Wildlife, 1151 Punchbowl St., Honolulu, HI 96813.

Lester W. Bryan, late retired Deputy State Forester, supported and encouraged the project in various ways. His lists of Champion Trees of Hawaii have been included here.

Joyce Davis Jacobson of Hilo, formerly at the Bishop Museum, has reviewed the manuscript and added notes on traditional Hawaiian uses of various tree species.

The late Otto Degener, author of "Flora Hawaiiensis," made available his published drawings. His notes on many species, especially uses by the early Hawaiians, have been adapted with credit.

Numerous publications have served as a basis for this one. Several important references have been cited under Other Publications. Many others are listed under Selected References. Credit for the illustrations, both original and from published sources, is given under the topic Illustrations.

Grateful acknowledgment is made to specialists at parks, botanical gardens, arboreta, and other areas for supplying lists of tree species. These lists have served as a basis for the compilation of Special Areas, which list locations of tree species.

For their very helpful review of the manuscript, special credit is due F. Raymond Fosberg, Dept. of Botany, Smithsonian Institution; Derral R. Herbst, U.S. Fish and Wildlife Service; and Carolyn A. Corn, Hawaii Division of Forestry and Wildlife.

Finally, Warren L. Wagner, Derral R. Herbst, and Seymour H. Sohmer, authors of "Manual of the Flowering Plants of Hawaii" (1989 in press), generously have made available much valuable, unpublished, information, especially on classification, nomenclature, and distribution. Rebecca Nisley and Betty Toczek of the USDA Forest Service prepared the camera copy with desktop publishing.

TREE SPECIES, DESCRIPTIONS AND ILLUSTRATIONS

TREEFERN FAMILY (DICKSONIACEAE)

1. Hāpu'u-'i'i, Hawaiian treefern *Cibotium chamissoi* Kaulf.

Large treeferns are among the most distinctive and most beautiful plants of Hawaii's tropical rain forests. They are easily identified as ferns by their giant feathery or lacelike fern leaves (fronds) unrolling from a densely hairy coil and by the absence of flowers, fruits, and seeds. The unbranched trunks, leafy only at the top and evergreen, qualify as trees.

Six species of treeferns, all belonging to the genus *Cibotium*, have been distinguished as native in Hawaii. All have their powdery beadlike masses (sori) of microscopic spores in yellowish boxlike cups (indusia) in 2 rows on edges of under surface of leaf segments. Smaller trunked ferns, genus *Sadleria* with 6 or fewer native species, are treelike and have been included with an example, species No. 4, 'ama'uma'u, *Sadleria cyatheoides*. These smaller ferns bear their spores along midvein between 2 long black folds or lines.

This species, the native treefern of largest size, is recognized by the rough leafstalks covered with long stiff, shaggy blackish hairs and by the broad thickened leaf segments with prominent veins often branching.

A small tree with trunk usually less than 10 ft (3 m) high, rarely to 23 ft (7 m), and with several very large erect spreading leaves adding as much as 10 ft (3 m). The unbranched trunks are covered in lower part with compact masses of fibrous blackish air roots as much as 2--3 ft (0.6--0.9 m) in diameter, usually much smaller.

Leaves (fronds) several, erect and spreading, the oldest dying and bending down along trunk, very large, feathery, divided 3 times (pinnate), mostly 6--12 ft (1.8--3.7 m) long and 3--5 ft (0.9--1.5 m) wide. Leafstalk or axis (stipe) to 3 ft (0.9 m) long, stout, green, rough, and warty, the enlarged base shaggy with long glossy reddish brown narrow flattened hairlike scales, beyond with coarse, long stiff black hairs or bristles often shaggy and forming dense cover. The main axis is rough and bears many pairs of branches (pinnae) to 2 1/2 ft (0.8 m) long, further divided. Branches (pinnules) many pairs, narrowly lance-shaped, 5--10 in (13--25 cm) long and 1--1 1/2 in (2.5--4 cm) wide, ending in very long narrow point, with very short stalk at base, divided or lobed 3/4 to midvein. Segments or lobes many, nearly paired, oblong rounded, to 1/2 in (13 mm) long and 1/4 in (6 mm) wide, with edges turned under, slightly thickened, upper surface green and hairless with prominent veins often forked, lower surface paler and usually hairless.

Older leaf segments bear, on lower surface at ends of veins and on edges, 2 rows of 7 or fewer yellowish boxlike cups (indusia), each containing masses of dark brown spore cases (sporangia), which shed masses of microscopic spores abundantly. These boxlike cups are 1/16 in (1.5 mm) wide, larger than in related species.

In cross section the trunk has a thick outer mass of compacted fibrous blackish air roots, next the small stem with hard outer wall. Within is the whitish soft pith containing scattered woody strands or leaf traces. Trunks of ferns are not divided into bark and wood and increase in diameter only slightly.

Fortunately, these treeferns are hardy. They withstand damage by cattle and when uprooted by wild hogs continue to grow, so long as the growing tip or "fiddlehead" is not destroyed. A felled trunk can form many shoots. However, a few treeferns may be killed by shading of larger trees. Young plants of species No. 116, 'ōhi'a lehua, *Metrosideros polymorpha*, often begin growth from seeds that lodge and germinate among the moist leaf bases and air roots in the top of a treefern. Then roots are sent to the ground. Thus, some Hawaiians believed that the treefern was the parent.

This species is common and widespread in wet forests at low and middle altitudes of 800--6,000 ft (244--1,829 m) through the 6 large islands. It is a characteristic understory plant in 'ōhi'a forests and with other treeferns often forms dense undergrowth. The largest treeferns and best displays are on the Island of Hawaii, such as in the forests of Puna, Hilo, and Kohala Mountains.

Special areas--Waimea Arboretum, Wahiawa, Volcanoes, Waiakea.

Champion--Height 35 ft (10.7 m), c.b.h. 15.8 ft (4.8 m), spread 24 ft (7.3 m). Honaunau Forest Reserve, Kailua-Kona, Hawaii (1968).

Range--Hawaiian Islands only.

Other common names--hei'i, hāpu'u ii.

Botanical synonym--*Cibotium menziesii* Hook.

This largest native treefern for many years was known as *Cibotium menziesii* Hook. At the same time the name *Cibotium chamissoi* was misapplied to the species now called *Cibotium splendens* (Gaud.) Krajina. This species honors Ludolf Adalbert von Chamisso (1781--1838), German naturalist and explorer.

Treeferns were utilized for their trunks in the 1960's (Nelson and Hornibrook 1962). The outer part, like bark,

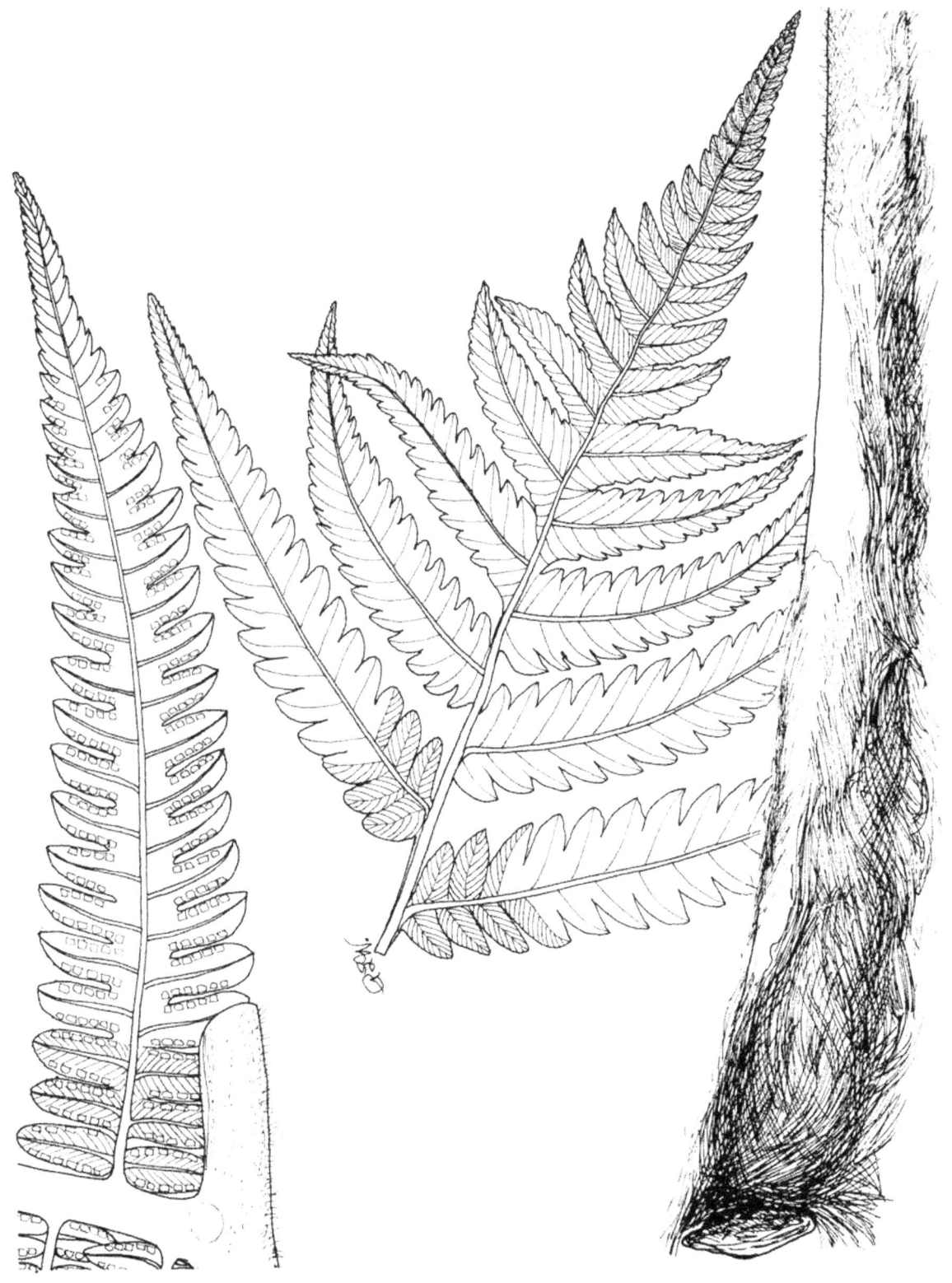

1. Hāpu'u-i'i, Hawaiian treefern   *Cibotium chamissoi* Kaulf.
Upper leaf surface (center) and lower leaf surface (left), 2/3 X, and base of axis (right), 1/6 X.

consisting of tightly packed fibrous blackish air roots, was sawed in narrow widths like slabs of bark or lumber. These sawed pieces of fibrous roots (called "poles") were shipped to the mainland for use as media or bases for growing orchids and other air plants. Some cutting of treeferns for this purpose continues. Chunks of treefern trunk are widely used as a rooting medium, and "pots" made of short cross sections have been a mainstay of the vanda orchid industry. This logging does not destroy the resource so long as the tops are left on the ground to grow again. The tops survive for long periods while forming new root systems. This is the preferred species for "pole" manufacture because of its acidity and moderate density.

2. Hāpu'u-pulu, Hawaiian treefern

This treefern is widespread in wet forests and is the most common species on the island of Hawaii. It is distinguished by the smooth leafstalks with masses of shiny golden brown, soft hairlike scales on enlarged base and lower part. Also, the leaf branches (pinnules) are stalkless, the leaf segments divided to midvein and beneath whitish and hairless, and the lowest pair of leaf segments common with earlike lobe and overlapping the axis.

A shrub or sometimes small tree with slender unbranched trunk 3--10 ft (0.9--3 m) high, rarely to 16 ft (5 m), and 6--8 in (15--20 cm) in diameter, and with erect spreading leaves adding as much as 8 ft (2.4 m). The trunk is smooth, but stubs of leafstalks remain attached.

Leaves (fronds) several, erect and spreading, the oldest dying and bending down along trunk, very large, feathery, divided 3 times (pinnate), mostly 6--8 ft (1.8--2.4 m) long. Leafstalk or axis (stipe) stout, slightly flattened, the enlarged base with masses of shiny golden brown hairlike scales to 2 in (5 cm) long, beyond light green and hairless. The main axis bears many pairs of branches (pinnae) to 2 1/2 ft (0.8 m) long, further divided. Branches (pinnules) many pairs, narrowly lance-shaped, 5--6 in (13--15 cm) long and 1/2--3/4 in (13--20 mm) wide, stalkless, divided to midvein. Segments many, nearly paired, long, narrow, rounded, slightly curved, to 3/8 in (10 mm) long and less than 3/16 in (5 mm) wide, with edges turned under and slightly wavy, thin, upper surface slightly shiny green with fine side veins slightly sunken, lower surface whitish and hairless.

3. Meu, Hawaiian treefern

This treefern, not distinguished in older references, is limited to wet forests of the Island of Hawaii. It is recognized by the slender smooth trunk, by the leafstalks appearing hairless but covered with soft matted dull brown hairs with masses of pale, dull gray brown,

The durable cut trunks of treeferns have served in corduroy trails across swamps, as water bars across mountain trails for drainage, and also for posts. The horizontal trunks usually sprout at one end to form a hedge of young leaves.

Wild hogs, which were introduced by the early Hawaiians, are enemies of treeferns. They uproot the plants and feed upon the starchy trunks. Degener reported that before the National Park was established, large tree trunks in the Kilauea region were dumped into steam crevices for cooking by volcanic heat, then fed to fatten hogs. In times of famine, the early Hawaiians cooked and ate the starchy trunks and young leaves. Actually, many people today eat peeled and boiled treefern growing tips and consider them a delicacy.

Cibotium glaucum (Sm.) Hook. & Arn.

Older leaf segments bear, on lower surface at ends of veins and on edges, 2 rows of 7 or fewer yellowish boxlike cups (indusia) more than 1/32 in (1 mm) wide, each containing masses of dark brown spore cases (sporangia), which shed masses of microscopic spores abundantly.

The main product of this treefern was *pulu*, the dense soft cover of glossy, silky scales of hairs at the base of young coiled leaves. *Pulu* served the Hawaiians as an absorbent surgical dressing and in embalming the dead. This "wool" was harvested, often by cutting down the larger plants, and exported mainly to California for stuffing mattresses and pillows. According to Degener (1930), the harvest was extensive by 1859, rose to the maximum export of almost 623,000 pounds in 1869, and ended by 1885. Because it becomes matted, breaks into powder, and absorbs moisture, *pulu* was replaced by other fibers. About 1920, treeferns were cut for the starch in their trunks, which was used for laundry and cooking. However, that industry was short-lived.

Widespread but apparently uncommon in wet forests at low and middle altitudes of 800--6,000 ft (244--1,829 m).

Special area--Volcanoes.

Range--Larger Hawaiian Islands except Maui.

This species propagates by trunk buds as well as spores. It grows well in shady, damp gardens at sea level and is often used as an ornamental. The other species do not survive at sea level.

Cibotium hawaiiense Nakai & Ogura

narrow flattened hairlike scales on enlarged base and lower part; and by the leaf segments light green beneath.

A shrub or sometimes small tree with slender unbranched trunk to 20 ft (6 m), and 4--6 in (10--15 cm) in

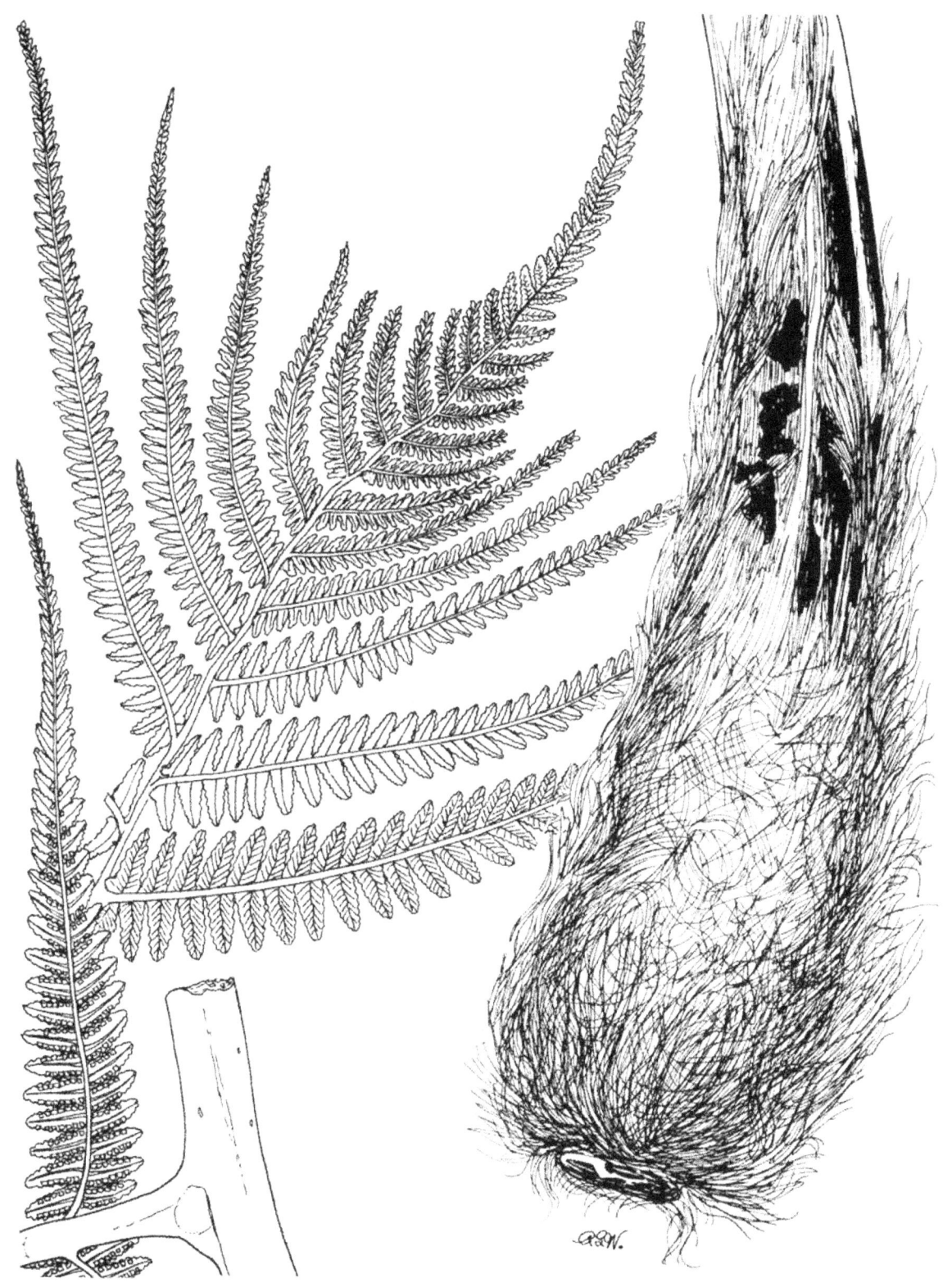

2. Hāpu'u-pulu, Hawaiian treefern *Cibotium glaucum* (Sm.) Hook. & Arn.
Upper leaf surface (above), lower leaf surface (lower left), base of axis (right), 2/3 X.

diameter, and with erect and nearly horizontal spreading leaves adding as much as 6 ft (1.8 m). Lower part of trunk is smooth, and upper part bears dead broken leaves.

Leaves (fronds) several, erect and spreading, the oldest dying and bending down along trunk and breaking off cleanly, very large, feathery, divided 3 times (pinnate), mostly 5--6 ft (1.5--1.8 m) long. Leafstalk or axis (stipe) stout, flattened above, the enlarged base and lower part with masses of pale dull brown hairlike scales 1 1/4--2 in (3--5 cm), beyond green and hairless. The main axis bears many pairs of branches (pinnae) to 2 1/2 ft (0.8 m) long, further divided. Branches (pinnules) many pairs, narrowly lance-shaped, 5--6 in (13--15 cm) long and 3/4 in (2 cm) wide, mostly with very short stalk at base, further divided or lobed almost to midvein. Segments or lobes many, nearly paired, oblong narrow rounded, to 3/8 in (10 mm) long and 3/16 in (5 mm) wide, with edges turned under and finely wavy, thin, upper surface light green with inconspicuous veins, lower surface dull light green with tiny cobwebby hairs.

Older leaf segments bear, on lower surface at ends of veins and on edges, 2 rows of 7 or fewer yellowish boxlike cups (indusia) more than 1/32 in (1 mm) wide, each containing masses of dark brown spore cases (sporangia), which shed masses of microscopic spores abundantly.

The uses are similar to those described for the first species of treefern, but the stem is rarely large enough in diameter to be sawed or even bucked into "pots."

Widespread in wet forests at low and middle altitudes of 800--6,000 ft (244--1,829 m) on the Island of Hawaii. An understory plant in 'ōhi'a forests, often abundant in thickets.

Special area--Hawaii Volcanoes National Park, uncommon except in Olaa tract.

Range--Island of Hawaii only.

Other common name--hāpu'u.

Closely related to *Cibotium splendens* (Gaud.) Krajina (earlier called *C. chamissoi*) and not distinguished in most references. Perhaps only a variety of that species, which is more common.

BLECHNUM FERN FAMILY (BLECHNACEAE)

4. 'Ama'u, sadleria *Sadleria cyatheoides* Kaulf.

This common trunked fern growing on open lava flows and in wet forests at middle altitudes through the Hawaiian Islands resembles a small treefern and becomes treelike. This species will serve as an example for its endemic genus of 5 or 6 species.

A shrub with slender unbranched trunk 3--5 ft (0.9--1.5 m) high, reported by Hillebrand (1888) as sometimes more than twice as tall, and 3--4 in (7.5--10 cm) in diameter. A tree with a trunk more than 16 ft (5 m) long to the first leaf scar was cut in the late 1950's near Glenwood, Hawaii, and displayed for many years at Hawaiian Fern-Wood Ltd. in Hilo. Tall trunks were also observed in Kona, Hawaii, and Waimea, Kauai. Trunk brown, upper part with old dead leaves hanging down on broken stalks, upper part with fibrous roots.

Leaves several, large, feathery, spreading from top of trunk, developing from large coil with long golden brown narrow flattened scales, 3--5 ft (0.9--1.5 m) long. Leafstalk or axis (stipe) 1--2 ft (0.3--0.6 m) long, light green, flattened above, densely covered at base with long golden brown narrow flattened hairlike scales to 2 in (5 cm) long, hairless above. Blade (frond) 2--3 ft (0.6--0.9 m) long, oblong, nearly hairless, twice divided (bipinnate). Branches (pinnae) 30--40 on a side, 6--10 in (15--20 cm) long and 1/2--1 in (1.3--2.5 cm) wide, divided almost to base into many narrow oblong curved segments, blunt at apex and turned down on edges, hairless, slightly thickened, upper surface slightly shiny green with fine sunken midvein and inconspicuous side veins, lower surface dull whitish green.

Older leaves bear, on lower leaf surfaces along midvein, 2 long black folds or lines (indusia) and between them a mass (sorus) of numerous dark brown spore cases (sporangia) that shed powdery masses of microscopic spores abundantly. Under favorable conditions spores, like seeds, develop into new plants.

Common and widespread in wet forests and open areas to 5,500 ft (1,676 m) altitude through the Hawaiian Islands.

Special areas--Haleakala, Volcanoes.

Other common names--'ama'uma'u, ma'uma'u, ma'u.

The trunked ferns of *Sadleria* are widespread within Hawaii. They grow under extreme conditions, from near the sea to mountain peaks, from rain forests to barren new lava flows, where they are among the pioneers or first plants to become established.

These ferns serve as nurse plants and soil for No. 116, 'ōhi'a lehua, *Metrosideros polymorpha*. The tiny windborne tree seeds lodge in the scaly and leafy fern stems and germinate there. Then the young trees, deriving moisture and nutrients from the fern leaves, crowd and replace the ferns. A weevil that bores into the stems

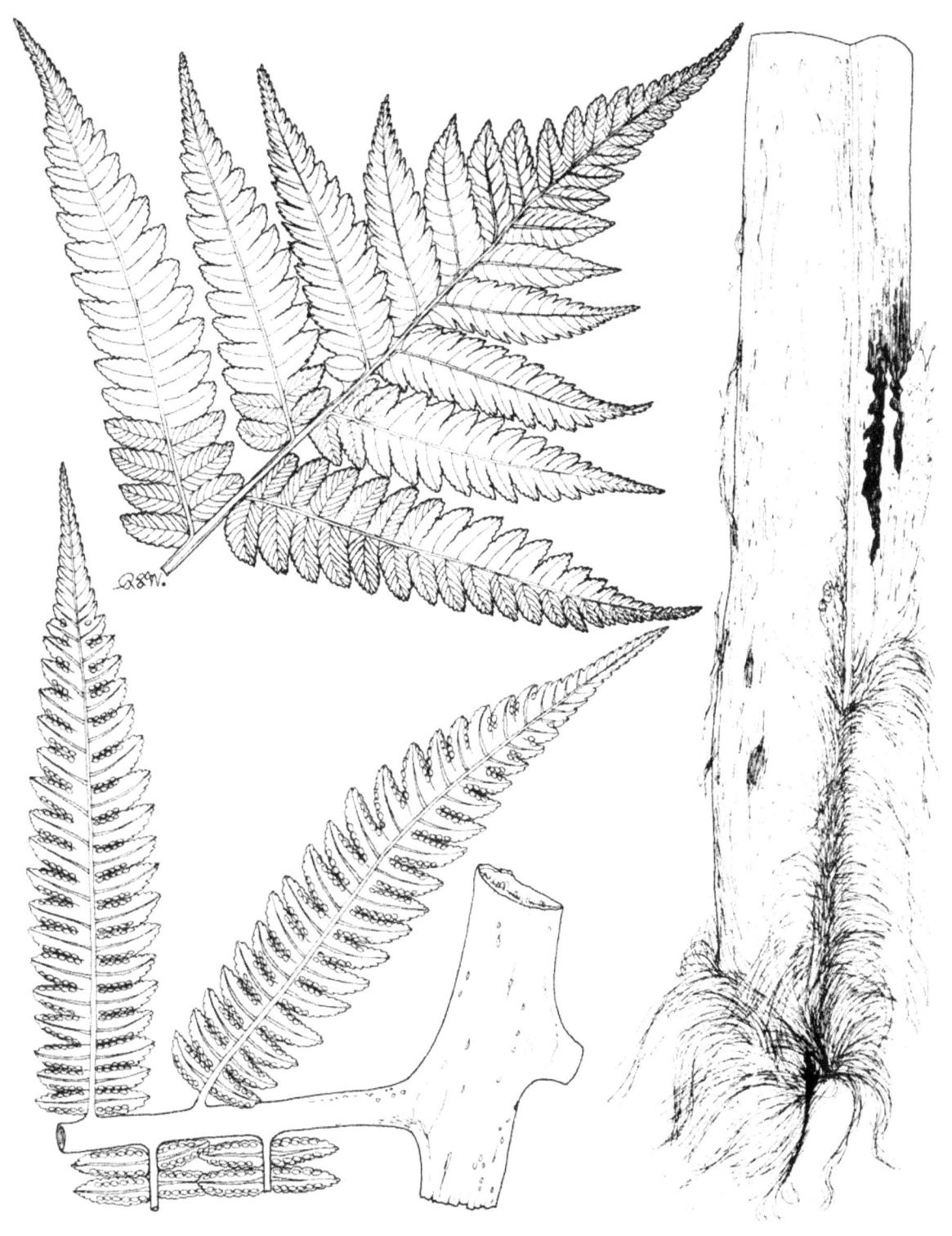

3. Meu, Hawaiian treefern *Cibotium hawaiiense* Nakai & Ogura
Upper leaf surface (above), lower leaf surface (below), base of axis (right), 2/3 X.

and damages the leaves has become a pest locally in recent years.

Various uses by the Hawaiians have been summarized by Degener (1930). Formerly, the soft hairlike scales of the young leaves were gathered commercially with those of the larger treeferns for stuffing mattresses and pillows. Leafstalks of this fern were beaten and used as sizing with bark in making tapa, or bark cloth. The large leaves served as thatch for the houses, especially at corners and ridges. For dry-land vegetable crops the fronds served as ground cover or mulch. The starchy, almost tasteless inner part of the trunks was cooked in the ground and eaten, mainly in times of famine. The young leaves, which taste like asparagus, were cooked also. A red dye for bark cloth was extracted from the outer part of the trunk by mashing the fibers and then squeezing the sap into a vessel. This liquid was then boiled with hot stones added to protect the container.

Halemaumau, the fire pit within Kilauea Crater, means "the house of 'ama'uma'u fern."

This species is the type of the genus *Sadleria*. The name commemorates Joseph Sadler (1791--1849), Hungarian botanist and physician.

ARAUCARIA FAMILY (ARAUCARIACEAE*)

5. Parana-pine

Araucaria angustifolia (Bert.) Kuntze*

Large evergreen tree with straight axis, horizontal branches usually in tiers or rings of 4--8, and spreading lance-shaped sharp-pointed leaves. Introduced about 1955 and relatively slow growing in Hawaii, it has not reached large size. At two different sites, it grew to 24 ft (7 m) in height in 6 years.

Leaves borne singly or sometimes paired, mostly 1 1/4--2 1/4 in (3--6 cm) long and to 1/4 in (6 mm) wide, green or whitish green, long-pointed, stiff and leathery, with keel and whitish lines beneath. Upper leaves on cone-bearing branches shorter and crowded in spirals.

Male and female cones usually on different trees, the male crowded near bases of leaves, cylindric, 3--4 in (7.5--10 cm) long and 1/2--3/4 in (1.3--2 cm) wide. Mature female cones large, rounded, slightly flattened, 5 in (13 cm) long and 6 1/2 in (16.5 cm) in diameter, narrowed from middle upwards. Cone-scales many, overlapping, ending in stiff points curved backward. Seeds large, shiny brown, oblong, to 2 in (5 cm) long, 3/4 in (2 cm) wide, and 3/8 in (1 cm) thick.

The wood of trees grown in Brazil has sapwood and heartwood of various shades of brown, sometimes streaked with red. It has a specific gravity of 0.55 and is a general purpose softwood used for construction, millwork, boxes, and pulp.

Formerly this species was an important timber species of southern Brazil. However, after cutting, the natural forests are being replaced by plantations of more rapidly growing pines.

This tropical species has not been extensively planted in Hawaii. It has been tried in a few plantations and has done well in two locations at 2,100 ft elevation (640 m). It may be seen at the Kalopa section of the Hamakua Forest Reserve on the Island of Hawaii.

Special areas--Keahua, Wahiawa, Kula.

Range--Native of southern Brazil and northeastern Argentina.

Other common names--candelabra-tree, Parana araucaria.

Botanical synonym--*Araucaria brasiliana* A. Rich.

The generic name is derived from Arauco, the province in Chile of the type species, monkey-puzzle araucaria, *Araucaria araucana* (Molina) K. Koch.

6. Columnar araucaria

Araucaria columnaris (G. Forst.) Hook.*

Large introduced evergreen tree with straight axis, branches horizontal or slightly drooping, in rings, narrow columnar crown, and both awlshaped and scalelike leaves. To 130 ft (39 m) or more in height and 3 ft (0.9 m) in trunk diameter. Bark gray, rough, thick, with horizontal cracks, resinous. Branches regular, with twigs spreading in one plane. Twigs long and very slender, shedding.

Leaves on young trees and branches narrow, awlshaped, to 1/2 in (13 mm) long, pointed; leaves on older branches broadly triangular, up to 1/4 in (6 mm) long, closely overlapping and curving inward, like a rope.

Male and female cones usually on different trees, the male cone about 2 in (5 cm) long, 1/2--3/4 in (1.3--2 cm) wide, borne single in large numbers at ends of twigs. Mature female cones large, elliptical, to 6 in (15 cm) long and 3--4 in (7.5--10 cm) in diameter. Cone-scales many, about 1 1/4 in (3 cm) wide. Seed 1 at base of cone-scale, with broad wings. Seedling with 4 narrow cotyledons about 1/16 in (1.5 mm) wide.

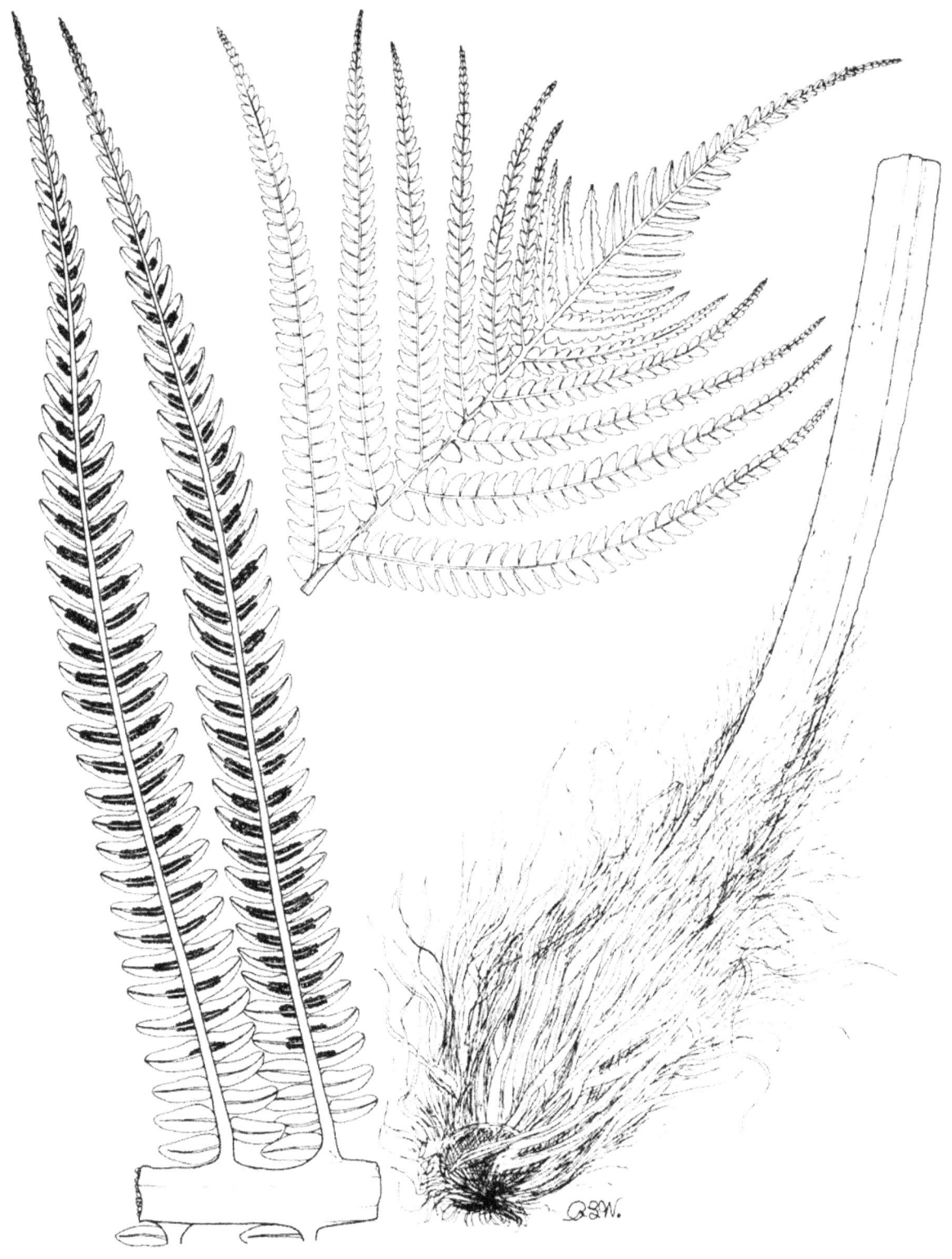

4. 'Ama'u, sadleria   *Sadleria cyatheoides* Kaulf.
Upper leaf surface (above), lower leaf surface (left), base of axis (lower right), 1 X.

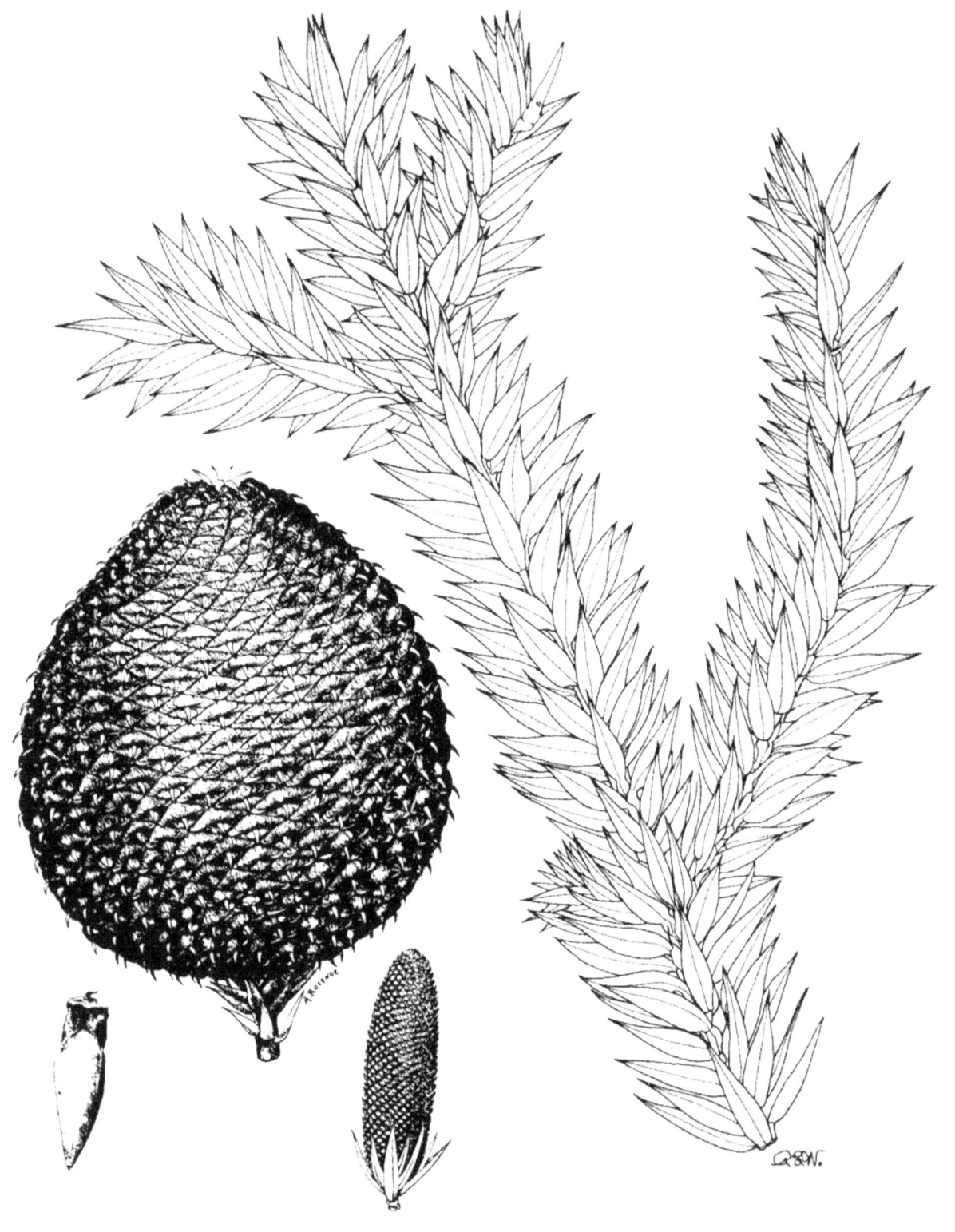

5. Parana-pine *Araucaria angustifolia* (Bert.) Kuntze*
Leafy twig, 1 X. Cone, cone-scale with seed, and male cone (lower left), 1 X (Barrett).

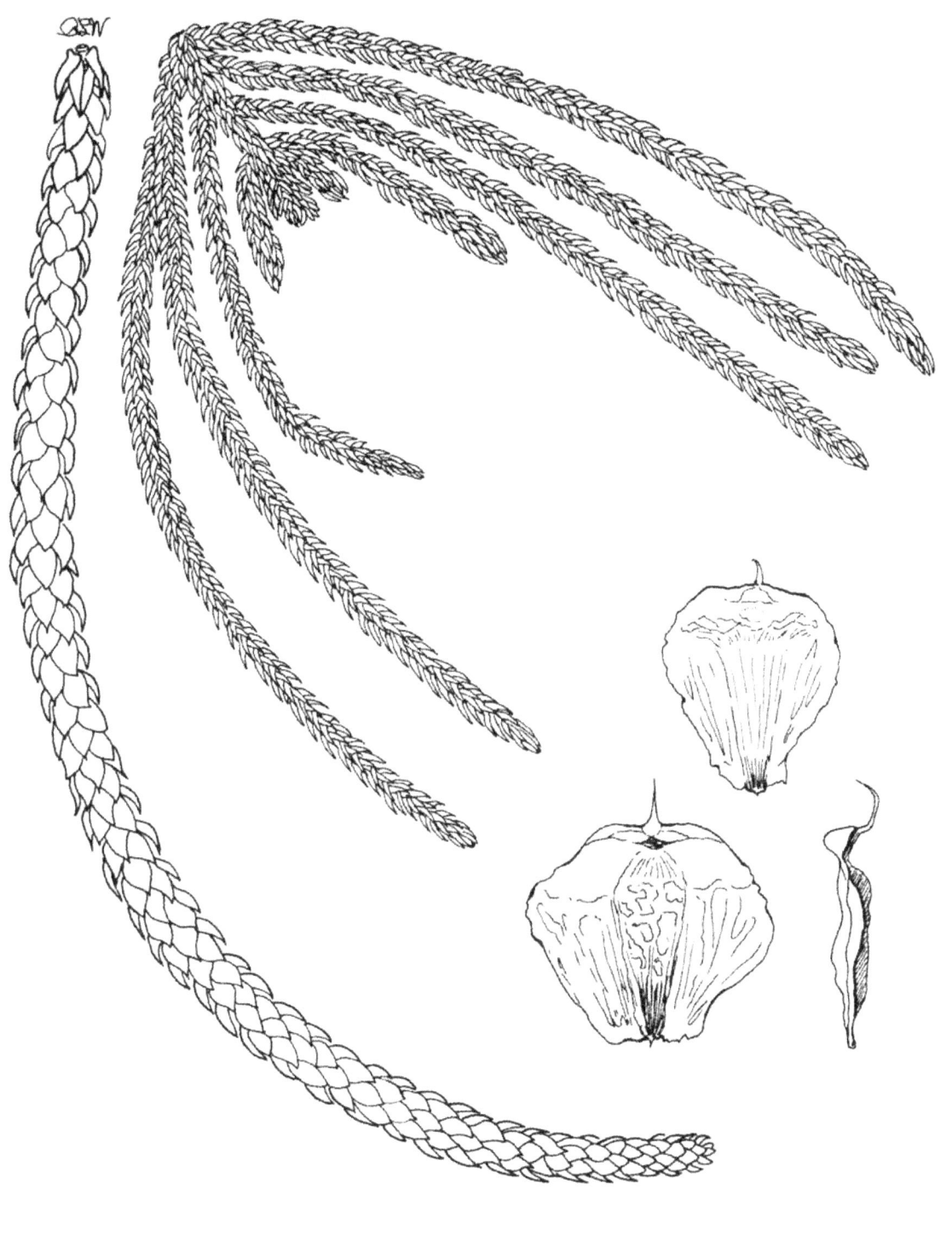

6. Columnar araucaria *Araucaria columnaris* (G. Forst.) Hook.*
Twig with scalelike leaves (left), twig with needlelike leaves (center), and seeds with cone-scale (lower right), 1 X.

The wood is a lustrous straw color. Tests of clear wood at the USDA Forest Service, Forest Products Laboratory have indicated that it has a strength approximately equal to that of Rocky Mountain Douglas-fir (*Pseudotsuga menziesii*). The lightweight wood (sp. gr. 0.44) is rarely available knot-free. It is used for attractive knotty pine paneling, turned bowls, and bracelets. For a short period when Hawaii had a veneer plant, it was used as veneer.

This tree is present on all islands, commonly known as "Norfolk-Island-pine" in Hawaii (see below). It may be seen above St. Louis Heights in Honolulu, at Schofield Barracks, and in Waiahole Valley on Oahu, at Lanai City, at Mahinahina near Lahaina, Maui, and at many other locations. Nearly 2 million board feet of this species have been cut on Hawaii and Maui for lumber and veneer production. It is also planted extensively in Christmas tree plantations and was in the recent past exported to the West Coast as a novelty Christmas tree.

Special areas--Aiea, Foster, Wahiawa.

Champion--Height 109 ft (33.2 m), c.b.h. 10.2 ft (3.1 m), spread 40 ft (12.2 m). Kukuihaele, Hamakua, Hawaii (1968).

Range--Native of New Caledonia and Isle of Pines.

Other common names--Cook araucaria, Cook-pine.

Botanical synonym--*Araucaria cookii* R. Br. ex Lindl.

Hawaii is a leader in supplying seed of "Norfolk-Island-pine" to other parts of the world. The seed supplied is *A. columnaris* rather than *A. heterophylla*, although possibly a little may be hybridized. Hybridization is unlikely because the two species shed pollen 6 months apart. Seed and foliage samples sent to Kew Gardens were all identified as *A. columnaris*. The very high percentage of albinism in the Hawaiian seed indicates serious inbreeding.

Identification of the seed is important. *Araucaria columnaris* does not grow with the erect habit that it attains in Hawaii and in its native New Caledonia when it is used as an ornamental in parts of Florida and Queensland, Australia. At those places, and perhaps others, it produced trees with very crooked, undesirable stems, and has a poor reputation for that reason.

It is said that, when discovered by Capt. James Cook in New Caledonia, the tall trees resembled pillars or columns of basalt from a distance. In its native habitat, but rarely in Hawaii, the lower branches shed and are replaced by short twigs forming a dense green column that widens abruptly near apex of the narrow crown.

In 1983, a hurricane hit Kauai and stripped all the branches from the narrow-crowned *A. columnaris* and the wider-crowned suspected *A. heterophylla*. When the branches sprouted back, they grew to the same length from top to bottom of the trees, producing columnar crowns regardless of the original crown shape. Thus, crown shape may be a result of wind damage.

Norfolk-Island-pine, *Araucaria heterophylla* (Salisb.) Franco* (*A. excelsa* (Lam.) R. Br.) is a closely related species very similar in appearance. However, it has a wider crown of coarser foliage and when open grown tends to have much longer branches and a more pyramidal appearance. Mature cones are rounded and often broader than long, 3--4 in (7.5--10 cm) long and 3 1/2--4 1/2 in (9--11 cm) in diameter. Seeds are larger, with swollen body to 3/8 in (1 cm) thick. Seedlings have 4 broad cotyledons 1/8--3/16 in (3--5 mm) wide. The species is native to Norfolk Island, which lies between New Zealand and New Caledonia in the South Pacific and is extensively cultivated in subtropical and tropical climates of the world and grown indoors in temperate regions.

Most trees called Norfolk-Island-pine in Hawaii, as well as seeds distributed from Hawaii under that name, apparently are columnar araucaria. Seeds of the two species are readily distinguished by size and the seedlings by cotyledon width.

7. Hoop-pine

Araucaria cunninghamii D. Don*

Hoop-pine or Moreton-Bay-pine, is a large introduced ornamental evergreen tree with straight axis, with a conical crown that becomes irregular and rounded with age. The branches spread widely in rings about 2 ft (0.6 m) or more apart, the lower horizontal and upper pointing upward, and the needlelike leaves are of 2 kinds. To 80 ft (24 m) and 1 1/2 ft (0.5 m) in trunk diameter, somewhat larger in age and in native home 100--200 ft (30--61 m) high and 2--6 ft (0.6--1.8 m) in diameter. Bark dark gray brown, smoothish, with horizontal lines suggesting hoops or bands, slightly scaly, peeling off in thin layers like birch, on small trunks gray and very smooth. Inner bark red, very hard, thick. Branches regular, several at a ring. Twigs long and very slender, often crowded and clustered or tufted, green, hairless, branching and spreading horizontally, shedding with attached leaves.

Leaves alternate and crowded, hairless, mostly spreading almost at right angle in 2 rows, needlelike, about 1/2 in (13 mm) or more in length or near forks becoming shorter and less than half as long, flattened and nearly 1/16 in (1.5 mm) wide, stiff, ending in very narrow point, extending down twig at base, shiny green above, shiny light green beneath. Leaves on uppermost branches and cone-bearing twigs spreading on all sides of twig, curving inward, shorter and 1/4--3/8 in (6--10 mm) long, short-pointed.

Male and female cones usually on different trees (dioecious). Male cones cylindrical, 1 1/2--3 in (4--7.5 cm) long and 1/2 in (13 mm) in diameter, at ends

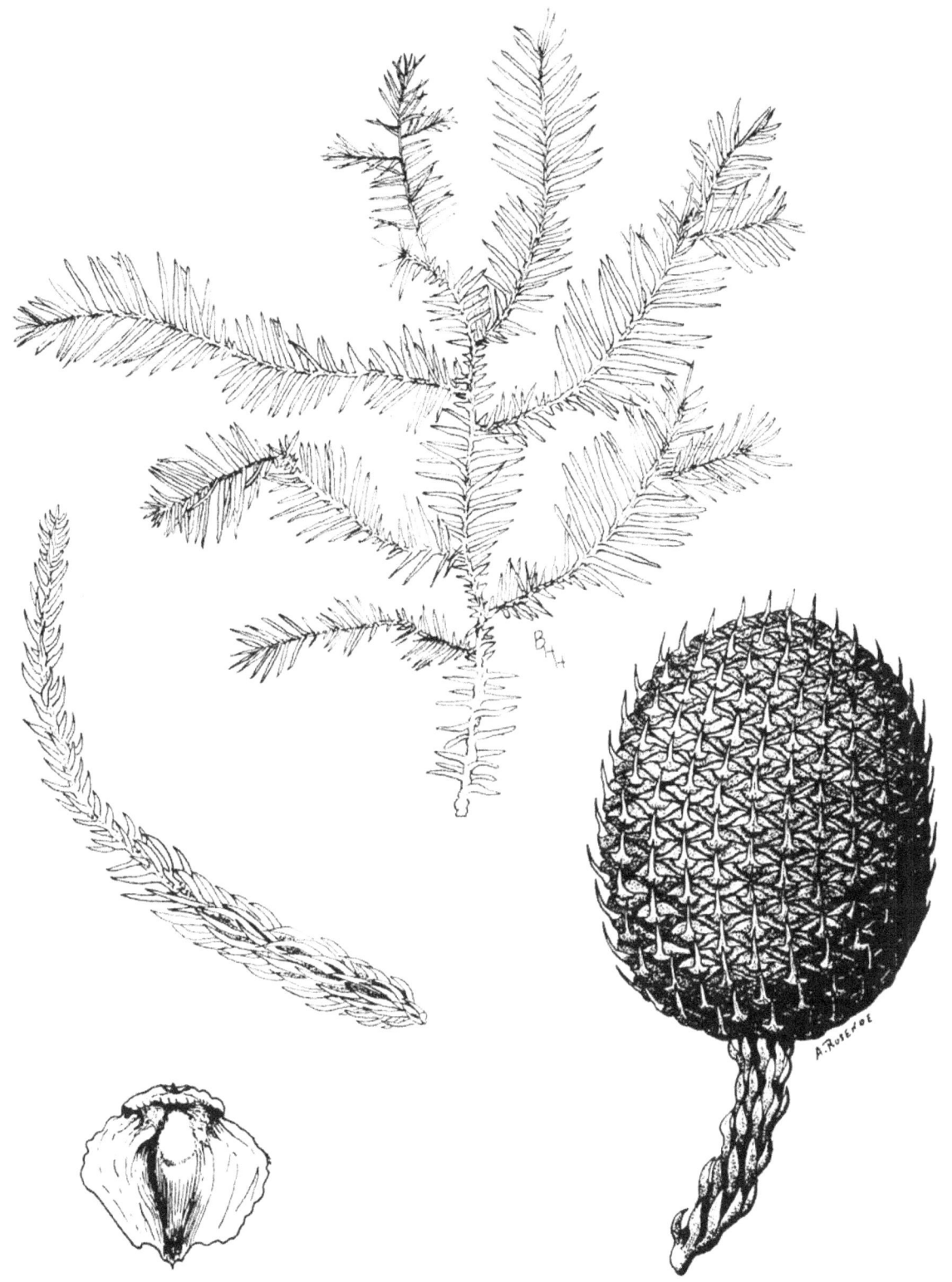

7. Hoop-pine   *Araucaria cunninghamii* D. Don*
Twig with needlelike leaves, twig with scalelike leaves (left). Cone-scale with seed (lower left), and cone (lower right), 1 X (Barrett).

of twigs. Female cones egg-shaped or rounded, 3--4 in (7.5--10 cm) long and 2--3 in (5--7.5 cm) in diameter, composed of many overlapping winglike cone-scales, ending in narrow stiff point curved backward, falling apart at maturity. Seed 1, about 3/8 in (1 cm long), attached to a cone-scale 1--1 1/4 in (2.5--3 cm) long and shedding together like a 2-winged seed.

The wood is pale yellow or whitish, lightweight (sp. gr. 0.42) fairly soft, straight-grained, easily worked, and readily stained. It is not resistant to decay or termites. In Australia, it is used for interior construction, including flooring and molding, furniture, veneers and plywood, and boxes and crates. It is not used in Hawaii because of the limited supply. A transparent whitish to yellowish resin is produced from wounds.

Introduced to Hawaii about 1880 as an ornamental and later used in forestry tests. A large individual on the grounds of Iolani Palace in Honolulu serves as the community Christmas tree. The Division of Forestry has planted 8,600 trees on the forest reserves on all islands but mostly on Kauai. This species is one of the most important native softwoods of Australia.

Special areas--Foster, Pepeekeo.

Champion--Height 101 ft (30.8 m), c.b.h. 7.6 ft (2.3 m), spread 28 ft (8.5 m). Foster Gardens, Honolulu, Oahu (1968).

Range--Native of eastern Australia. Scattered in wet forest of coastal ranges, subtropical with warm humid summers and mild winters. Planted as a forest tree in Australia and South Africa and as an ornamental through the tropics.

Other common names--Moreton-Bay-pine, Cunningham araucaria, Hawaiian-star-pine.

This species honors Allan Cunningham (1791--1839), British-born botanist who explored Australia.

PINE FAMILY (PINACEAE*)

8. Slash pine

Pinus elliottii Engelm.*

Large introduced narrow-leaf or needle-leaf evergreen tree of forest plantations. Trunk long and straight, to 90 ft (27 m) in height and 2 ft (0.6 m) in diameter, with branches in horizontal tiers or rings, more than 1 annually. Bark gray, very thick, cracking into long narrow plates and peeling off, exposing dark brown layer, becoming purplish brown with large flat scaly plates. Inner bark whitish beneath dead orange brown outer layer, resinous. Twigs stout, brown, rough, and scaly. Winter buds cylindrical, pointed, reddish brown, with fringed spreading scales.

Leaves or needles 2 or 3 in cluster, stiff, spreading, 6--12 in (15--30 cm) long, slightly shiny dark green, with many fine whitish lines on each surface, finely toothed edges, and sharp point. Sheath at base of leaves about 1/2 in (13 mm) long.

Cones few, conical or narrowly egg-shaped, symmetrical, 3--6 in (7.5--15 cm) long, maturing in 2 years and shedding, leaving a few basal scales attached. Cone-scales at exposed end shiny dark brown, 4-sided with stout short prickle. Seeds paired and exposed at base of cone scales, with egg-shaped, blackish mottled body 1/4 in (6 mm) long and detachable wing about 1 in (2.5 cm) long.

This species and longleaf pine (*Pinus palustris*) are leading producers of naval stores or oleoresins. The lumber of both serves elsewhere for miscellaneous factory and construction uses, flooring, railroad-cars, and ships. Other uses are poles, piling, and pulpwood. Some thinnings of the Hawaiian plantations have been used for fenceposts.

The wood of one tree grown at Lalakea, Hawaii, was tested at the Forest Products Laboratory. Also, an increment core survey was made of the wood specific gravity of several stands (Skolmen 1963). These studies indicated that the wood produced by the species in Hawaii is considerably less dense (sp. gr. 0.42) than is normal for the species in its native habitat (sp. gr. 0.54). Thus, the wood may have strength properties and pulp yields similar to those of western hemlock (*Tsuga heterophylla*), rather than those of the very strong southern-grown slash pine. Such strength properties would almost certainly be adequate for construction lumber.

Slash pine grows well in plantations in the Hawaiian Islands. It has been planted on eroded lands on Kauai and Molokai in mixture with other pines to control erosion. There are 1,100 acres (446 ha) of pine, mostly slash, in the Puu Ka Pele Forest Reserve on Kauai and 700 acres (283 ha) on the south-facing ridges of Molokai. In addition, this species has been planted at the Waiahou Spring Forest Reserve and Kula Forest Reserve on Maui and at several locations on the Island of Hawaii. It attains its best growth at 3,000--4,000 ft (914--1,219 m) elevation where the rainfall is about 60 in (1,524 mm), but it is also growing very well in two locations on stony organic muck soil in 200 in (5,080 mm) rainfall. On Molokai, it has been damaged by the fungus *Diplodia pinea*.

Range--Southeastern continental United States, Coastal Plain from southern South Carolina to southern Florida, also Lower Florida Keys, and west to southeastern Louisiana.

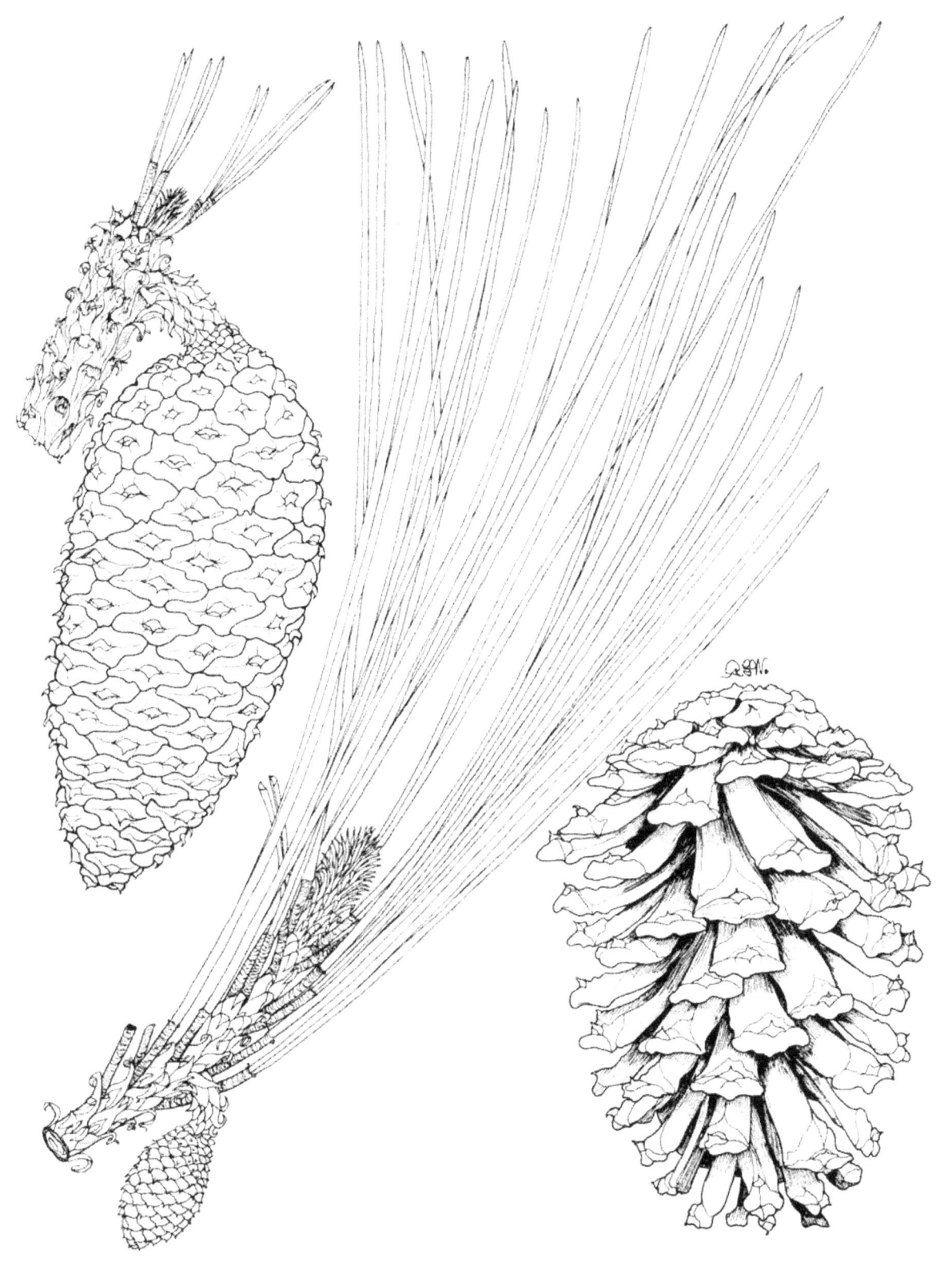

8. Slash pine *Pinus elliottii* Engelm.*
Closed cone (upper left), leafy twig with year-old cone, and open cone (lower right), 1 X.

9. Jelecote pine

Pinus patula Schiede & Deppe*

Large introduced narrow-leaf or needle-leaf evergreen tree of graceful shape, in forest plantations on moist slopes, with long slender drooping or "weeping" needles, many clustered cones remaining attached, and reddish bark on upper trunk and branches. To 95 ft (29 m) or more in height, with straight or forked trunk 1 1/2 ft (0.5 m) in diameter. Bark with scattered warts, smoothish, papery reddish brown, becoming thick, rough, furrowed, and gray. Inner bark light brown, fibrous, spreading, resinous. Branches long and spreading, more than 1 tier or ring annually. Young twigs slender, with whitish bloom, hairless, becoming reddish brown. Winter buds cylindrical, long-pointed, 1/2--1 in (1.3--2.5 cm) long, with long-pointed fringed spreading scales.

Leaves or needles usually 3 in cluster, sometimes 4--5, shiny green, yellow green when young, very slender, drooping, 4--8 in (10--20 cm) long, only 1/32 in (1 mm) wide, with many fine whitish lines on each surface, finely toothed edges. Sheath at base of leaves 1 in (2.5 cm) long.

Cones crowded in clusters of 5--10, narrowly conical, 3--4 1/2 in (7.5--11 cm) long, shiny dark brown, 1-sided or oblique at base, stalkless or nearly so, bent down, maturing in 2 years, remaining attached with scales closed. Cone-scales with exposed end flattened and with tiny shedding prickle. Seeds paired and exposed at base of cone-scales, with triangular mottled gray black body less than 1/4 in (6 mm) long and wing 1/2--3/4 in (13--19 mm) long.

The specific gravity of the wood of this species was found to be 0.44 in an increment core survey (Skolmen 1963). This was denser than trees grown in New Zealand but less dense than trees grown in South Africa. The species is very important for pulp and lumber in South Africa and Swaziland but has not been utilized in Hawaii.

This subtropical pine has been planted at Waihou Spring Forest Reserve, altitude 3,800 ft (1,158 m) and the Kula Forest Reserve at 5,500 ft (1,676 m) on Maui, and at several locations on the Island of Hawaii. Trees from all the provenances tried have grown with a very steep branch angle and usually with multiple leaders. Because of its generally poor form as a timber tree, this species has not been used in recent plantings.

Champion--Height 108 ft (32.9 m), c.b.h. 6.4 ft (2.0 m), spread 27 ft (8.2 m). Kula Forest Reserve, Maui (1968).

Range--Native of mountains of eastern Mexico (Tamaulipas to Oaxaca).

Planted as an ornamental in subtropical regions, it is also so used on Maui in upland gardens.

Other common name--Mexican weeping pine.

10. Cluster pine

Pinus pinaster Ait.*

Large introduced narrow-leaf or needle-leaf evergreen tree of forest plantations, with paired long stout needles, clustered large brown cones remaining attached, and thick rough bark. To 80 ft (24 m) or more in height with long straight trunk 2 1/2 ft (0.8 m) in trunk diameter, usually without branches except near top. Bark very thick, rough, deeply furrowed into long narrow ridges, blackish, in upper part or on smaller trees smoothish and gray. Branches horizontal, usually 1 tier or ring annually. Twigs light brown, hairless, becoming ridged and rough from bases of scale leaves. Winter buds large, 3/4--1 in (2--2.5 cm) long, long-pointed, with whitish-brown fringed spreading scales.

Leaves or needles 2 in cluster, dull gray green, stout, stiff, 4--8 in (10--20 cm) long, flattened and more than 1/16 in (1.5 mm) wide, with many fine whitish lines on each surface, finely toothed edges, and blunt point. Sheath at base of leaves 1 in (2.5 cm) long.

Cones 1 to several clustered in ring, egg-shaped, 3--5 in (7.5--13 cm) long, shiny brown, symmetrical or nearly so, nearly stalkless, spreading or bent down, maturing in 2 years, opening or sometimes closed, remaining attached. Cone-scales with exposed end 4-angled and bearing a stout blunt point. Seeds paired and exposed at base of cone-scales, with elliptical body 3/8 in (1 cm) long and wing 1--1 1/2 in (2.5--4 cm).

The wood is moderately heavy and hard, coarse-textured, and resinous. An increment core survey of wood density indicated that Hawaii-grown wood had a specific gravity of 0.49. It produces true annual rings in Hawaii. Sapwood pale yellow to whitish; heartwood reddish brown. Uses elsewhere include construction, boxes, mine timbers, utility poles, and railway crossties. Chiefly in western France, large forest areas are managed for production of oleoresins or naval stores by tapping operations. Resin is extracted also from stumps and waste wood by destructive distillation.

The trees thrive in sandy soils in warm climates and have served to reclaim large areas of sand dunes along coasts of France and Portugal.

This pine has been planted at Waihou Spring Forest Reserve, altitude 3,800 ft (1,158 m) and at the Kula Forest Reserve at 6,200 ft (1,890 m) on Maui. It has also been planted on the leeward slopes of Molokai and Kauai as well as several locations on Hawaii. Trees have been severely damaged by the Eurasian pine adelgid ("aphid") (*Pineus pini*) on Molokai, Maui, and the Island of Hawaii, and are also very susceptible to a fungus

9. Jelecote pine　　*Pinus patula* Schiede & Deppe*
Leafy twig with male cones and twig with cone (lower left), 1 X.

(*Diplodia pinea*), which has killed many trees on Molokai. Because of these attacks and the relatively slow growth rate as compared to the southern pines, cluster pine will not be used in future plantings.

Champion--Height 90 ft (27.4 m), c.b.h. 9.9 ft (3.0 m), spread 54 ft (16.5 m). Waihou Springs Forest Reserve, Olinda, Maui (1968).

Range--Native of western Mediterranean region from Portugal, Spain, and Morocco east to southern France, Corsica, and eastern Italy, greatly extended beyond by cultivation.

Other common names--maritime pine, seaside pine.

11. Monterey pine

Pinus radiata D. Don*

Large introduced narrow-leaf evergreen tree of forest plantations. Trunk straight, with irregular branches in 2 or more tiers or rings annually and growing angled slightly to strongly upward. To 150 ft (46 m) in height and 4 ft (1.2 m) in diameter. Bark gray, when young with prominent horizontal lines (lenticels); furrowed into flat ridges, the branches smoothish. Inner bark with reddish-brown and light brown layers, resinous. Twigs hairless. Winter buds 1/2--3/4 in (13--19 mm) long, short-pointed, brown, resinous, with closely pressed scales.

Leaves or needles 3 in cluster, crowded, slender, 4--6 in (10--15 cm) long, shiny green, with many fine whitish lines on each surface, finely toothed edges, sharp-pointed. Sheath at base of leaves 3/8--1/2 in (10--13 mm) long.

Cones usually 3--5 around branch in 1 or more clusters annually, egg-shaped, pointed at apex, very 1-sided or oblique at base, almost stalkless, turned back, 3--5 1/2 in (7.5--14 cm) long, shiny light brown, maturing in 2 years and remaining attached and usually closed on tree many years. Cone-scales with exposed end thick, hard, raised and rounded, those on outer side very large, bearing minute prickle. Seeds paired and exposed at base of cone-scales, with elliptical blackish body 1/4 in (6 mm) long and wing about 1 in (2.5 cm).

The wood is pale brown, slightly hard, and resinous, usually with very indistinct growth rings. The specific gravity based on an increment core survey was 0.46, the same as that of trees grown in California. This density indicates that the wood should be suitable for most structural purposes. The generally steep branch angle and poor form would produce low lumber grade yield if the species were sawn.

This subtropical pine has been used in many forest plantations on all the islands of Hawaii except Oahu. Planted at 5,200 ft (1,585 m) at Kulani Honor Camp southwest of Hilo, Hawaii, it has done quite poorly because of frequent defoliation from volcanic sulfur fumes. Near Waikii, Hawaii, it is doing well except for occasional damage by the Eurasian pine adelgid. At the Kula Forest Reserve at 6,000 ft (1,829 m) on Maui, it is presently doing well after being set back by the Eurasian pine adelgid, which is now believed to be controlled by an introduced insect parasite. On Molokai, most of the Monterey pine has been killed by a fungus (*Diplodia pinea*). Because of its high susceptibility to this disease, to wind damage, and to adelgid damage, it has been deleted from future planting plans. These problems are unfortunate because the tree grows better over a wider elevational range than any other pine planted in Hawaii, 1,500--7,500 ft (457--2,286 m). It also has a tendency to foxtail or form very long unbranched leaders that have slow diameter growth. In at least one area, Kialialinui, Maui, growth equal to the best in New Zealand was measured in 14-year-old trees (Skolmen 1963).

Special areas--Waihou, Kula.

Range--Very rare and local at 3 localities on coast of central California (San Mateo, Santa Cruz, Monterey, and San Luis Obispo Counties). Also a var. on Guadalupe Island, Mex.

Other common name--insignis pine.

Botanical synonym--*Pinus insignis* Dougl. ex Loud.

The most common pine in forest plantations through the southern hemisphere for pulpwood and lumber, particularly in Australia, New Zealand, and South Africa. Grown also for shade and ornament.

12. Loblolly pine

Pinus taeda L.*

Large introduced narrow-leaf or needle-leaf evergreen tree of forest plantations on moist slopes, 70 ft (21 m) or more in height; with long straight trunk 1 ft (0.3 m) or more in diameter. Bark gray, thick, rough, deeply furrowed into long narrow scaly plates. Inner bark within thick dead outer layer, light brown, fibrous, resinous. Branches horizontal, usually 2 or more tiers or rings annually. Twigs hairless, whitish, becoming brown and ridged. Winter buds conical, 1/4--1/2 in (6--13 mm) long, light brown, with fringed spreading scales.

Leaves or needles 3 in cluster, stiff, spreading, 6--9 in (15--23 cm) long, with many fine whitish lines on each surface, finely toothed edges, and sharp point. Sheath at base of leaves nearly 1 in (2.5 cm) long.

Cones several in 2 or more groups, conical, symmetrical nearly stalkless, 3--5 in (7.5--13 cm) long, sym-

10. Cluster pine *Pinus pinaster* Ait.*
Leafy twig (left), twig with male cones (right), and cone (below), 1 X.

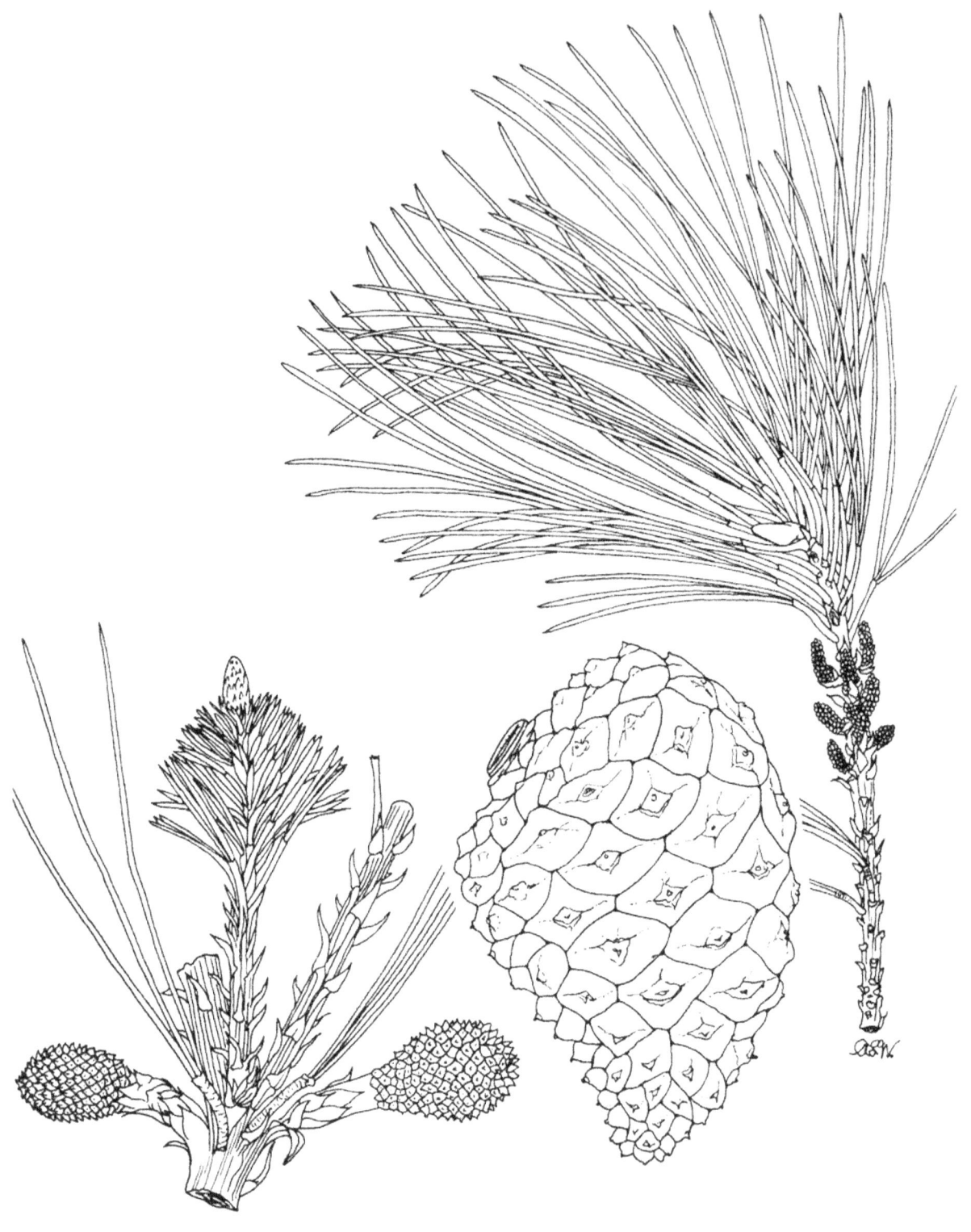

11. Monterey pine   *Pinus radiata* D. Don*
Foliage and male cones (upper right), year-old cones (lower left), and mature cone (lower right), 1 X.

12. Loblolly pine   *Pinus taeda* L.*
Leafy twig and cone (lower right), 1 X.

metrical, dull brown, maturing in 2 years and shedding. Cone-scales 4-sided at exposed end, with ridge and stout curved spine. Seeds paired and exposed at base of cone-scales, with mottled dark brown body 1/4 in (6 mm) long and wing about 1 in (2.5 cm) long.

The wood grown in Hawaii has very broad bands of summerwood and because of its fast growth, wide growth rings. Sapwood yellowish, heartwood light brown. Its specific gravity (0.42) found by increment core survey is somewhat less than that of wood grown in the southeastern states, where it is the principal softwood, but still adequate for structural strength. Used as lumber on the mainland for building material including millwork, boxes and crates, agricultural implements, and low-grade furniture. It is also an important plywood species. Veneer is used for containers. This and other southern pines are the leading native pulpwoods and leading woods in production of slack cooperage.

This pine is being tested in a large spacing study at Waihou Spring Forest Reserve, altitude 3,800 ft (1,158 m) on Maui. It has also been planted along with *Pinus elliottii* on Kauai, Molokai, and Hawaii.

Range--Southeastern continental United States, Coastal Plain, and Piedmont from southern New Jersey and Delaware south to central Florida and west to eastern Texas, and north in Mississippi Valley to extreme southeastern Oklahoma, central Arkansas, and southern Tennessee.

Other common names--oldfield pine, shortleaf pine.

REDWOOD FAMILY (TAXODIACEAE*)

13. Sugi, cryptomeria

Cryptomeria japonica (L. f.) D. Don*

Introduced aromatic evergreen tree with small awl-shaped or needlelike leaves. In Hawaii, it grows to 80 ft (24 m) in height and 1.5 ft (0.5 m) in diameter. Crown cone- or pyramid-shaped or irregular, narrow. Trunk straight, tapering from enlarged base. Branches spreading horizontally. Bark reddish brown, with long fissures, smoothish to slightly shaggy, fibrous, peeling off in long shreds. Inner bark light pink within red-brown outer layer, resinous. Twigs very slender, dull green, hairless, spreading or drooping, mostly shedding with leaves after a few years. Buds minute.

Leaves borne singly (alternate) and spreading on all sides of twig, awl-shaped or needlelike, 1/4--5/8 in (6--15 mm) long, curved forward and inward, slightly stiff and blunt-pointed, slightly flattened and 4-angled, dull blue green, hairless, base extending down twig. Flushes of new growth in winter bright orange.

Males cones in clusters of 20 or more at leaf bases near ends of twigs, oblong, 1/4--3/8 in (6--10 mm) long, orange or reddish. Female cones on same tree, single on short leafy stalks, rounded, 1/2--3/4 in (13--19 mm) in diameter, dull brown, opening at maturity the first year, but remaining attached many many years. Cone-scales 20--30, outer surface flattened and angled with 3-5 spinelike points at top and another curved down from center. Seeds 2--5 at base of cone-scale, slightly 3-angled, less than 1/4 in (6 mm) long, dark brown, with narrow wings on edges. Sometimes a leafy twig grows from apex of the slightly pointed cone.

The wood is soft, lightweight, aromatic, easily worked, and durable. The sapwood is whitish or yellow, and heartwood reddish brown. Wood grown in Hawaii has a specific gravity of 0.41 and thus should have strength properties similar to western redcedar (*Thuja plicata*). One of the most important timbers of Japan, it is used for construction, paneling, furniture, and boxes there. In Hawaii, where there are 2.5 million board feet, the wood has so far been used only for fence posts.

Cryptomeria is the national tree of Japan, where it is important in forestry plantations, and is widely planted as a tree and ornamental, and is found about temples. Alone in its genus, this species has many horticultural varieties.

Forest plantations have been established in moist middle altitudes of Hawaii, mostly at 2,500--6,000 (762--1,829 m) on Kauai, Maui, and the Island of Hawaii. Trees may be seen at Kokee on Kauai, on the Kula and Waihou Spring Forest Reserves on Maui, and along the old Volcano Road at Volcano Village, Hawaii. The best stand in the State of Hawaii is on the land of Papa in South Kona, where the trees were planted in the late 1880's. About 500,000 trees were planted by the Division of Forestry at various locations on the Forest Reserves between 1910 and 1960. Grown also as an ornamental and windbreak.

Champion--Height 105 ft (32.0 m), c.b.h. 8.8 ft (2.7 m), spread 18 ft (5.5 m). Hoomau Ranch, Honomolino, Hawaii (1968).

Range--Japan and China. Widely planted in warm temperate regions, including Europe, eastern United States north along coast to New York City and Boston, and in Pacific States.

Other common name--Japanese-cedar.

13. Sugi, cryptomeria *Cryptomeria japonica* (L. f.) D. Don*
Twig with cone and male cones, 1/2 X; cone (lower left), 2 X; and male cone (lower right), 5 X (Degener).

14. Redwood
Sequoia sempervirens (D. Don) Endl.*

Large to very large introduced aromatic evergreen tree with short needlelike leaves, in Hawaii reaching about 110 ft (34 m) in height. Trunk straight, becoming enlarged at base, to 3.5 ft (1 m) in diameter, with narrow irregular crown of slender, slightly drooping branches. Bark rough, very thick, fibrous, corky or spongy, dark brown, deeply furrowed into long narrow ridges, exposing reddish brown layers. Inner bark light brown, fibrous, slightly bitter. Twigs very slender, angled, light green, turning brown, ending in scaly brown buds 1/8 in (3 mm) long, mostly shedding with leaves.

Leaves spreading in 2 rows, 1/4–1/2 in (6–13 mm) long flattened and nearly 1/8 in (3 mm) wide, stalkless with twisted base onto twig, short-pointed, with sunken midvein, slightly shiny dark green above and with whitish lines beneath, gradually shorter toward both ends of twig. Leaves on leading and cone-bearing twigs slightly spreading around twig, scalelike and shorter, only 1/4 in (6 mm) long, sharp-pointed.

Seed-bearing or female cones drooping at ends of slender twigs, oblong, or egg-shaped, 3/4–1 in (2–2 1/2 cm) long, hard and woody, reddish brown, maturing and opening the first year, remaining attached. Cone-scales 14–20, the ends 4-sided with horizontal ridge and raised point in center. Seeds 3–7 exposed at base of each cone-scale, about 1/16 in (1.5 mm) long, light brown, with narrow wings. Male or pollen cones on same tree, oblong, 5/16 in (8 mm) long, borne singly at end of leafy twig, short-stalked.

The heartwood is reddish, and the sapwood light yellow. The wood is lightweight (sp. gr. 0.36), soft, fine-textured, and easily worked. When grown in its natural habitat, it is very durable in the soil and resistant to termites. Wood grown in Hawaii is very fast-grown and lacks resistance to decay and termites. The Hawaii-grown wood has been tested by the Forest Products Laboratory and found to be similar to California-grown second growth redwood in most properties (Youngs 1960). Redwood trees in Hawaii are quite limby, thus supplying only low-grade lumber suited for construction. In Hawaii, redwood is traditionally used as clear, all-heart lumber for house siding. The locally grown trees will not supply this product, nor will they supply wood with the durability for which redwood is noted. Construction lumber and fencing will be the main uses for Hawaii-grown redwood.

Redwood has been planted in many Forest Reserves. Maui has more than 280 acres (113 ha) with about 7 million board feet in the Kula Forest Reserve at 5,500 ft (1,676 m). The tree may be seen at Kokee State Park on Kauai, Waihou Spring Forest Reserve on Maui, and near Volcano Village on Hawaii. It is also planted at about 5,500 ft (1,676 m) on the land of Piha in the Hilo Forest Reserve and at 3,000 ft (914 m) in the Honaunau Forest Reserve. Except in the Kula Forest Reserve on Maui, where it attains a good form, redwood in Hawaii is very strongly tapered in stem form.

Redwood is planted widely as an ornamental and shade tree in moist tropical regions. Pieces of burls placed in water will sprout and produce house plants for temperate regions.

This species is the world's tallest tree. The tallest, in northwestern California, attains a height of about 368 ft (112 m). California redwood, also including giant sequoia, or Sierra redwood, (*Sequoiadendron giganteum* (Lindl.) Buchholz), is the State tree of California.

Range--Pacific Coast region from extreme southwestern Oregon south to central California.

Other common names--coast redwood, California redwood.

CYPRESS FAMILY (CUPRESSACEAE*)

15. Arizona cypress
Cupressus arizonica Greene*

Medium-sized introduced aromatic evergreen tree with small scalelike leaves. Trunk straight, to 40 ft (12 m) high and 1 ft (0.3 m) in diameter, with conical or rounded crown of spreading branches. Bark commonly reddish brown, smoothish or scaly, peeling off in thin scale plates; in one variety rough, thick, furrowed, gray or blackish. Twigs numerous, 4-angled, branching nearly at right angles in all directions, green, becoming dark brown.

Leaves paired, scalelike, overlapping in 4 rows against twig, blunt-pointed, 1/16 in (15 mm) long, pale blue green, often with gland dot and drop of resin on back.

Seed-bearing (female) cones short-stalked, rounded, 3/4–1 in (2–2.5 cm) in diameter, hard and woody, gray brown, remaining attached several years. Cone-scales 6 or 8, shield-shaped, with raised point in center. Seeds about 100 per cone, oblong or 3-angled, brown, about 1/8 in (3 mm) long, with narrow wing. Pollen cones (male) on same tree, more than 1/8 in (3 mm) long, yellow.

Wood moderately soft and lightweight; heartwood, light brown, fragrantly scented, and very durable. Else-

14. Redwood *Sequoia sempervirens* (D. Don) Endl.*
Leafy twig, 1 X; and cone (lower right), 2/3 X.

where used for fenceposts. In Hawaii, wood not widely known or distinguished from other cypresses and not used.

This handsome tree with blue green foliage and scaly reddish brown bark has been planted elsewhere in the world as an ornamental and in shelterbelts. It is cultivated also as a Christmas tree in southeastern States. In Hawaii, a total of 10,300 trees were planted between 1910 and 1960. L. W. Bryan (1947) reported that it was growing well at low and middle elevations in Hawaii. It may be seen at Waihou Spring Forest Reserve on Maui, where 80 trees were planted in 1936. In 1939, the largest planting of this species, 3,200 trees, was in the Kula Forest Reserve on Maui.

Range--Southwestern continental United States from trans-Pecos Texas to Arizona and southern California and in northern Mexico (5 varieties or closely related species).

Other common names--Arizona rough cypress, Arizona smooth cypress.

The commonly cultivated variation with reddish brown scaly bark is Arizona smooth cypress (var. *glabra* (Sudw.) Little).

16. Mexican cypress

Cupressus lusitanica Mill.*

Mexican cypress is an introduced conifer of gardens and old farm lots and also of forest plantations, where it has been tested. The paired scalelike leaves on 4-angled twigs have an inconspicuous gland dot on back, and the rounded, hard, brown cones are about 5/8 in (15 mm) in diameter.

A medium-sized evergreen tree, resinous and aromatic, becoming 40 ft (12 m) high and 1 1/2 ft (0.5 m) in trunk diameter, with straight erect axis and dense regular or narrow crown of green to dark green foliage. Bark reddish brown, smooth to fissured, and sometimes with a few scales or shreddy. Inner bark whitish, slightly fibrous, and slightly resinous. Leafy twigs alternate, numerous, spreading, branching regularly in 4 rows, less than 1/16 in (1.5 mm) in diameter. Older twigs reddish brown, rough and scaly, with dead leaves persistent.

Leaves crowded, opposite in 4 rows, mostly dark green, scalelike, pressed against twig, less than 1/16 in (1.5 mm) long, short-pointed, angled or keeled, with inconspicuous gland dot, covering twig and shedding together. The foliage has a resinous odor and taste.

Pollen and seeds are borne on the same tree (monoecious). Male cones numerous toward the apex of short branches, cylindric, 3/16 in (5 mm) long and less than 1/8 in (3 mm) broad, greenish yellow, the scales in 4 rows bearing pollen sacs and pollen. Female cones (strobili) begin as a few inconspicuous green scales less than 1/8 in (3 mm) across, with naked ovules, at the end of short twigs. At maturity the second year, the hard woody cone, about 5/8 in (15 mm) in diameter, changes from whitish green to dull brown. It is composed mostly of 8 rounded but angular cone-scales, pressed together at edges, each with a stout central raised point about 1/8 in (3 mm) high. Later, the cone opens to free numerous brown seeds more than 1/8 in (3 mm) long, irregularly flattened with borders slightly winged.

The sapwood is pale brown and the heartwood pinkish brown, not distinct from the sapwood. The wood is a moderately lightweight softwood (sp. gr. 0.40), fine-textured and fragrant, but not durable (Scott 1953). Used for construction and furniture elsewhere, but not in Hawaii.

Occasionally, Mexican cypress has been planted in the Forest Reserves. These plantings total only 3,700 trees for the entire State. The Division of Forestry in 1929 planted 62 trees on Kapapa Island off Oahu and 100 trees at Kahuku also on Oahu. Another planting site was Waihou Spring Forest Reserve on Maui. This species is suited primarily to lower elevations and is reported by Bryan (1947) to grow very rapidly. It is used primarily as an ornamental and windbreak in Hawaii. Trees seen in old homesteads often were planted by Portuguese settlers along with *Cupressus macrocarpa* to remind them of Portugal and the Azores.

For ornament, the plants can be pruned in different shapes or trimmed as living hedges. Small symmetric plants 3--6 ft (0.9--1.8 m) high would serve as attractive Christmas trees and could be grown in plantations for this purpose. The leafy branches also serve for decorations and wreaths. The trees are subject to windthrow on poorly drained soils.

Champion--Height 45 ft (13.7 m, c.b.h. 3.3 ft (1.0 m), spread 32 ft (10.0 m). Honaunau Forest Reserve, Kailua-Kona, Hawaii (1968).

Range--Native in mountains of Mexico, Guatemala, El Salvador, and Honduras. Widely spread in cultivation and naturalized southward in mountains of Central America to Costa Rica, in Andes from Colombia and Venezuela to Argentina and Chile and in the Old World. Introduced in northern Florida. In Kenya it is grown extensively and has become an important timber tree. Common in South Africa.

Botanical synonym--*Cupressus benthamii* Endl., *C. lindleyi* Klotzsch.

This variable species consists of several forms. One has drooping or weeping branches and another a very long and narrow columnar crown. The scientific name, meaning "of Lusitania," an old name of Portugal, was based on planted trees in that country before the origin was definitely known.

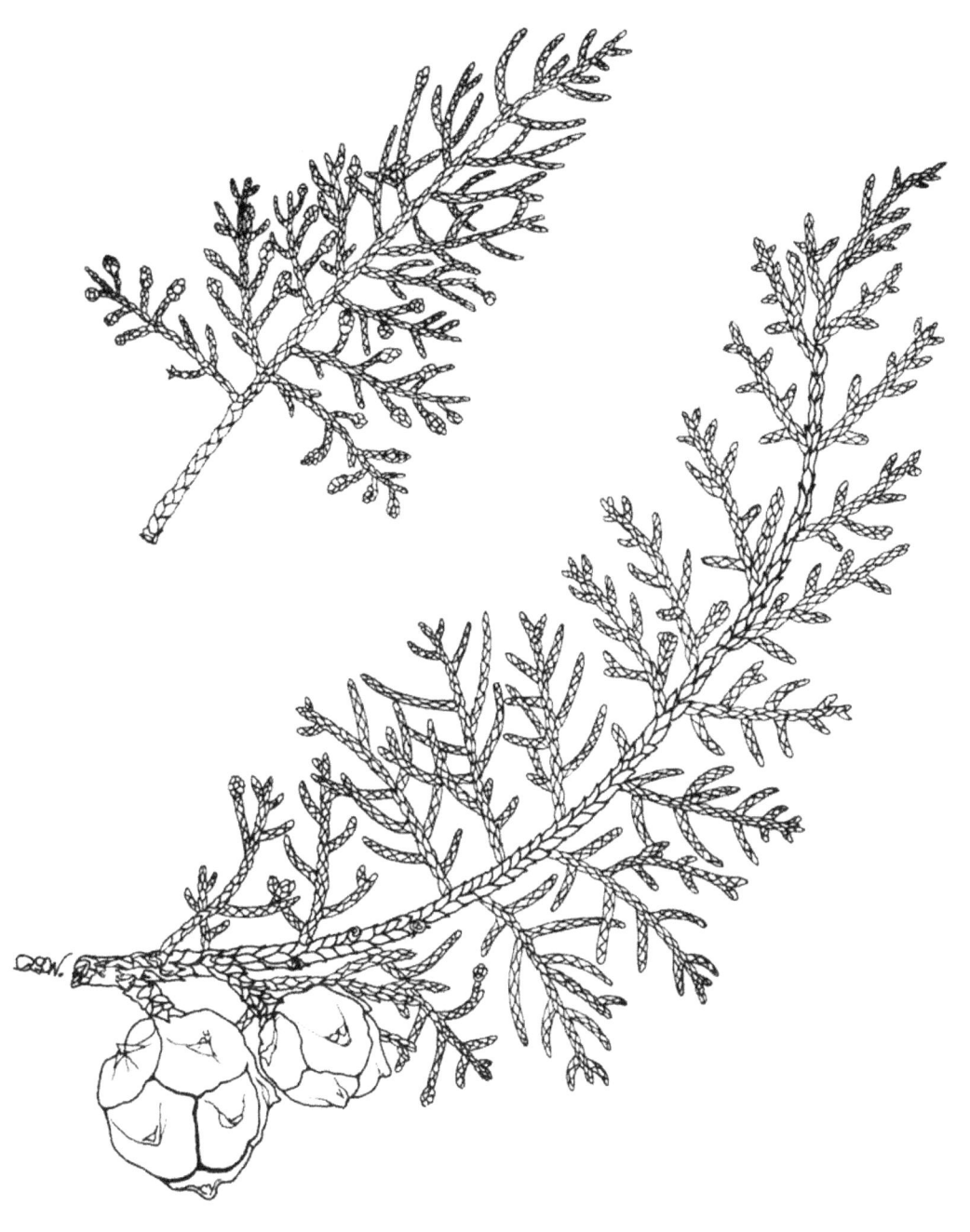

15. Arizona cypress *Cupressus arizonica* Greene*
Twig with male cones (upper left) and twig with cones, 1 X.

17. Monterey cypress

Cupressus macrocarpa Hartw.*

Medium-sized to large introduced aromatic evergreen tree with small scalelike leaves. Trunk straight, to 110 ft (34 m) high and commonly 2 1/2 ft (0.8 m) in diameter, with conical or spreading crown. Bark dark gray, rough, thick, furrowed into flat ridges. Inner bark light brown with outer dark brown layer, slightly bitter, resinous. Twigs slender, 4-angled.

Leaves paired, scalelike, overlapping in 4 rows against twig, blunt-pointed, 1/16--1/8 in (1.5--3 mm) long, dark green, without gland.

Female (seed-bearing) cones clustered on short stalks, rounded, 1--1 1/2 in (2.5--4 cm) in diameter, brownish, maturing second year and remaining attached. Cone-scales 8--14, shield-shaped, rounded at edges, with raised point in center. Seeds about 140 per cone, 3/16 in (5 mm) long, light brown, angled, with several gland-dots and with narrow wing. Male (pollen) cones on same tree, oblong, 1/8 in (3 mm) long, yellow.

Heartwood, yellowish brown, is not readily distinguished from the sapwood. The fine-textured, slightly aromatic wood is moderately lightweight (sp. gr. 0.40). Strong and suited to most types of construction, but reported to lack durability in ground contact (Streets 1962). A small amount of lumber has been sawn in Hawaii, and the wood was used for general construction. Butt logs of large trees are relatively knot-free, but are deeply fluted, making lumber grade yield quite poor.

This species, distinguished by the larger cones, is the most common cypress in Hawaii and is most frequently seen as a large tree at old homestead sites in the uplands. Widely planted in the Forest Reserves on all islands: more than 216,000 trees were planted between 1910 and 1960. It grows well at 1,500--5,000 ft (457--1,524 m) altitude in most areas and may be seen at Kokee, Kauai, Waihou Spring (Olinda), Maui, and at Waimea and Volcano Village on Hawaii.

Elsewhere, the trees have been planted widely for ornament, hedges, and windbreaks, not only in the Pacific States, but in warm temperate parts of Europe, South America, Australia, and New Zealand. However, this species is no longer recommended in the Pacific States because of a fungus disease.

Champion--Height 85 ft (25.9 m) c.b.h. 27.6 ft (9.4 m), spread 81 ft (24.7 m). Waikoekoe, Waimea Village, Hawaii (1968).

Range--Very rare and local on Pacific coast near Monterey, California.

18. Italian cypress

Cupressus sempervirens L.*

Italian cypress is a handsome introduced conifer occasionally planted for ornament and in forestry tests. Columnar Italian cypress (var. *stricta* Ait.), the common variety, is readily identified by the very narrow columnar crown. The paired scalelike leaves in 4-angled twigs have a gland dot on back, and the rounded hard gray or brown cones are relatively large, 1--1 1/4 in (2.5--3 cm) in diameter.

Medium-sized coniferous evergreen tree, resinous and aromatic, becoming 60 ft (18 m) tall and 1 1/2 ft (0.5 m) in trunk diameter. In its native home this is a large tree with stout trunk and spreading branches. The variety commonly cultivated has short erect branches. Bark brown, smoothish, becoming rough and fissured. Inner bark light brown with outer reddish layer, fibrous, resinous and bitter. Leafy twigs alternate, numerous, crowded and spreading, much branched, slender, less than 1/16 in (1.5 mm) in diameter.

Leaves scalelike, paired in 4 rows, closely pressed against twig, blunt, with gland dot in groove on back, about 1/32 in (1 mm) long, dull green, covering twig and shedding together.

Male and female cones borne on the same tree (monoecious), the male elliptical, more than 1/8 in (3 mm) long, yellowish. Female cones short-stalked, woody, composed of 8-14 flat cone-scales irregularly 5--6 sided, with a short point or knob in center. Seeds 8--20 at base of each cone-scale, 3/16 in (5 mm) long, elliptical, flattened, angled, and slightly winged, brown.

The wood is described as yellow or light brown, moderately hard, fine-textured, aromatic, easily worked, and durable. Elsewhere, it has been used for construction, furniture, and clothes chests. It has not been used in Hawaii.

This species is very common in gardens throughout Hawaii. It has also been tested on a small scale in the Forest Reserves. A total of 3,000 trees were planted in the forests between 1923 and 1955. Examples may be seen at Kokee, Kauai, and in the Olaa Forest Parks along the Volcano Road on Hawaii.

This classic cypress of ancient Greece and Rome is conspicuous in formal gardens and cemeteries of southern Europe. It is hardy in subtropical and warm temperature climates, such as southern continental United States from Florida to Arizona and California. Because of its narrow shape, it is used for borders as well as formal planting. Hedges can be formed by clipping.

Champion--Height 82 ft (25.0 m), c.b.h. 8.4 ft (2.6 m), spread 20 ft (6.1 m). Ulupalakua Ranch, Ulupalakua, Maui (1968).

Range--Native in the eastern Mediterranean region of southern Europe in Syria, Cilicia, Greece, and the islands of Rhodes, Crete, and Cyprus and in the mountains of northern Iran in western Asia.

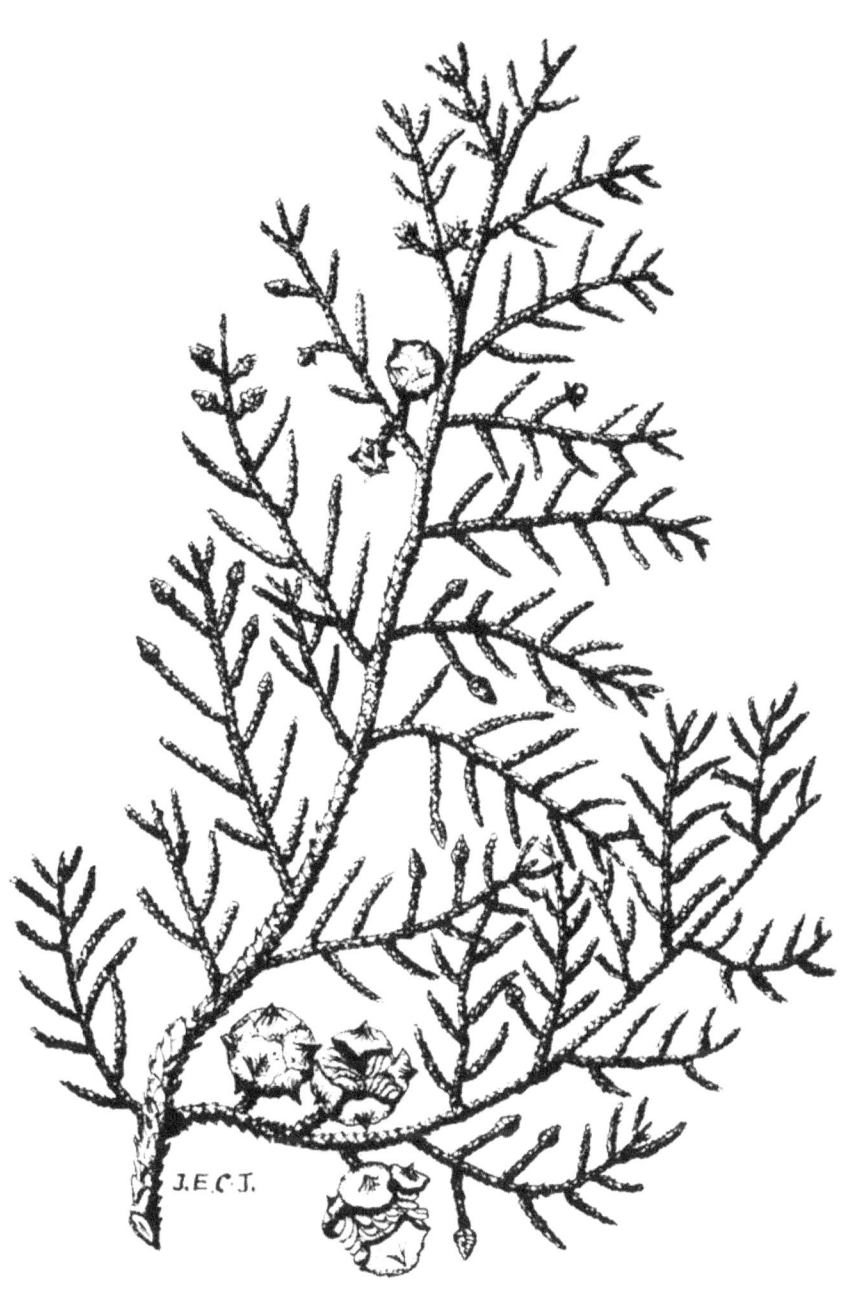

16. Mexican cypress *Cupressus lusitanica* Mill.*
Twig with cones and male cones, 1 X.

17. Monterey cypress   *Cupressus macrocarpa* Hartw.*
Twig with young female cones (upper left), twig with male cones (upper right), and twig with cones (below), 1 X.

18. Italian cypress *Cupressus sempervirens* L.*
Leafy twig (above), twig with cones (lower right), 1 X (P. R. v. 2).

19. Bermuda juniper *Juniperus bermudiana* L.*

Small to medium-sized introduced evergreen tree with small scalelike leaves. Trunk straight, with many branches in compact cone-shaped crown to 40 ft (12 m) high. Bark brown, furrowed into scaly ridges and peeling. Inner bark with dead brown outer layer, whitish, fibrous, resinous. Twigs 4-angled, stout, more than 1/16 in (1.5 mm) wide.

Leaves paired, scalelike, overlapping in 4 rows against twig, blunt-pointed, 1/16 in (1.5 mm) or more in length, with groove along back, gray green. Leaves on young plants in groups of 3, awl-shaped, to 1/2 in (13 mm) long, those on older twigs in groups of 2, 3-angled, pointed, to 3/16 in (5 mm) long.

Trees male and female, the male (pollen-bearing) cones cylindrical, about 3/16 in (5 mm) long. Female (seed-bearing) cones berrylike, rounded or slightly 2-lobed, about 1/4 in (6 mm) long, whitish blue, composed of few united scales, maturing in 1 year. Seeds 2--3, egg-shaped, pointed, shiny brown, grooved.

Sapwood is yellowish white, and the heartwood reddish brown. This wood is fine-textured, aromatic, of low density, and durable. Used for furniture, cabinetmaking, and shipbuilding in other parts of the world, but not used in Hawaii.

In Hawaii, Bermuda juniper is grown as an ornamental and in hedges. It is also planted in the Forest Reserves on a small scale. Between 1921 and 1953, 6,500 trees were planted in the forests on all islands. Examples may be seen at Waiahole, Oahu, Waiahou Spring Reserve (Olinda), Maui, and Pepeekeo Arboretum, Hawaii. This subtropical species is not hardy northward in temperate regions.

Champion--Height 72 ft (21.9 m), c.b.h. 6.9 ft (2.1 m), spread 56 ft (17.1 m). Kainaliu, North Kona, Hawaii (1968).

Range--Native only to the island of Bermuda. Formerly abundant there, but now rare because of a disease as well as cutting.

Other common names--Bermuda redcedar, Bermuda cedar.

SCREWPINE FAMILY (PANDANACEAE)

20. Hala, screwpine *Pandanus tectorius* Parkins.

Picturesque evergreen tree of coasts and lowlands recognized by the many large prop roots around the short, smooth light gray trunks, by the few widely forking stout branches, ending in a cluster of many crowded, spirally arranged large strap-shaped leaves with saw-toothed edges, and by the fruits resembling a pineapple.

Small tree 10--33 ft (3--10 m) high, the trunk and branches with irregular ring scars from fallen leaves. The straight cylindrical prop roots or stilt roots covered with small spines or prickles support the main trunk and spreading branches. Branches divide regularly into 2 equal widely spreading forks.

Leaves long and very narrow, thick and leathery, about 3 ft (0.9 m) long and 2 in (5 cm) wide or to twice that size, alternate but crowded, with broad clasping base, parallel saw-toothed edges, midvein and many inconspicuous parallel side veins, and ending in a long tapered drooping point. The upper surface is shiny green, the lower surface dull light green with spines along midvein. Dead brown leaves hang down and gradually fall away.

Two kinds of trees are distinguished not only by their flowers but by their trunks. Flowers are male and female on different plants (dioecious), small, simple, and without calyx and corolla. Male flowers are very numerous in drooping clusters 1--2 ft (0.3--0.6 m) long from the center of a cluster of leaves, and very fragrant. They consist of many stamens 1/8--1/4 in (3--6 mm) long, crowded on threadlike branching stalks along an axis (raceme) with several spiny-edged pale yellow, very fragrant bracts ending in a long very narrow point. Female flowers in terminal compact greenish heads have pistils densely crowded with colored scales.

The multiple fruit (syncarp) borne singly on a long stalk, is a large hard heavy ball 4--8 in (10--20 cm) in diameter, composed of 40--80 fruits (drupes). Each fruit is 1 1/2--2 3/4 in (4.7 cm) long and 3/8--3/4 in (1--2 cm) wide, angled and slightly flattened, shiny pale yellow to orange to red, hard and fibrous, containing usually 5--11 seeds or empty cells. These bright-colored fragrant fruits are scattered by animals that eat the sweetish pulp. The soft orange pulp was also used as food by Hawaiians in times of famine. The old dried fruits are spongy and probably float to other islands.

Trunks of male trees are hard and solid throughout and have wood that is yellow with dark brown fiber bundles, very strong, but brash when subjected to a sudden load and difficult to split. Those of female trees are very hard in the outer part, but soft and fibrous or juicy within. Elsewhere, trunks of female trees have served as water pipes after removal of the pith.

Hawaiians utilized most portions of the plant for various purposes. The leaves (lau) served as thatch in

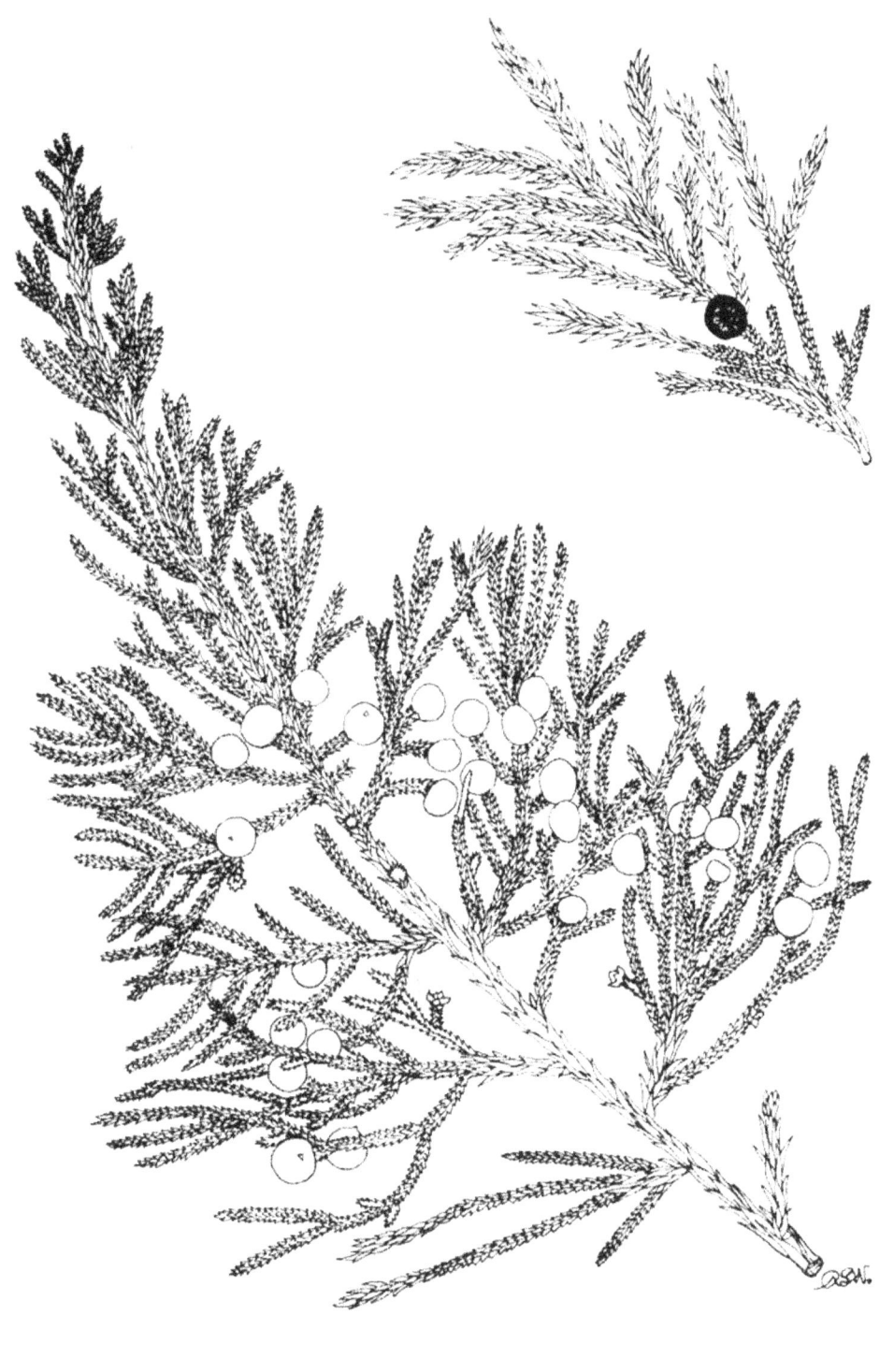

19. Bermuda juniper *Juniperus bermudiana* L.*
Twigs with cones, 1 X.

houses and were plaited into baskets and mats, or floor coverings. A roof of lauhala leaves is said to last about 15 years in low rainfall coastal areas while one of coconut leaves will last only 3 years. Finely divided lauhala is used to make hats. The dry weeds with fibrous end were employed as brushes in dying bark cloth. Parts of the fruit were sliced into pieces that were strung into leis or garlands with leaves of other plants intertwined.

Hala is common through the Hawaiian Islands in lowlands, especially windward sides along coasts and from sea-level to 2,000 ft (610 m).

Special areas--Keahua, Waimea Arboretum, Foster, Volcanoes.

Range--Through the Hawaiian Islands and southwestward in the South Pacific Islands to northern Australia, New Guinea, west to Philippine Islands, Moluccas, and Java.

This is the only species of Pandanus native in Hawaii, according to a conservative classification. It was formerly united under P. odoratissimus L. f. of the South Pacific region.

Other common names--puhala, lauhala, pandanus; kafu (Guam, N. Marianas); ongor (Palau); fach (Yap); fach (Truk); kipar (Pohnpei); moen (Kosrae); bop (Marshalls); fala (Am. Samoa).

Also, fruit color varieties: hala (yellow), hala'ula (orange), hala lihilihi'ula (red yellowish below), halapia (pale yellow).

The name screw-pine is suggested by the leaves arranged spirally like a screw and by the ball-like fruits similar to a pine cone.

Several species have been introduced as ornamentals, illustrating white-bordered or striped leaves, short leaves, and large fruits.

Champion--Height 35 ft (10.7 m), c.b.h. 4.5 ft (13.7 m), spread 40 ft (12.2 m). Keaau, Hilo, Hawaii (1968).

GRASS FAMILY (GRAMINEAE)

21. 'Ohe, common bamboo

Bambusa vulgaris Schrad. ex Wendl.*

Bamboos are giant introduced evergreen grasses with clustered jointed hollow stems and feathery foliage. In size, they may be classed as trees, though they grow in clumps of several stems, like many shrubs. Their large treelike size and usefulness justify their inclusion here.

This common species was one of the first introduced bamboos and is one of the largest established in Hawaii. The description here is from the Puerto Rican tree handbook. Neal (1965, p. 60-64) published information about bamboos with a key to 8 genera.

It is difficult to distinguish between some of the different species, as the flowers needed for positive identification are rarely produced. However, as a group, bamboos are easily recognized by: (1) clusters of several to many slender, tapering, slightly curved stems, in the species of largest size, 2--4 in (5--10 cm) in diameter, dark green to orange, with swollen rings or joints 8--18 in (20--46 cm) apart; (2) several very slender branches spreading horizontally and regularly at the joints; and (3) grass leaves in 2 rows, consisting of a basal sheath around the slender twig and long-pointed blade with many lateral veins parallel to midrib.

The stems (culms) attain 30--50 ft (9--15 m) in height and toward the top diverge from the center. The smooth surface, green to dark green, becomes orange or yellow in age. From a distance the plants appear like clumps of giant ferns. The slender side branches about 1/4 in (6 mm) in diameter are nearly horizontal and bear wirelike yellow green twigs. Spines are absent in this species.

The light green leaf sheaths are 1 1/2--2 1/2 in (4--6 cm) long, closely fitting the twig. Blades are 6--10 in (15--25 cm) long and 3/4--1 3/4 in (2--4.5 cm) wide, or as short as 2 in (5 cm) at base of twig, with rough edges, long-pointed at apex and short-pointed where narrowed and jointed into sheath. The upper surface of the flat thin blade is green and slightly shiny, the lower surface pale blue green.

Large bamboos bloom only once. After many years, many plants growing together flower simultaneously, produce seeds, and then die. Like most other grasses, bamboos have inconspicuous flowers, usually light brown or straw-colored. The flower cluster (panicle) of this species is composed of slender branches bearing bracted clusters of 3--15 or more stalkless spikelets 1/2--3/4 in (13--19 mm) long, oblong and pointed, each with several to many flowers (florets) about 3/8 in (10 mm) long. The flower has 2 narrow scales, 6 stamens with purple protruding anthers, and pistil, producing an oblong grain.

Not divided into bark and wood, the stem is hollow except at nodes, lightweight, hard, and strong. The very hard outer wall is about 1/2 in (13 mm) thick, whitish brown, and fibrous.

A new stem completes its height growth from the clustered roots at base in about 3 months, elongating very rapidly, up to 8 in (20 cm) daily. Nor does it expand in diameter after formation.

The young growing shoot at the outside of a clump is readily distinguished by the absence of branches and by the presence at each node or joint of a large leaf with

20. Hala, screwpine *Pandanus tectorius* Parkins.
Cluster of old female flowers in head (left), 1/2 X; male flower clusters (right), 1/4 X; male flower (top center), 2 X; fruits (below), 1/2 X (Degener).

by the presence at each node or joint of a large leaf with triangular spreading blade. These clasping leaves along the main axis have a very large gray green sheath 6--12 in (15--30 cm) long, extending nearly to the next node and bearing many brown needlelike hairs that stick in the flesh when touched, and a short triangular pointed yellow green blade 2--3 in (5--7.5 cm) long and broad, also with a few brown hairs. Toward the apex of the elongating stem the leaves are closer together and overlapping.

Bamboos of this and other species have many uses besides ornament. Their masses of intertwining roots and accumulations of leaf litter check erosion on roadside banks and slopes. Poles of various kinds for construction, fences, fenceposts, ladders, tool handles, flagpoles, and stakes are easily made from bamboo. The stems will serve as temporary water pipes after opening them on one side at each node and removing the partitions. Short pieces are used as pots for seedlings to be transplanted later. Bamboo boards can be prepared by slitting, splitting, and spreading open the stems. The split pieces are woven into baskets. Bamboo stems have been utilized in the manufacture of various articles, including furniture, lattices, fishing rods, picture frames, lampshades, mats, and flower vases. This species is suitable for paper pulp because the stems have relatively long wood fibers.

With outer scales removed, the tender growing tips of bamboo shoots can be eaten by boiling for about a half hour and changing the water once or twice to remove any bitter taste. There is no distinct flavor except for a slight suggestion of young corn. Bamboo shoots are added to meat stews, salads, and other ways. Young shoots are gathered under stands in the forest by all ethnic groups in Hawaii, even though such use had its origin in China and Japan.

This species of bamboo was introduced into Hawaii probably from China in the early part of the 19th century, according to Hillebrand. He observed that Hawaiians used bamboo for fishing poles and outriggers for canoes. The Hawaiians, however, introduced and used another species, *Schizostachyum glaucifolium* (Rupr.) Munro. It has thin-walled culm with long internodes and was used primarily to make musical instruments such as nose flutes. Neal mentioned additional products made by the Hawaiians. Bamboo was used for fans, mats, bellows, straight edges, knives, and pens and stamps for marking tapa bark cloth. Elsewhere, many other uses have been developed, such as construction of houses and manufacture of tools.

Through the Hawaiian Islands, this bamboo is cultivated commonly as an ornamental and soil binder along roadsides and streams and grows as if wild. Bamboo has been planted in large stands in the forest, particularly on Oahu and Maui. A typical stand is near Reservoir No. 3 in Nuuanu Valley on Oahu. Others may be seen from the Hana Road on Maui. Natural vegetative propagation by breaking and rooting of the fragile short branches has been observed elsewhere. However, normal establishment of seedlings is rare because of very infrequent production of seed. The plant is considered noxious by foresters in Hawaii because it takes over and shades out all other vegetation in the wet gulches where it grows best. Fortunately, entire clones will flower and die periodically, affording an opportunity for other vegetation to recapture the bamboo stands.

Special areas--Waimea Arboretum, Wahiawa.

Range--Native of tropical Asia, perhaps India, but widely planted through the tropics. Common in Puerto Rico and Virgin Islands and grown also in southern Florida.

Other common names--feathery bamboo; bambú (Puerto Rico, Spanish); pi'ao palao'an (Guam).

Golden bamboo (var. *aureo-variegata* Beadle) is a handsome ornamental variety characterized by variegated stems with yellow and green vertical stripes.

This is one of the most widely cultivated bamboos in tropical and subtropical regions through the world, because of its feathery foliage and large size. It is easily propagated by division of clumps and stem cuttings. This species will endure light frosts but usually is killed to the ground by an infrequent freeze at 28 degrees Fahrenheit (- 2 degrees centigrade).

PALM FAMILY (PALMAE)

22. Coconut, niu *Cocos nucifera* L.**

This familiar symbol of the tropics and especially sandy shores scarcely needs a description. Distinctive characters of the world's best known palm include the slender, often leaning trunk enlarged at base, smoothish with horizontal rings, and many very long spreading feathery or pinnate leaves, and the familiar coconut fruits within a large fibrous husk. The coconut was the official tree of the Territory of Hawaii before statehood.

Medium-sized palm, usually 30--60 ft (9--18 m) high, sometimes taller and rarely almost 100 ft (30 m). Trunk 16--20 in (0.4--0.5 m) in diameter at enlarged base and 8--12 in (0.2--0.3 m) above, gray or brown, slightly fissured, bent slightly perhaps by the constant coastal winds. At the apex is the relatively broad evergreen crown of erect spreading and drooping leaves.

21. 'Ohe, common bamboo *Bambusa vulgaris* Schrad. ex Wendl.*
Plants (left), much reduced; leafy twig (right), 1 X (P. R. v. 1).

Leaves many, alternate, 12--20 ft (3.7--6 m) long, with basal sheath of coarse brown fibers, long leafstalk, and numerous segments. Basal sheath nearly 2 ft (0.6 m) high on sides of leafstalk, surrounds trunk and breaks as younger leaves expand. Leafstalk stout, 3--5 ft (0.9--1.5 m) long, yellowish and slightly concave. Blade 9--15 ft (2.7--4.6 m) long and 3--5 ft (0.9--1.5 m) wide, composed of numerous segments spreading regularly in one plane on both sides of axis. Segments or leaflets very narrow (linear), 2--3 1/2 ft (0.5--1.1 m) long and 2 in (5 cm) wide, shorter toward apex, long-pointed, leathery, parallel-veined, shiny yellow green above, and dull light green beneath. The lowest dead leaves hang down against trunk, eventually shedding and forming a smooth ring scar. Generally, coconuts produce an average of 12 leaves per year, and the age of a tree may be estimated by counting the ring scars and dividing by 12. Stem diameter is greatly influenced by vigor of the tree and in recently developed areas in Hawaii trees with a conspicuous constriction in their upper stems may be seen. These constrictions resulted from reduced diameter growth for a period after the full-sized trees were transplanted.

Flower clusters (panicles) 3--4 ft (0.9--1.2 m) long at leaf bases rise from 2 long narrow long-pointed sheaths (spathes), the inner about 4 ft (1.2 m) long, and bear many slightly fragrant stalkless flowers. A branch about 1 ft (0.3 m) long has numerous small male flowers and near base 1 or few much large female flowers, which open later (monoecious). Male flowers 3/8--1/2 in (10--13 mm) long and broad have 3 small pointed whitish sepals 1/8 in (3 mm) long, 3 oblong petals nearly 1/2 in (13 mm) long, 6 widely spreading stamens, and sterile pistil with 3 styles. Female flowers about 1 1/4 in (3 cm) long and broad, rounded or 3-angled, have 2 broad scales at base, 3 broad round sepals 3/4--1 in (2--2.5 cm) long, 3 rounded whitish or light yellow rounded petals 1--1 1/4 in (2.5--3 cm) long, and light green pistil 1 1/4 in (3 cm) long with 3-celled ovary and 3 minute stigmas.

The fruit has an egg-shaped or elliptical light brown fibrous husk 8--12 in (20--30 cm) long, bluntly 3-angled and not splitting open. The elliptical or nearly round inner brown fruit with 3 round spots near end is essentially a very large hollow seed covered with a hairy outer shell. Inside is a whitish layer of slightly sweet and oily stored food 3/8 in (1 cm) thick and a large central cavity containing watery or milky liquid. Flowering and fruiting continuously through the year.

Ranking among the 10 most useful tree species to mankind in the world, coconut is the most important of cultivated palms through the tropics. The fruits are eaten raw, prepared into candies, or shredded with pastries. When immature, the soft jellylike flesh can be eaten with a spoon. The watery liquid of green fruits and the milky juice of mature ones are pure, nutritious, cool, and refreshing drinks. These green fruits are sold on city streets in the tropics. Under the name copra the dried white oily part of ripe fruits is marketed in large quantities for the manufacture of soaps and coconut oil. The latter is used in the preparation of margarine and other foods, and for cooking. The white suspension of grated coconut is an essential sauce in most Polynesian cuisine. Coconut is also classed as a honey plant. The sugary sap collected from cut unopened flower clusters is a fresh beverage known as toddy, and a source of alcohol.

The trunks serve for posts. Walking sticks have been made from the outer layer or ring of the trunk. The inner part is a very soft, light brown pith with scattered reddish brown bundles. The wood of coconut palms from Fiji was tested in New Zealand and found to be marginally useful if handled properly (Alston 1973). Only the wood in the outer part of the stem at the butt is sufficiently dense and strong to be useful. The rest of the stem should be discarded. The wood is difficult to season without collapse and is extremely dulling to saws. It is used only rarely for rafters and house posts on islands where more suitable woods are not available. Two pieces of coconut wood are in the decorative carving on the door behind the thrones in Iolani Palace, Honolulu.

Many large coconut plantations or orchards have been established along coast lines of many tropical regions, especially in southern Asia and South Pacific islands. Also, the trees thrive inland where soil moisture is ample, and are hardy in dry climates if irrigated. Trees begin to bear at about 6 years and annually produce 40 or more nuts, which mature in 9 or 10 months. In Hawaii, copra plantations were attempted in several places during the late 1800's. One such plantation is at the Coco Palms Hotel on Kauai. It was found that the trees did not bear as well at Hawaii's latitude as elsewhere in the Pacific, so the ventures were given up. Dwarf varieties are recommended for ornamental planting around homes, as falling coconuts may be dangerous. Trees in parks and along streets must be kept trimmed of young fruits for this reason. People have been killed by falling nuts in Honolulu. Falling fronds have also caused serious injury.

Early Hawaiians managed groves of coconuts and had many uses for the various parts, as reported by Neal (1965). The leaves served for thatching, baskets, fans, brooms, string, etc. From the fibrous material at leaf bases, sandals and strainers were made. The husks provided sennit, cordage used in fishing, house construction, etc. The terminal leaf buds and pith of trunks were eaten raw. Mature and immature nuts provided food and drink. Coconut oil was used for light, ointment, and hair oil. The shells were made into utensils and implements. The hard outer part of the trunk served as wood for posts, furniture, construction, and spears. The sweet sap of flowering buds was tapped as a source of sugar, wine, and vinegar.

22. Coconut, niu   *Cocos nucifera* L.**
Male flowers and female flower (left), 2/3 X, and palm, much reduced (P. R. v. 1).

Throughout the Hawaiian Islands, as elsewhere through the tropics, coconut palms line sandy beaches and have been planted along streets and in parks. The pre-European variety used by the Hawaiians is distinctly more slender-stemmed for its height than the more recently introduced varieties. This variety is still common in such places as Keanae and Hana, Maui, and may also be seen on all islands in older settlements.

Special areas--Waimea Arboretum, Foster, Iolani.

Champion--Height 92 ft (28 m), c.b.h. 50 ft (1.4 m), spread 28 ft (8.5 m). Hilo, Hawaii (1979).

Range--Native land apparently Indian Ocean region. Now thoroughly naturalized on tropical shores of the world. Growing naturally and in cultivation through the Hawaiian Islands, in southern Florida, and in Puerto Rico and Virgin Islands.

Other common names--coco-palm, coconut-palm; coco, cocotero (Puerto Rico, Spanish); niyog (Guam); nizok (N. Marianas); iru (Palau); lu (Yap); nu (Truk); ni (Pohnpei and Marshalls); nu (Kosrae); niu (American Samoa).

Wherever its place of origin, coconut apparently has spread naturally by means of the large lightweight husks and waterproof shells. The seeds are able to germinate after floating in salt water a few months. Whether the coconut arrived in Hawaii before the first human inhabitants is uncertain. If not native, this species has become thoroughly naturalized as though wild.

Reports that this species was native or pre-Columbian in the New World, such as on the Pacific coast of Costa Rica and Panama, have been rejected. Columbus did not find coconut, and early Spanish writers in the New World did not mention it. However, this valuable palm apparently reached Puerto Rico within a century after discovery in 1493.

Both the common and scientific names apparently are derived from Portuguese coco, a word adopted in European literature. It is said that the meaning refers to the resemblance of the fruit with its three spots to an ape's face.

Coconut is one of the largest seeds known. It is surpassed only by the 1-seeded 2-lobed fruit weighing up to 50 pounds (22.7 kg) of the double-coconut (*Lodoicea maldivica* (Gmel.) Pers.), a tall fan palm native to the Seychelles Islands in the Indian Ocean, which may be seen at Foster Botanic Garden.

23. Loulu, pritchardia *Pritchardia* spp.

Hawaii has only a single group of native palms, medium-sized fanpalms of the genus *Pritchardia*. They are found in moist or dry forests from the coast almost to 4,000 ft (1,219 m). This genus is characterized by fan-shaped or wedge-shaped leaves, smoothish trunk with horizontal lines, and absence of spines. A generalized description of the group follows.

Trunk straight, unbranched, 13--30 ft (1--9 m), sometimes much taller, mostly stout and about 1 ft (0.3 m) in diameter, smooth, gray, finely fissured, with horizontal lines from leaf scars. Height varies in different species from about 5 ft (1.5 m) to a maximum of 75 ft (23 m) or more.

Leaves crowded and spreading at top of trunk, very large, coarse, evergreen, about 15--30, composed of stout, spineless gray green leafstalk about 3 ft (0.9 m) long and fan-shaped or wedge-shaped blade about 3 ft (0.9 m) in diameter, folded and divided toward edges into many narrow, thick and leathery segments, with many fine longitudinal lines of parallel veins. The upper surface is green and the lower surface varying in different species from green and almost hairless to whitish and waxy or silvery and densely hairy. Older dead leaves hang down and are often persistent.

Flower clusters (panicles) single at leaf bases, about 2 ft (0.6 m) long, composed of a long stout stalk covered with large gray or straw-colored scales and many light green to yellowish branches, hairless or densely hairy. Flowers many, with slight odor of cheese, stalkless, small, about 1/2 in (13 mm) long and broad, composed of greenish cuplike base with 3-toothed calyx, 3 narrow petals that fall upon opening, 6 spreading stamens united at base into cup, and pistil with 3-celled ovary and short style.

Fruits (drupes) many, nearly stalkless, hanging down, round or elliptical, varying in size in different species, 1/2--2 in (13 mm--5 cm) long, with point from style at apex, shiny green and turning to black, sometimes yellowish, with fibrous outer wall and hard inner wall. Seed 1, round or oblong, brown, 1 in (2.5 cm) or less in length, white within, soft and edible when immature, becoming hard.

The main use of these native Hawaiian palms is as ornamentals. A few species are in cultivation here and elsewhere in the tropics. Formerly, the Hawaiians made hats from the young leaves and used the mature leaves for thatching and fans. The immature soft seeds, which have a taste slightly like coconut, were eaten. Domestic animals and rats feed on the seeds, and wild hogs also dig up and devour young plants. Formerly, the trunks were used to make fences.

The native plams of this genus are mostly distributed in the wet forests of the six large islands chiefly at about 2,000--3,000 ft (610--914 m) altitude, sometimes higher, and in dry forest at sea level. One species is found on Nihoa, a very small island 150 miles (241 km) northwest of Kauai. An extinct one is recorded from Laysan.

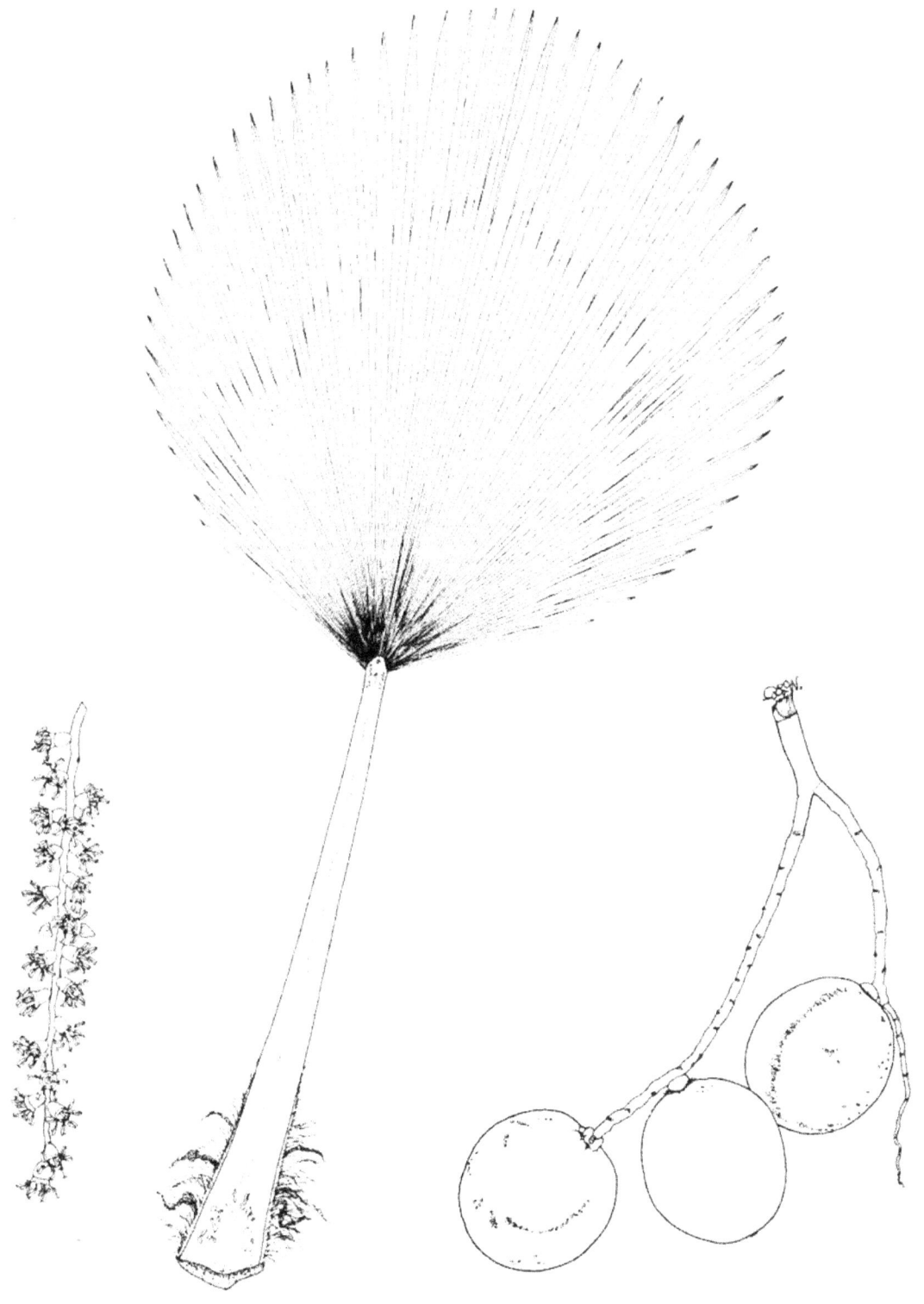

23. Loulu, pritchardia *Pritchardia martii* (Gaud.) H. Wendl.
Leaf, about 1/18 X, and flowers (lower left) and fruits (lower right), 2/3 X.

The genus *Pritchardia*, according to an early monograph (Beccari and Rock 1921), had about 31 named species of fanpalms, nearly all confined to Hawaii except for about 5 in Fiji and other South Pacific islands. Many very localized forms in Hawaii probably are varieties of very few species, according to Fosberg and Herbst (1975). Robert W. Read (pers. comm.) recognizes 18 species in Hawaii as a result of a revision for the "Manual of Flowering Plants of Hawaii." A few are rare and have been proposed as endangered. Most native species have a local range within a single island. Variation is great, even within one plant in different years.

The name honors William T. Pritchard, author of Polynesian Reminiscences and British Consul in Fiji, where the first species was named in 1861. That species, *Pritchardia pacifica* Seem. & H. Wendl., Fiji fanpalm, is an introduced ornamental in Hawaii.

The drawing is from a plant of *Pritchardia martii* (Gaud.) H. Wendl. of Oahu, growing in the Lyon Arboretum.

Special areas--Kokee, Waimea Arboretum, Foster.

Champion (*Pritchardia beccariana* Rock)--Height 55 ft (16.8 m), c.b.h. 4.5 ft (1.4 m), spread 19 ft (5.8 m). Waiakea Forest Reserve, Hilo, Hawaii (1968). Taller plants are at Hoomau Ranch on the same island.

Other common names--hawane, pritchardia.

AGAVE FAMILY (AGAVACEAE)

24. Ti, common dracaena

Cordyline fruticosa (L.) Chev.**

Ti (pronounced "tea") is an evergreen shrubby plant apparently introduced by the early Hawaiians for its many uses. It is related to No. 25, halapepe or golden dracaena, *Pleomele aurea*, which it resembles slightly. Characterized by a slender trunk, few wide-spreading branches (none when small), and clusters of many very large, longstalked narrowly oblong or lance-shaped leaves crowded in spiral.

A shrub or small tree to 15 ft (4.6 m) high with light gray smoothish trunk to 3 in (7.5 cm) in diameter, becoming warty and slightly cracked, with horizontal rings, not divided into bark and wood. Within the thin brown outer layer, the trunk is whitish, soft, and bitter.

Leaves are alternate but very crowded in a spiral at end of erect stout hairless branch, with stout grooved greenish leafstalk of 2--4 in (5--10 cm), hairless. Blades narrowly oblong, 7--18 in (18--45 cm) long and 2--4 in (5--10 cm) wide, broadest near middle and gradually narrowed to long-pointed ends, not toothed on edges, thin and flexible, with many long fine parallel veins, shiny green on both surfaces, leaving a ring scar.

Flower clusters (panicles) large, arising from center of cluster of leaves, 12--15 in (30--38 cm) long, curved and branched. Flowers many, stalkless on slender drooping branches, from narrow whitish buds 1/2 in (13 mm) long, tinged with purple, composed of narrow calyx whitish tube with 6 pointed lobes curled back, 6 yellow spreading stamens inserted in throat, and white pistil with 3-celled ovary and slender style.

Fruits (berries) rarely formed, about 1/4 in (6 mm) in diameter, yellow, turning to bright red. Seeds few, shiny black.

This species served the original inhabitants in many ways. The papery leaves provided food wrapping, plates, drinking cups, bandages, leis, religious symbols, and thatch for roofs and walls of houses. They were made into raincoats and sandals. In the late 1800's, King Kalakaua introduced from the Gilbert Islands a technique of shredding the leaves into skirts for dancing the hula. Whistles and flute-like musical instruments were made from rolled strips of fresh leaf. The leaves provided fodder for livestock also. The entire leafy stem was and still is used as a sled for "ti leaf sliding." One sits on the leaf, grasps the stalk, then toboggans down wet hillsides.

The plants have a large tuberous root that weighs as much as 300 pounds (136 kilos) and stores sugar. The edible baked roots have a taste like molasses candy. From the Europeans, the Hawaiians learned to distill brandy and whisky from the baked fermented roots. This beverage called okolehao is sold in stores at present.

Ti plants are favorite ornamentals and are easily propagated from cuttings. Numerous cultivated varieties have been developed with purple, pink, and striped leaves, also red flowers. A table decoration of sprouting leaves and roots is made by placing short stem cuttings in water.

This useful species was the subject of various stories and legends. A hedge of these plants around a house was believed to ward off evil spirits and bring good luck.

Common through the islands at low altitudes, persistent around houses and in open areas bordering wet forests.

Special areas--Waimea Arboretum, Foster, Haleakala, City, Volcanoes, Kipuka Puaulu, Hawaii.

Range--Widespread in Pacific Islands, Australia, tropical Asia, and planted through the tropics, the original native range uncertain.

Other common names--ki, la-i, "dracaena," "common dracaena."

Botanical synonym--*Cordyline terminalis* (L.) Kunth.

24. Ti, common dracaena *Cordyline fruticosa* (L.) Chev.**
Flowering twig, 1/12 X; bud and flowers (below), 2 X (Degener).

25. Halapepe, golden dracaena

Pleomele aurea (Mann) N. E. Br.

Halapepe (genus *Pleomele* or *Dracaena*) is a distinctive small evergreen tree resembling the related mainland genus of *Yucca*, with trunk and very few stout nearly erect branches, ending in a cluster of many crowded large sword-shaped or strap-shaped leaves, not forming a compact crown. Plants are scattered in dry areas on middle slopes of the 6 large Hawaiian Islands.

In the broad sense, halapepe is described here as a variable species under its oldest scientific name, *Pleomole aurea* (Mann) N. E. Br., *sens. lat.* Altogether, 9 species have been named from the Hawaiian Islands, each limited to 1 or 2 of the 6 large islands (St. John 1985). The segregates are separated mainly by differences in measurements of flowers and flower parts and of leaves.

Height about 15--25 ft (4.6--7.6 m), sometimes to 40 ft (12 m). The straight trunk 1--3 ft (0.3--0.9 m) in diameter, light gray, smooth, slightly scaly or fissured, not divided into bark and wood. The outer dead part reddish brown and inner part whitish or light yellow with reddish streaks, fibrous, and almost tasteless or slightly bitter. Smallest branches about 3/4--1 1/4 in (2--3 cm) in diameter, light gray, smooth with many crowded nearly horizontal lines or ridges of leaf scars, forming irregular rings.

Leaves sword-shaped or strap-shaped, spreading and drooping, 8--20 in (20--51 cm) long and 3/8--1 1/4 in (1--3 cm) wide, with expanded whitish slightly clasping base, curved, leathery, hairless, ending in long blunt point, slightly shiny green on both surfaces, without midvein, not toothed on edges.

Flower clusters (panicles) large, massive, about 2 ft (60 cm) long at end of leafy branch, curved downward, with stout persistent woody axis, much branched, with leaflike scales. Flowers many, crowded, 1-3 together on slender stalk, narrowly bell-shaped, golden yellow or varying from greenish yellow to orange, about 1 1/2--2 in (4--5 cm) long, persistent, composed of narrow tube 1 1/2 in (4 cm) long and 3/8 in (1 cm) wide, 6 narrow spreading lobes 5/8 in (1.5 cm) long, 6 slender stamens attached near base of lobes and about the same length, and pistil inside enlarged base of tube, composed of rounded greenish ovary, 1/4 in (6 mm) long, slender long style, and slightly 3-lobed stigma. Flowering in early spring or, in moist areas, summer.

Berries many on curved stalks, round or slightly 2--3-lobed, 3/8--5/8 in (1--1.5 cm) in diameter, bright red, dark brown when dry. Seeds mostly 1, sometimes 2 or 3, elliptical or rounded or slightly angled, 1/4 in (6 mm) long, whitish or light brown.

The inner portion of trunk corresponding to wood is whitish or light yellow mottled with reddish streaks, fine-textured, extremely soft and easily cut. Hawaiians formerly carved religious statues from the soft trunks. The branches served for decorating their altars, including that of Laka, the goddess of hula.

Halapepe is common in dry areas, especially the aa (rough) lava fields through the Hawaiian Islands, usually at 600--2,000 ft (183--610 m) altitude). Cultivation of this attractive plant should be encouraged.

Special areas--Kokee, Waimea Arboretum, Wahiawa, Bishop Museum.

Champion--Height 20 ft (6.1 m), c.b.h. 3.5 ft (1.1 m), spread 14 ft (4.3 m). Kaupulehu, Kailua-Kona, Hawaii (1968).

Range--Native only in Hawaiian Islands.

This genus is placed here in the Agave Family (Agavaceae), along with related genera of similar habit, such as *Cordyline*, *Yucca*, and *Agave*.

The common name halapepe, meaning baby hala, refers to the resemblance to the larger plant No. 20, hala or screwpine, *Pandanus tectorius* Parkins.

Botanical synonym--*Dracaena aurea* Mann.

The segregate species with distribution by islands are: *Pleomele auwahiensis* St. John (*P. rockii* St. John), Molokai and Maui; *P. fernaldii* St. John, Lanai; *P. forbesii* Deg., Oahu; *P. halapepe* St. John, Oahu; *P. hawaiiensis* Deg. & I. Deg. (*P. kaupulehuensis* St. John, *P. konaensis* St. John), Hawaii. The drawing here is that of the segregate *P. halapepe* (Degener 1933-1986).

CASUARINA FAMILY (CASUARINACEAE*)

26. River-oak casuarina

Casuarina cunninghamiana Miq.*

Three of the 11 species of *Casuarina* introduced from Australia and Pacific islands for forestry and other purposes are described here. These evergreen trees are known by the many drooping, very slender green twigs like pine needles or wires, each with many joints bearing tiny scalelike pointed gray or brown leaves in a ring, and by the rounded hard warty fruits like a ball or cone with many minute winged seeds. The green twigs manufacture food in the absence of leaves and are shed gradually when old. The trees are important nitrogen

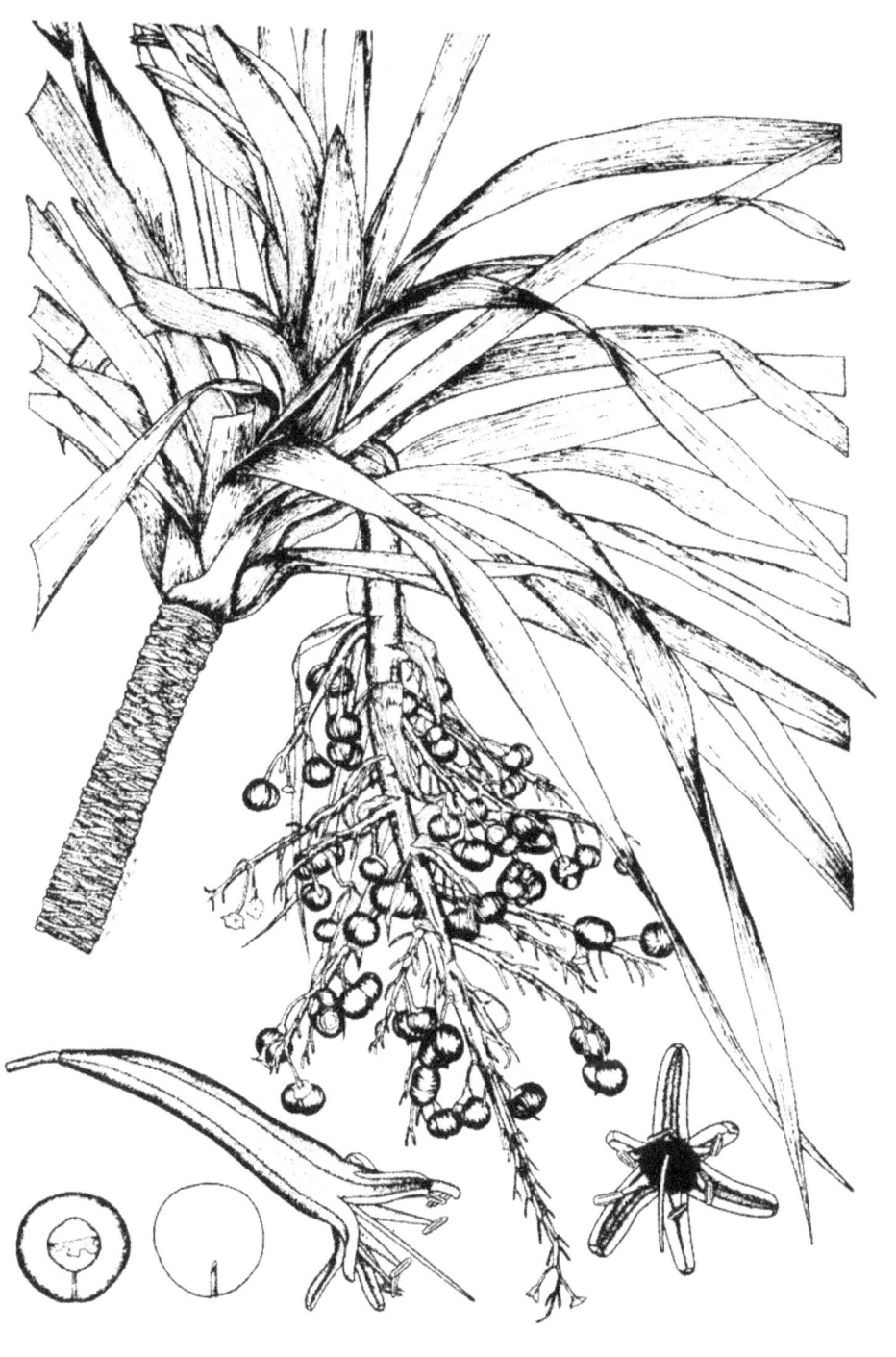

25. Halapepe, golden dracaena *Pleomele aurea* (Mann) N. E. Brown
Twig with fruits, 1/3 X; flowers and seeds (below), 1 X (Degener).

fixers in a symbiotic relationship with fungi (Actinomycetes) of the genus *Frankia*, analogous to that of legumes with *Rhizobia*. They are thus able to thrive on depleted, very eroded sites where other plants cannot survive.

The name is from the Malay word *kasuari*, cassowary, because of the fancied resemblance of the twigs to the plumage of that bird. Trees of this genus have been known also as beefwood from the reddish color of the wood. The term "she-oak," widely applied in Australia, according to Maiden (1907) refers to the wood being like oak (with broad rays) but not as strong. Other common names are "Australian-pine," from the foliage and fruits, and in Hawaii "ironwood," from the hard wood. The last names are not recommended, because this family is not related to pines or other conifers and because "ironwood" has been applied to many different trees with confusion.

Three common species are described and illustrated here. No. 26, River-oak casuarina, *Casuarina cunninghamiana* Miq., has the shortest and thinnest drooping wirelike twigs, mostly 3--7 in (7.5--18 cm) long and less than 1/32 in (1 mm) wide, with scale leaves 8-10 in a ring, and small conelike fruits about 3/8 in (10 mm) in diameter. No. 27, horsetail casuarina, *Casuarina equisetifolia* L. ex J. R. & G. Forst., has longer twigs about 1/32 in (1 mm) wide, with scale leaves 6-8 in a ring, and larger fruits 1/2--3/4 in (13--19 mm) in diameter. No. 28, longleaf casuarina, *Casuarina glauca* Sieber ex. Spreng., has long thicker twigs nearly 1/16 in (1.5 mm) wide, with scale leaves 12-16 in a ring, and fruits about 1/2 in (13 mm) in diameter. Although the 3 species described here can be recognized in Hawaii, many if not most trees show characteristics of more than 1 species and are believed to be hybrids.

River-oak casuarina is a medium-sized evergreen tree to 80 ft (24 m) high and 2 ft (0.6 m) in trunk diameter, with thin irregular crown of drooping twigs. Bark gray brown, smoothish but becoming rough, thick, and furrowed into narrow ridges. Inner bark brown and dark red within, gritty or slightly bitter. Wiry gray green drooping twigs mostly 3--7 in (7.5--18 cm) long and less than 1/32 in (1 mm) wide, with 8-10 long fine lines or ridges ending in scale leaves, shedding gradually. A few main twigs, finely hairy and pale green when young, develop into rough or smooth brownish branches.

Scale leaves less than 1/64 in (0.4 mm) long, 8-10 in a ring (whorled) at joints or nodes less than 1/4 in (6 mm) apart. Leaves on main twigs in rings as close as 1/16 in (1.5 mm), to 1/8 in (3 mm) long and curved back.

Flower clusters inconspicuous, light brown, male and female on different trees (dioecious). Male flower clusters (like spikes or catkins) terminal, narrowly cylindrical, 1/4--3/4 in (6--19 mm) long and less than 1/8 in (3 mm) wide. Tiny male flowers crowded in rings within grayish scales consist of 1 exposed brown stamen less than 1/8 in (3 mm) long with 2 tiny brown sepal scales at base. Female flower clusters are short-stalked lateral balls (heads) more than 1/4 in (6 mm) across spreading styles, consisting of pistil less than 1/4 in (6 mm) long with small ovary and long threadlike dark red style.

The multiple fruit is a small brown or gray hard warty ball about 3/8 in (10 mm) in diameter, often longer than broad and slightly cylindrical, composed of long broad hard points of 1/8 in (3 mm) long and broad, each from a flower. An individual fruit nearly 1/4 in (6 mm) long splits open in 2 parts at maturity to release 1 winged light brown seed (nutlet) less than 1/4 in (6 mm) long.

The hard heavy wood is composed of light brown sapwood and reddish-brown heartwood. The wood has broad rays which form a pronounced ray fleck similar to oak on radial surfaces--thus the common name. It is heavy (sp. gr. 0.58) with a relatively large shrinkage in drying. It could be used for turnery, but at present, at least in Hawaii, is used only for fuelwood.

This species has been tested in windbreaks but is suitable also for ornament and shade, having very fast growth. It may be better adapted to mountains than the coast, but in Hawaii it has been planted mostly at lower elevations. Altogether, 13,000 trees are recorded as having been planted in the Forest Reserves. Most of these were planted on Oahu at the Honouliuli and Waimanalo Reserves. The trees may be seen at many locations. There are some among other casuarinas on Kalakaua Avenue near the Aquarium in Waikiki. Others are at Ualakaa Park (Round Top) on Oahu and Waiakea Arboretum near Hilo and Kalopa State Park, both on the Island of Hawaii.

In Australia, this species attains the largest size in the genus and is also the most cold hardy. This species is adapted to subtropical mountains and is planted north to Florida, southern Arizona, and California, in the interior plateaus of Mexico, through the northern Andes, and south to Argentina and Chile. In central and south Florida, it is recommended for shade, shelter, and windbreaks, but not in cities, because of the large disruptive root systems.

Champion--Height 80 ft (24.4 m), c.b.h. 6.8 ft (2.1 m), spread 42 ft (12.8 m). Kohala Forest Reserve, Aamakao, Hawaii (1968).

Range--Native of Australia, but planted and naturalized in various tropical and subtropical regions.

Other common names--Cunningham casuarina, "ironwood," "Australian-pine," river she-oak, pino australiano, pino de Australia (Puerto Rico, Spanish).

26. River-oak casuarina *Casuarina cunninghamiana* Miq.*
Twig with female flowers and fruits, twig with male flowers (right), 1 X (Maiden).

27. Horsetail casuarina

Casuarina equisetifolia L. ex J. R. & G. Forst.*

Horsetail casuarina is the species most commonly planted in Hawaii and in other tropical and subtropical regions around the world, where it has become naturalized. A rapidly growing medium to large tree becoming 50--100 ft (15--30 m) tall and 1--1 1/2 ft (0.3--0.5 m) in trunk diameter, with thin crown of drooping twigs. The bark is light gray brown, smoothish on small trunks, becoming rough, thick, furrowed and shaggy, and splitting into thin strips and flakes exposing a reddish brown layer. Inner bark is reddish and bitter or astringent. The wiry drooping twigs mostly 9--15 in (23--38 cm) long, are dark green, becoming paler, with 6-8 long fine lines or ridges ending in scale leaves, shedding gradually like pine needles. A few main twigs, gray and finely hairy, become rough and stout and develop into brownish branches.

Scale leaves less than 1/32 in (1 mm) long, 6-8 in a ring (whorled) at joints or nodes 1/4--3/8 in (6--10 mm) apart. Leaves on main twigs in rings as close as 1/8 in (3 mm), to 1/8 in (3 mm) long and curved back.

Flower clusters inconspicuous, light brown, male and female on same tree (monoecious). Male flower clusters (like spikes or catkins) terminal, narrowly cylindrical, 3/8--3/4 in (10--19 mm) long and as much as 1/8 in (3 mm) across stamens, minute and crowded in rings among grayish scales, consisting of 1 protruding brownish stamen less than 1/8 in (3 mm) long with 2 minute brown sepal scales at base. Female flower clusters are short-stalked lateral balls (heads) less than 1/8 in (3 mm) in diameter or 5/16 in (8 mm) across spreading styles, consisting of pistil 3/16 in (5 mm) long including small ovary and long threadlike dark red style.

The multiple fruit is a light brown hard warty ball 1/2--3/4 in (13--19 mm) in diameter, often longer than broad and slightly cylindrical, composed of points less than 1/8 in (3 mm) long and broad, each from a flower. An individual fruit splits open in 2 parts at maturity to release 1 winged light brown seed (nutlet) 1/4 in (6 mm) long.

The sapwood is pinkish to light brown, the heartwood dark brown. The fine-textured wood is very hard, heavy (sp. gr. 0.81), and very susceptible to attack by dry-wood termites. Tests of the wood have been made in Puerto Rico. It is strong, tough, difficult to saw, but cracks and splits, and is not durable in the ground. Rate of air-seasoning is moderate, and amount of degrade is considerable. Machining characteristics are as follows: planing and turning are fair; and shaping, boring, mortising, sanding, and resistance to screw splitting are good. In Hawaii, the wood is used only as fuel. Elsewhere, the wood is used in the round. Uses include fenceposts and poles, beams (not underground), oxcart tongues, and charcoal. The bark has been employed in tanning, in medicine, and in the extraction of a red or blue-black dye. In southern Florida, the fruits have been made into novelties and Christmas decorations. Often propagated by cuttings for street, park, ornamental, and windbreak plantings, it can also be trimmed into hedges. It is used for reforestation because of its rapid growth and adaptability to degraded sites.

This tree grows rapidly, reportedly as much as 80 ft (24 m) in height in 10 years, and adapts to sandy seacoasts, where it becomes naturalized. It is very salt tolerant.

Common and naturalized along sandy coasts of Hawaii and up to more than 3,000 ft (914 m). It is used as windbreaks, such as along the Kohala Mountain Road, Hawaii; at Waimanalo, Oahu; and Hanalei, Kauai, near the pier. More than 70,000 trees were planted on the Forest Reserves and many others on private lands. The species was successfully established on severely eroded Kahoolawe where it was to be a windbreak for other tree species. However, goats broke through a fence and ate all the trees. The same system was used in the 1890's to plant the extremely windy Nuuanu Valley near the Pali.

Special areas--Waimea Arboretum, Kalopa.

Champion--Height 89 ft (27.2 m), c.b.h. 17.3 ft (5.3 m), spread 56 ft (17.1 m). Olowalu, Maui (1968).

Range--Native of tropical Asia and Australasia but planted and naturalized in various tropical and subtropical regions. Planted and naturalized in southern Florida and Puerto Rico and Virgin Islands.

Other common names--"shortleaf ironwood," "common ironwood" (Hawaii); "Australian-pine," beefwood, she-oak, toa; pino australiano, pino de Australia (Puerto Rico, Spanish).

28. Longleaf casuarina

Casuarina glauca Sieber ex Spreng.*

This introduced species differs from related species in the longer and thicker drooping dull green wiry twigs 12--16 in (30--40 cm) long and nearly 1/16 in (1.5 mm) in diameter, with scale leaves 12-16 in a ring. The cone-like fruits are about 1/2 in (13 mm) in diameter. Stands are easily recognized because they are dense or closely spaced due to the root suckering habit of the tree.

Medium-sized evergreen tree to 40--50 ft (12--15 m) high, with straight trunk to 1 1/2 ft (0.5 m) in diameter, becoming slightly enlarged at base, with thin crown of drooping twigs. Bark on erect branches gray brown, smoothish, on trunk becoming rough, thick, and furrowed. Inner bark light brown, slightly fibrous and astringent. Wiry drooping twigs with 12--16 long fine lines or

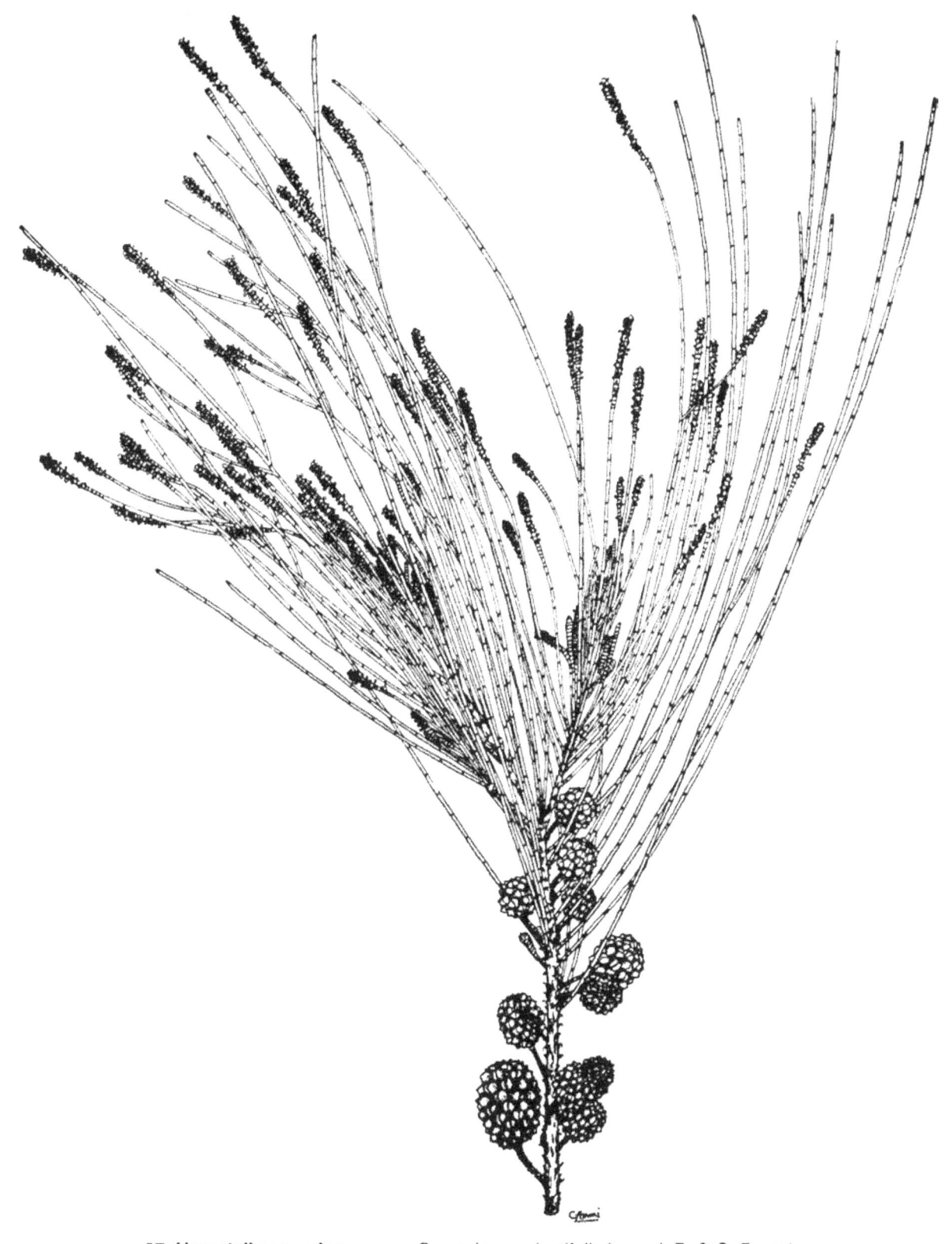

27. Horsetail casuarina — *Casuarina equisetifolia* L. ex J. R. & G. Forst.*
Twig with male flowers and fruits, 2/3 X (Degener).

ridges ending in scale leaves, the twigs shedding and forming rounded scar. A few main twigs become stout, brown, and rough.

Leaves consisting of 12--16 tiny pointed brown scales 1/32 in (1 mm) long, about 1/2--3/4 in (13--19 mm) apart. Main twigs with rings as close as 1/8 in (3 mm), with longer scale leaves to 3/16 in (5 mm) and curved back, becoming stout, brown, and rough.

Flower clusters light brown, male and female apparently on different trees (dioecious). Male flower clusters (like spikes or catkins) at ends of long twigs, narrowly cylindrical, mostly 3/4--1 1/2 in (2--4 cm) long and 5/16 in (8 mm) in diameter. Male flowers crowded in rings of narrow pointed scales, consist of 1 stamen less than 1/8 in (3 mm) long with 2 tiny sepal scales at base. Female flower clusters are short-stalked lateral balls (heads) about 1/4 in (6 mm) in diameter, consisting of pistil with small ovary and long threadlike style.

The multiple fruit is a gray hard warty ball composed of many finely hairy points about 1/8 in (3 mm) long and broad, each from a flower. An individual fruit splits open in 2 parts at maturity to release 1 winged light gray seed (nutlet) less than 1/4 in (6 mm) long.

Sapwood pale yellow and heartwood dark brown, often has a beautiful figure provided by the prominent oak-like rays. The wood is hard, heavy, fine-textured, strong, but not durable in ground contact. It is difficult to season and to work. It takes a good polish, but is said to be brittle. Uses in Australia include shingles, fence rails, staves, paneling, furniture, oxen yokes, and marine piling. In Hawaii, it is used only for fuel.

By far, this is the most common species of *Casuarina* in the Forest Reserves, with almost 1 million trees planted. Because of its suckering habit, the species was early recognized as excellent for recapturing erosion scars. It has come into disfavor, however, because it also takes over good land wherever planted. Most ranchers in Hawaii regard it as noxious. A typical stand may be seen near the beginning of the Aiea Loop trail on Oahu.

Special area--Kalopa.
Champion--Height 65 ft (20.9 m), c.b.h. 10.4 ft (3.2 m), spread 37 ft (11.3 m). Waikoekoe, Waimea Village, Hawaii (1968).
Range--Native of coasts of eastern and southern Australia.
Other common names--"longleaf ironwood," "salt-marsh ironwood" (Hawaii); swamp-oak, she-oak (Australia).

This species grows naturally in Australia along coasts in swampy margins of tidal areas and should be suitable for planting in saline soils.

BAYBERRY FAMILY (WAXMYRTLE FAMILY, MYRICACEAE*)

29. Firetree   *Myrica faya* Ait.*

Evergreen introduced shrub or small tree, with many narrowly elliptical leaves, twigs and leaf surfaces with brown or yellow dot scales visible under a lens, and many dark red to blackish fruits more than 1/4 in (6 mm) in diameter. Shrub or small tree to 40 feet (12 m) high and 10 in (25 cm) in trunk diameter. Bark brown to gray, smooth or becoming slightly fissured, inner bark greenish yellow, bitter. Twigs greenish, angled, with raised half-round leaf scars.

Leaves many, crowded, alternate, hairless, thin, narrowly elliptical, 1 1/4--4 1/2 in (3--11 cm) long, 3/8--3/4 in (1--2 cm) wide, short-pointed at apex, widest beyond middle, tapering to base and slender leafstalk about 1/4 in (6 mm) long, edges slightly turned down and often wavy toothed, above slightly shiny dark green, beneath dull green with veins slightly raised.

Flower clusters (spikes) at leaf bases, 1/2--1 1/4 in (1.3--3 cm) long, unbranched. Flowers many, stalkless, minute, male and female on same plant (monoecious), without calyx or corolla, each above a scale. Male flowers 3/16 in (5 mm) long, with 4 pinkish tinged stamens. Female flowers often joined together in groups of 3 above a scale.

Fruits (drupes) many, stalkless along slender unbranched axes 3/4--2 1/4 in (2--6 cm) mostly back of leaves, round, more than 1/4 in (6 mm) in diameter, turning from greenish to dark red to blackish; the surface with many tiny round beadlike scales, flesh reddish, almost tasteless. Seed 1, brown, rounded, more than 1/8 in (3 mm) diameter.

Introduced into Hawaii as an ornamental probably by Portuguese settlers, who made wine from the fruits, and now naturalized in moist areas throughout the islands. It is especially common in the pasture land above the town of Paauilo on the Hamakua Coast of Hawaii. Also present and spreading at Hawaii Volcanoes National Park and on the fringes of the Alakai Swamp, Kauai, where it was planted in 1927 and again in 1940. Almost all plantings were made by the Division of Forestry in 1926 and 1927. Since 1940, it has been considered one of Hawaii's most noxious plants. A nitrogen fixer, it has the ability to take over the best pasture land by forming dense thickets. An active eradication program is underway but is a continuing struggle with the present system of poisoning and uprooting.

Special areas--Kokee, Volcanoes.
Range--Native of Azores and Canary Islands.

28. Longleaf casuarina *Casuarina glauca* Sieber ex Spreng.*
Twig with female flowers, twig with fruits (lower left), twig with male flowers (right) 1 X (Maiden).

29. Firetree *Myrica faya* Ait.*
Twig with male flowers (upper left) and twig with fruits, 1 X.

30. Nepal alder *Alnus nepalensis* D. Don*
Twig with buds of female flowers and cones, 1 X.

BIRCH FAMILY (BETULACEAE*)

30. Nepal alder
Alnus nepalensis D. Don*

Medium-sized deciduous tree introduced in forest plantations in moist mountain areas, with elliptical wavy toothed leaves, becoming 30--90 ft (9--27 m) in height. Trunk straight, 1--2 ft (0.3--0.6 m) in diameter, becoming slightly enlarged at base. Crown spreading, irregular. Bark gray, smooth with horizontal corky ridges, becoming fissured. Inner bark pinkish or brown streaked, bitter or astringent. Twigs greenish, becoming brown, hairless, with raised half-round leaf scars, and pith triangular in cross section.

Leaves alternate, with paired narrow greenish stipules shedding early. Leafstalk short, less than 3/8 in (1 cm) long, light green. Blades elliptical, 2 1/2--5 in (6--13 cm) long and 1 1/2--3 in (4--7.5 cm), wide, short-pointed at both ends, finely wavy toothed on edges, thin, becoming hairless, above dull green with many slightly curved side veins, beneath paler with dotlike yellow brown scales and slightly raised veins.

Flowers male and female on different twigs, minute, crowded with scales. Male flowers in several stalked, long, narrow drooping clusters (catkins). Female flowers in conelike clusters on branching side twigs, from narrowly cylindric dark gray green buds 5/16 in (8 mm) long. Fruits many, conelike, 1/2--5/8 in (13--25 mm) long, becoming dark brown, with many spreading scales, remaining attached. Seeds (nutlets) minute, rounded and flat, light brown, more than 1/16 in (2 mm) long, with 2 broad wings.

The wood of 5 trees from the Kohala Forest Reserve was tested for some strength properties, machining, and veneer manufacture at the Forest Products Laboratory (Gerhards 1964; Lutz and Roessler 1964; Peters and Lutz 1966). It is pale brown or blond and marked with occasional broad rays, as is red alder (*Alnus rubra* Bong.). It is of low density (sp. gr. 0.34), similar in most properties to red alder but softer. It is easily seasoned and machines well, but is not resistant to decay or termite attack. The wood is suitable for plywood corestock, drawer sides and backs, and other interior uses in furniture manufacture but is not presently used in Hawaii. In India, it is reported that the bark serves for tanning and dyeing. It is a nitrogen fixer, as are other alders.

Within forest plantations in moist mountain areas of Hawaii, such as at 3,000 ft (914 m) altitude in Kohala Forest Reserve on Hawaii, where there are about 500,000 board feet of timber. Also at 3,500 ft (1,067 m) in Molokai Forest Reserve, Molokai. It is suited to very wet forests but is subject to windthrow and breakage in windy areas.

Champion--Height 75 ft (22.9 m), c.b.h. 11.3 ft (3.4 m), spread 38 ft (11.6 m). Pepeekeo Arboretum, Hilo, Hawaii (1968).

Range--Himalaya Mountains of Nepal and India. Reported to be common in temperate forests.

ELM FAMILY (ULMACEAE)

31. Gunpowder-tree
Trema orientalis (L.) Blume*

Medium-sized introduced evergreen weedy tree, recognized by the narrowly ovate unequal-sided very long-pointed leaves with finely toothed edges and 3 main veins from base, spreading in 2 rows on long slender twigs. To 60 ft (18 m) high and 2 ft (0.6 m) in trunk diameter. Bark light gray grown, smoothish, finely fissured. Inner bark pink, soft and fibrous, bitter. Twigs long, slender, unbranched, spreading, light green, finely hairy, turning brown.

Leaves borne singly (alternate) in 2 rows with slender leafstalks 3/8--3/4 in (1--2 cm) long, finely hairy. Blades narrowly ovate, 2 1/2--4 in (6--10 cm) long and 3/4--1 1/4 in (2--3 cm), wide, very long-pointed at apex, base slightly notched with sides unequal, edges finely toothed, thin, with 3 main veins from base. Upper surface light green, slightly shiny, slightly rough, with veins sunken, lower surface dull and paler, soft hairy, with prominent light yellow veins.

Flower clusters (cymes) at leaf bases, branched, 3/8--2 in (1--5 cm) long and broad. Flowers many, nearly stalkless, light green, 1/8 in (3 mm) long, male and female. Female flowers composed of 5 tiny sepals and pistil with 1-celled ovary and 2 whitish hairy spreading styles.

Fruits (drupes) round, nearly 3/16 in (5 mm) in diameter, pink to black, fleshy. Seed 1 round, brown, more than 1/16 in (1.5 mm) long.

Wood pale brown or buff, sapwood not distinct from heartwood. It is lightweight (sp. gr. 0.40), easy to work, and not resistant to decay or termites (Reyes 1938). The wood has not been used in Hawaii. It is said to make good charcoal for gunpowder, thus the common name.

A weed tree of rapid growth, extending into forest openings in moist lowland areas in Hawaii. It is particularly common in the vicinity of Hilo airport.

31. Gunpowder-tree   *Trema orientalis* (L.) Blume*
Twig with flowers and fruits, 1 X.

Champion--Height 72 ft (21.9 m), c.b.h. 20.5 ft (6.2 m), spread 108 ft (32.9 m). Pepeekeo, Hawaii (1968).

Range--Native from southeastern Asia through Malaysia.

Other common names--charcoal tree; banahl (N. Marianas); elodechoel (Palau).

Trema cannabina Lour. (*T. amboinensis* (Willd.) Blume) is a related native species, a small tree to 30 ft (9 m) high with smaller hairy black fruits about 1/8 in (3 mm) long, only slightly fleshy. Elsewhere, a fiber for fish nets has been made from the bark and medicine from other parts. In Hawaii, uncommon in lowland areas, recorded from the 5 largest islands. Native and more common in other Pacific Islands, such as Samoa, Fiji, and Tahiti, and southeastern Asia to China and India.

MULBERRY FAMILY (MORACEAE)

32. 'Ulu, breadfruit *Artocarpus altilis* (Parkins.) Fosberg**

Breadfruit, introduced by the early Hawaiians, is a handsome tree planted for its edible fruits and attractive foliage. It is easily recognized by the very large, usually deeply 7--11-lobed shiny dark green leaves, the yellowish green rounded or elliptical fruits 4--8 in (10--20 cm) long, and the milky sap that exudes from cuts. The commonly cultivated Hawaiian variety is seedless.

A medium-sized spreading evergreen tree 40--60 ft (12--18 m) high and 2 ft (0.6 m) or more in trunk diameter, with relatively few stout branches. Bark brown, smooth, with warty dots (lenticels). Inner bark whitish and almost tasteless, with white, slightly bitter sap or latex. Twigs very stout, 1/2--1 in (13--25 mm) in diameter, green and minutely hairy, with rings at nodes, ending in large, pointed, finely hairy bud 5 in (13 cm) or less in length, formed by a big scale (stipule) around developing leaf.

Leaves alternate on very stout green leafstalks of 1--2 in (2.5--5 cm). Blades elliptical in outline, about 15--20 in (38--51 cm) long and 8--12 in (20--30 cm) wide, sometimes larger, the pinnate lobes long-pointed, base short-pointed, thickened and leathery, upper surface nearly hairless except along veins, and lower surface lighter green and finely hairy at least on veins. Leaves of the seeded variety are less deeply lobed, usually have 9 or 11 lobes instead of 7, and are more hairy.

Flowers are very numerous and minute, male and female on same tree (monoecious) in separate thick, fleshy clusters (heads) borne singly at leaf bases on stalks of about 2 in (5 cm). Male cluster a cylindrical or club-shaped soft mass about 5--12 in (13--30 cm) long and 1 in (2.5 cm) in diameter, yellowish and turning brown. Male flowers 1/16 in (1.5 mm) long, consisting of 2-lobed calyx and 1 stamen, crowded on outside. Female flower cluster elliptical or rounded, about 2 1/2 in (6 cm) long and 1 1/2 in (4 cm) in diameter or larger, light green. Female flowers 3/8 in (10 mm) long and 1/16 in (1.5 mm) wide, composed of tubular conelike pointed, hairy calyx projecting 1/4 in (6 mm) and pistil with sunken 1-celled 1-ovuled ovary and 2-lobed style. Female flowers in seedless variety sterile and projecting only about 1/32 in (1 mm).

Fruits (multiple) are brownish, covered with individual fruits, and contain a whitish starchy pulp. Fruit surface in seeded variety is composed of greenish conical spinelike projections, each from 1 flower. Seeds several large brown edible. Seedless variety has smoothish surface honeycombed with individual fruits about 3/16 in (5 mm) across. Fruits mature mainly from June to August in Hawaii.

Sapwood light yellow to yellowish brown; heartwood golden colored, sometimes flecked with orange. Wood very soft, lightweight (sp. gr. 0.27), but relatively strong for its weight. It is very susceptible to attack by dry-wood termites. There are numerous large pores but no growth rings. Rate of air-seasoning and amount of degrade are moderate. Machining characteristics are as follows: planing is fair; shaping, turning, boring, and mortising are very poor; sanding is poor; and resistance to screw splitting is excellent. The wood is suitable for boxes, crates, light construction, and toys.

Because of its lightness, Hawaiians used breadfruit wood for surf boards and canoe hulls. It is said to have been the preferred wood for the platform or deck between the hulls of large double-hulled canoes. A large drum was hollowed from a section of the trunk. An inferior grade of tapa or bark cloth was made from the bark.

Fruits are gathered before maturity and roasted or boiled as a starchy vegetable. Young fruits can be sliced and fried, and the seeds boiled or roasted. A dessert and preserves can be made from the starchy male flower clusters. In Hawaii, breadfruit was not an important food but served partly to fatten hogs. This was probably because the Hawaiians had not introduced good varieties. In the Marquesas, whence the Hawaiians are believed to have emigrated, breadfruit was and still is the primary foodstuff of the people. There, it is made into a paste and fermented in pits into a cheese-like substance. Throughout Polynesia, it is eaten with a sauce made of coconut milk, seawater, and lime juice.

The trees are also attractive for ornament and shade. Elsewhere, cut foliage has served as forage for cattle during periods of drought. Rough leaves were used as sandpaper.

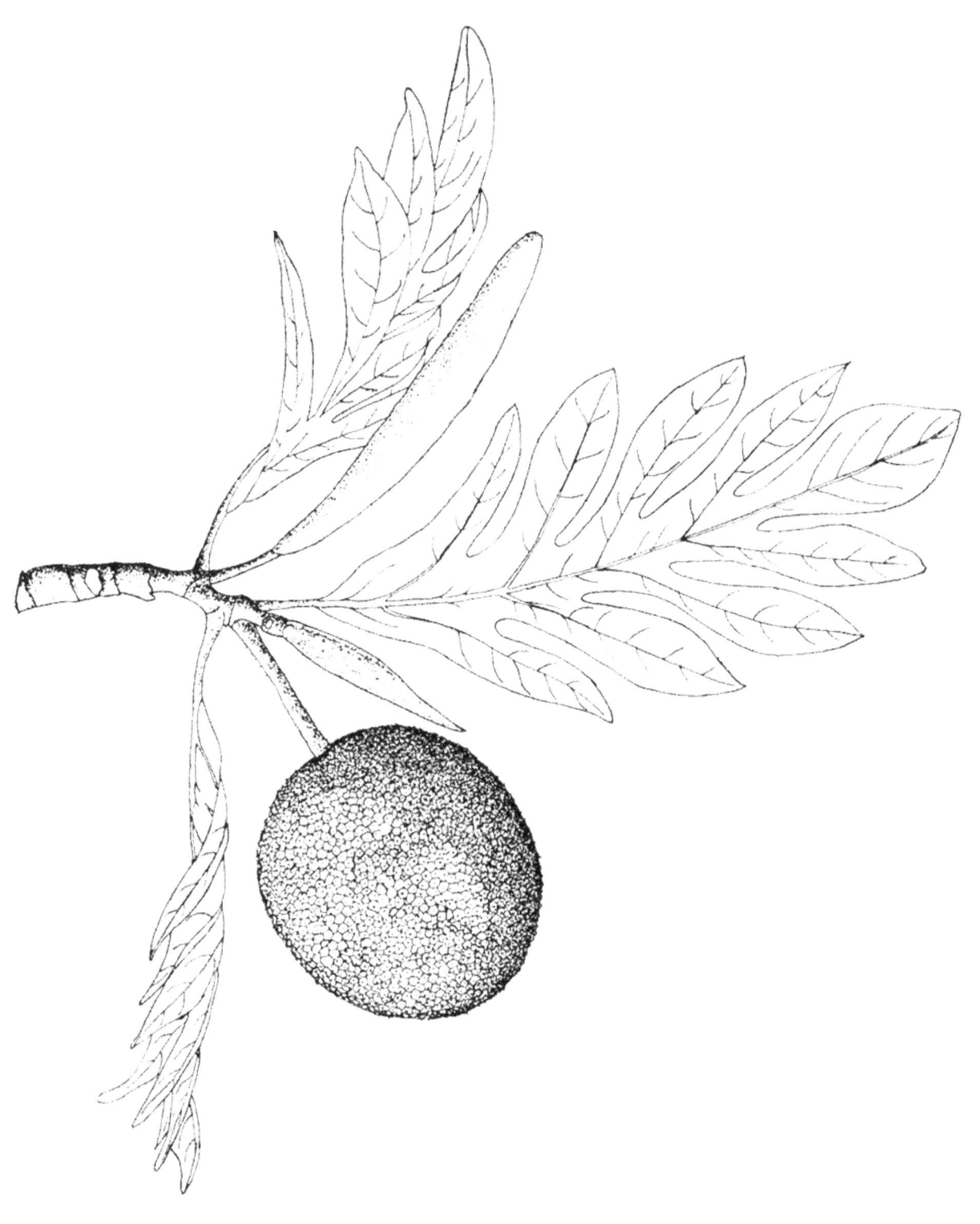

32. 'Ulu, breadfruit   *Artocarpus altilis* (Parkins.) Fosberg**
Twig with male flowers and immature fruits, 1/3 X (P. R. v. 1).

Special use of the sticky milky sap of breadfruit and certain other plants as bird lime has been told by Degener. Under instructions from King Kamehameha, the royal birdcatchers captured certain small birds for their feathers of special colors. These rare feather colors were needed to decorate the precious war cloaks, helmets, and images. The black mamo bird, now extinct, had a small tuft of yellow feathers under each wing. The glue was smeared on long baited poles which were placed where the birds would perch and become entangled. After the choice feathers were plucked and the feet of the birds carefully washed in oil or juice, the birds reportedly were released unharmed to grow more feathers!

According to a legend related by Degener, the chief Kahai brought the breadfruit tree to Hawaii from Samoa in the twelfth century and first planted it at Kualoa, Oahu. Only one variety was known in Hawaii, while more than 24 were distinguished by native names in the South Seas. Absence of the variety with seeds indicates that this species was not native. In Hawaii, the trees are found near dwellings or in lowland valleys at former homesites.

This tree was introduced by the British into the West Indies in 1793 from Tahiti of the South Sea Islands to provide cheap food for slaves. It was claimed that three or four mature trees could provide starchy food to support a person throughout the year. Captain William Bligh in the ship Providence chartered by the British Government brought plants to St. Vincent and Jamaica. This special expedition was undertaken to transport potted plants of the seedless variety the great distance. An earlier attempt with a cargo of plants on board the ship *Bounty* failed because of the famous mutiny against Captain Bligh in 1789. About the same time the French brought a few breadfruit trees to other islands of the West Indies.

Propagation is by root cuttings, suckers, or layering and in the seeded variety by seeds. Growth is rapid.

Common as a fruit tree through the tropics, the seeded variety escaping from cultivation. In Puerto Rico and Virgin Islands, it is planted around homes and escapes occasionally. Rare in southern Florida, fruiting only at Key West.

In Hawaii, planted and persistent in moist lowlands.

Special areas--Waimea Arboretum, Foster, City.

Champion--Height 54 ft (16.5 m), c.b.h. 10.3 ft (3.1 m), spread 59 ft (18.0 m). Hilo Hotel, Hilo, Hawaii (1968).

Range--Native in South Pacific Islands, probably New Guinea, but planted through the tropics.

Other common names--panapén, árbol de pan, palo de pan, pan, pana (Puerto Rico, Spanish); dogdog (Guam); lemai (N. Marianas); arudo (Palau); maa (Yap); mai (Truk, Pohnpei); mohs (Kosrae); ma (Marshalls); 'ulu (Am. Samoa).

Botanical synonym--*Artocarpus communis* J. R. & G. Forst., *A. incisus* (Thunb.) L. f.

33. Trumpet-tree

Cecropia obtusifolia Bertol.*

The trumpet-tree, introduced as watershed cover, is easily recognized by the very open crown of few stout hollow branches and few very large long-stalked umbrellalike leaves deeply divided into 9--15 oblong lobes, whitish beneath. A medium-sized evergreen tree to 50 ft (15 m) high, with branches arising high and curving upward. The trunk to 8 in (20 cm) in diameter has a few prop roots at base. Bark light gray, smooth and warty, with prominent rings. Inner bark whitish, fibrous, almost tasteless, its sap turning black on exposure. Branches few, very stout, hairy, greenish gray, smooth with prominent rings and dots, and hollow. End buds large, short-pointed, about 4 in (10 cm) long, covered by a large dark red hairy scale (stipule), which falls early, leaving a ring scar.

The few leaves alternate at ends of branches have very large round reddish hairy leafstalks of 6--16 in (15--40 cm). The leathery rounded blades 10--20 in (25--51 cm) in diameter have the leafstalk inserted near the middle (peltate), with 9--15 long oblong lobes often broadest beyond middle, with apex short-pointed or rounded, each lobe with dark red midvein and many straight parallel side veins. Upper surface dull dark green and rough, lower surface whitish green and finely hairy.

Flower clusters (groups of spikes) paired at leaf bases, develop inside a large pointed pinkish hairy bud scale, and bear very numerous tiny flowers, male and female on different trees (dioecious). The male flower cluster is composed of 12--15 yellow ropelike branches 3 1/4--5 1/2 in (8--14 cm) long and more than 1/8 in (3 mm) in diameter, hanging from a stalk of 1 1/2--4 in (4--10 cm). The tiny yellow male flowers 1/32 in (1 mm) long have a tubular calyx and 2 stamens. Female flower cluster consists of 3--6 greenish or whitish ropelike branches 6--14 in (15--36 cm) long and 3/16 in (5 mm) in diameter, hanging from a stalk to 6 in (15 cm) long. The numerous tiny female flowers in a mass of whitish hairs have tubular calyx and pistil with 1-celled ovary, 1 ovule, short style, and enlarged hairy stigma.

The fruiting branches (multiple fruits) consist of 3-6 long greenish or gray ropelike branches more than 1/4 in (6 mm) thick and slightly fleshy, hairy when dry, covered with brown dots. These are the individual 1-seeded fruits (achenes), less than 1/16 in (1.5 mm) long, dark brown.

The wood is yellowish and soft, with large pith, hollow in branches. According to tests elsewhere, it is suitable for pulpwood. The fibrous bark has been made into ropes. The wood of a closely related species, *Cecropia*

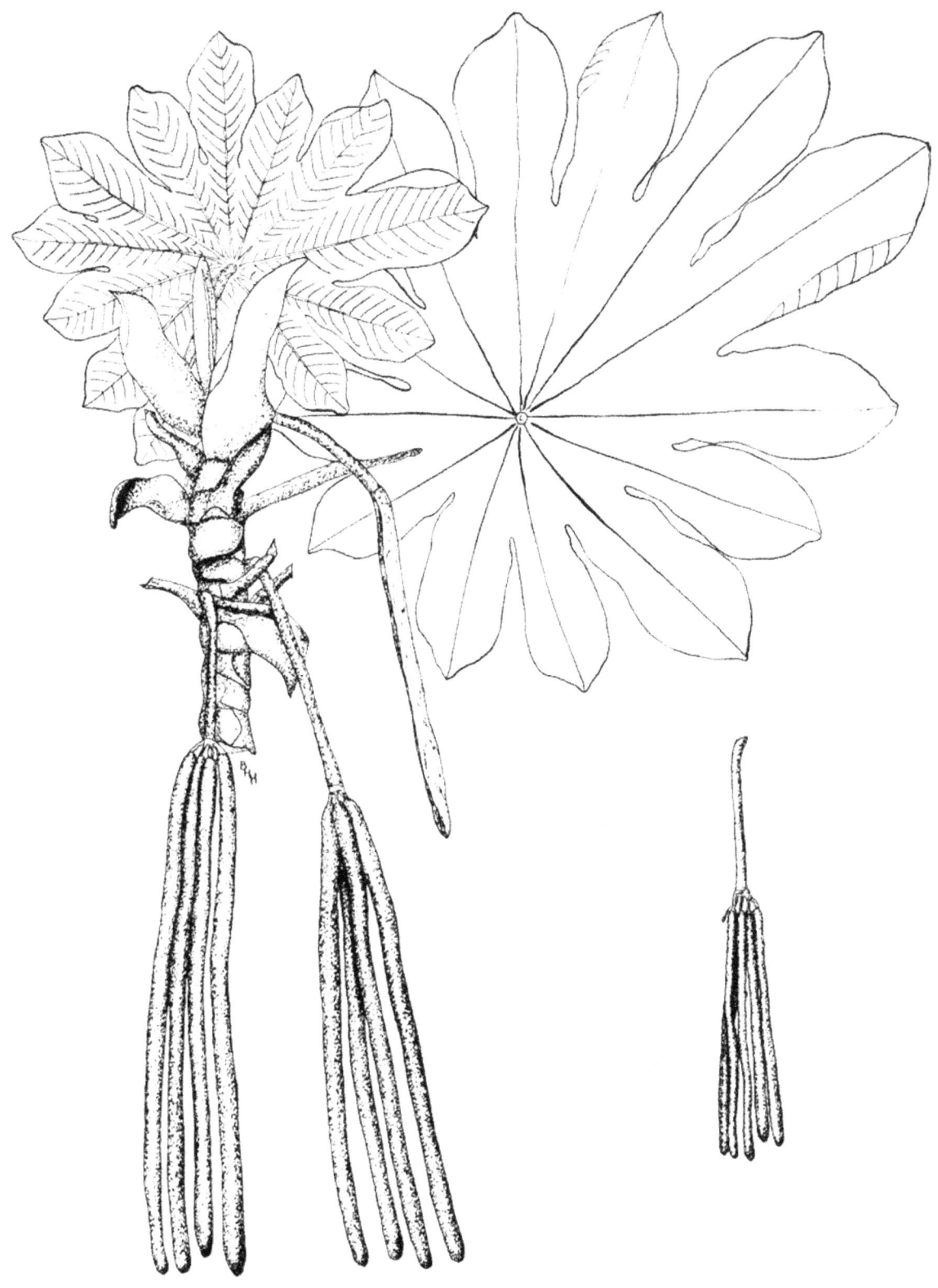

33. Trumpet-tree *Cecropia obtusifolia* Bertol.*
Twig with leaves and fruit clusters, and male flower cluster (lower right), 1/2 X.

peltata L., from Puerto Rico was tested at the Forest Products Laboratory and found to be suitable for certain low value uses as lumber or veneer. The wood is similar to that of black cottonwood in density and strength, but has a very high tangential shrinkage, which results in severe drying degrade (Bendtson 1964).

Sparingly introduced in Hawaii. It was aerially seeded in the Panaewa Forest Reserve near Hilo in 1928 and has become naturalized there and in the lower Waiakea Forest Reserve. It is also common near Kilauea, Kauai, where it has naturalized, and at the end of Manoa Road on Oahu. A total of only 752 trees are recorded as having been planted by the Division of Forestry, mostly in 1927 and 1928. It is a pioneer and weed tree of very rapid growth in clearings, other open areas, and secondary forests.

Special areas--Waimea Arboretum, Waiakea.

Champion--Height 70 ft (21.3 m, c.b.h. 9 ft (2.7 m), spread 72 ft (21.9 m). Paradise Park, Honolulu, Oahu (1968).

Range--Native from Mexico and Belize through Central America to Panama and northern South America.

Other common names--trumpet; guarumo (Spanish).

Botanical synonym--*Cecropia mexicana* Hemsl.

In Hawaii, this species apparently was introduced under the name of a related West Indian trumpet-tree, *Cecropia peltata* L., which was also planted. The latter, which is common in Puerto Rico, differs in leaves less deeply divided into short rounded lobes and the 2--5 short stout fruiting branches 2 1/2--4 in (6--10 cm) long and 3/8--1/2 in (10--13 mm) in diameter.

The English common name refers to a use of the hollow branches and leafstalks for trumpets or other musical instruments.

Hollow branches of this and related species in their native homes are inhabited by biting ants that bore holes to the interior. Early naturalists supposed that the ants repaid the tree by driving away other insects. However, the related trees in Puerto Rico have no ant dwellers and are abundant. On the Hawaiian trees, few small harmless ants, also introduced, were noted.

34. Chinese banyan

Ficus microcarpa L. f.*

This large ornamental tree of the fig genus, planted in parks and gardens, is distinguished by its short trunk and very dense broad rounded crown, by small dark green elliptical leaves with very small figlike fruits paired at base, by numerous aerial roots about trunk or hanging hairlike from lower branches, and by the milky juice or white latex exuding from cuts.

Large evergreen introduced tree to 65 ft (20 m) high, the trunk to 3 ft (0.9 m) and often with buttresses at base, crown often broader than tall, hairless throughout. Bark smooth, gray. Inner bark whitish and tasteless but containing slightly bitter latex. Twigs slender, gray, ending in long-pointed green scale (stipule) 3/8 in (1 cm) or less in length, which forms bud and leaves ring scar at leaf base upon shedding.

Leaves alternate on leafstalks of 1/4--3/8 in (6--10 mm). Blades 1 1/2--3 in (4--7.5 cm) long and 5/8--1 1/2 in (1.5--4 cm) wide, short-pointed at both ends and often nearly diamond-shaped (rhomboidal), with midvein and 2 main side veins from base along toothless margin, thick and leathery, upper surface dark green and slightly shiny, lower surface paler.

As flowers in this genus of figs are not visible, it appears that the trees have fruits but no flowers. The figlike multiple fruit (syconium), actually a compound fruit, corresponds to an enlarged overgrown flower stalk bearing on inner walls numerous tiny male and female flowers (monoecious) and the small seeds, each technically a fruit from a single flower. This species has small rounded figlike fruits paired and stalkless at leaf bases, about 5/16 in (8 mm) in diameter, with tiny pointed opening at apex and 3 pointed scales (bracts) 1/16 in (1.5 mm) long at base, green, turning yellow or reddish at maturity. Inside the fruit are many tiny male and female flowers (monoecious) and seeds. Fruiting probably through the year.

Sapwood whitish and heartwood light brown. Wood is moderately heavy (sp. gr. 0.50), marked by terminal parenchyma, with growth rings, soft and easy to work, very susceptible to attack by dry-wood termites.

This is a common species of the genus grown as an ornamental in lowlands of Hawaii. It has escaped from cultivation and can be found occasionally in the forest. Near homes, it seeds prolifically in drain pipes and gutters where small deposits of silt permit rooting. Actually, only 469 trees of this species are reported as having been planted in the Forest Reserves. It does, however, serve as a representative of the genus, of which at least 33 other species have also been planted in the forests (about 60 spp. have been introduced). The 3 most common in the forests are Port Jackson fig, *Ficus rubiginosa* Desf. (40,000 trees planted), Moreton Bay fig, *Ficus macrophylla* Desf. (36,000 trees), and rough-leaf fig, *Ficus nota* Merr. (25,000 trees).

The dense crowns are frequently trimmed into rounded shapes. Rooting of cuttings is uncertain, but sometimes successful. Better results have been obtained by air layering or marcottage, in which a fairly large branch can be used. In some places this tree is considered objectionable because of its size, the litter of the numerous fruits, or because a thrips insect may

34. Chinese banyan *Ficus microcarpa* L. f.*
Twig with fruits, 1 X (P. R. v. 1).

deform the foliage and may irritate the eyes of persons beneath the tree.

Special area--Foster.

Champion--Height 104 ft (31.7 m), c.b.h. 90.1 ft (27.5 m), spread 195 ft (59.4 m). Keaau Village, Puna, Hawaii (1968). This banyan with trunks and air roots grown together has the greatest trunk circumference and greatest crown spread of all Hawaiian champions.

Range--Native of India and Malaysia but widely planted in tropical regions.

35. A'ia'i, Hawaiian false-mulberry

Small slender evergreen tree native through the islands in wet and dry forests. Characterized by milky sap, shiny, finely toothed leaves alternate in 2 rows, and rounded brownish purple fleshy fruits about 3/8 in (1 cm) long. Usually small tree or shrub, sometimes to 40 ft (12 m) high, with trunk to 2 ft (0.6 m) in diameter, slightly enlarged at base. Bark light gray, smooth. Inner bark yellowish within green outer layer, fibrous, almost tasteless, with bitter white sap or latex. Twigs becoming brown and nearly hairless, with stipules forming bud 3/16 in (5 mm) long and leaving ring scars at leaf bases.

Leaves alternate in 2 rows, with paired pointed greenish stipules soon falling and with short finely hairy leafstalk about 3/16 in (5 mm) long. Blades oblong to lance-shaped, short-pointed at apex, rounded or slightly notched at base, finely toothed on edges, thin, with 10-12 straight veins on each side, above dark green and hairless, beneath light green with raised yellowish veins slightly hairy.

Flowers male and female mostly on different trees (dioecious). Male flowers in long narrow clusters (spikes) to 4 1/2 in (11 cm) long and 1/4 in (6 mm) wide, very numerous, crowded, whitish to purplish, about 3/16 in (5 mm) long, consisting of 4-lobed finely hairy calyx

Planted in southern Florida, Puerto Rico, and the Virgin Islands and elsewhere in tropical America for ornament and shade. It is a popular tree of plazas or town squares.

Other common names--Malayan banyan, India-laurel fig; laurel de la India (Spanish); jagüey (Puerto Rico); fig (Virgin Islands); nunu (N. Marianas); lulk (Palau).

Botanical synonym--*Ficus nitida* Thunb., not Blume. Formerly referred to as *F. retusa* L.

Streblus pendulinus (Endl.) F. Muell.

and 4 spreading stamens. Female flowers few in short cluster (spike) less than 1/2 in (13 mm) long, composed of 4-lobed calyx and pistil with green ovary and 2 spreading stigmas.

Fruits (drupes) few, rounded, brownish purple, about 3/8 in (1 cm) long, slightly flattened and pointed, hairless, shiny with calyx at base, juicy sweetish white flesh, and rounded stone 1/4 in (6 mm) long.

The wood is described as light brown, fine-textured, hard, and tough. That of a related species in Australia was used for boomerangs.

Uncommon in wet and dry forests from near sea level to about 5,300 ft (1,676 m) altitude. Recorded from Kauai, Oahu, Lanai, Molokai, Maui, and Hawaii.

Special area--Kokee.

Champion--Height 38 ft (11.6 m), c.b.h. 4 ft (1.2 m), spread 20 ft (6.1 m). Kaupulehu, Kailua-Kona, Hawaii (1968).

Range--New Guinea to Micronesia, s. to Norfolk Island., e. Australia, New Hebrides, Fiji, Rapa, and Hawaii.

Botanical synonyms--*Pseudomorus sandwicensis* Deg., *P. brunoniana* var. *sandwicensis* (Deg.) Skottsb., *Streblus sandwicensis* (Deg.) St. John.

NETTLE FAMILY (URTICACEAE)

36. Mamaki

Pipturus albidus (Hook. & Arn.) Gray

This variable species (in the broad sense including closely related species) is characterized by ovate leaves with wavy toothed edges, 3 main veins from base, and under surface light gray or brown and finely hairy. The fibrous bark was an important source of tapa or paper cloth.

A small tree to 30 ft (9 m) high and 1 ft (0.3 m) in trunk diameter, or a shrub, with long drooping branches. Bark light brown, smooth, with scattered raised dots. Inner bark streaked green, fibrous, mucilaginous, and almost tasteless. Twigs finely gray hairy, slightly enlarged at nodes and often slightly zigzag.

Leaves alternate, varying greatly in shape, size, and hairiness, with slender finely hairy leafstalks of 1--3 in

(2.5--7.5 cm). Blades ovate or elliptical, 2 1/2--8 in (6--20 cm) long and 1 1/4--6 in (3--15 cm) wide, thin or slightly thickened, long-pointed at apex, blunt or rounded at base, with wavy toothed edges, with 3 main veins from base slightly sunken and often reddish, upper surface green and slightly rough, under surface mostly light gray and finely hairy. Microscopic mineral growths (cystoliths) like crystals are present.

Flower clusters (heads) stalkless at leaf bases, rounded, 1/4--1/2 in (6--13 mm) in diameter, gray hairy. Flowers male and female mostly in different clusters of same plant (monoecious), many, stalkless, without corolla. Male flowers less than 1/8 in (3 mm) long, composed of cup-shaped 4-lobed finely hairy calyx and 4

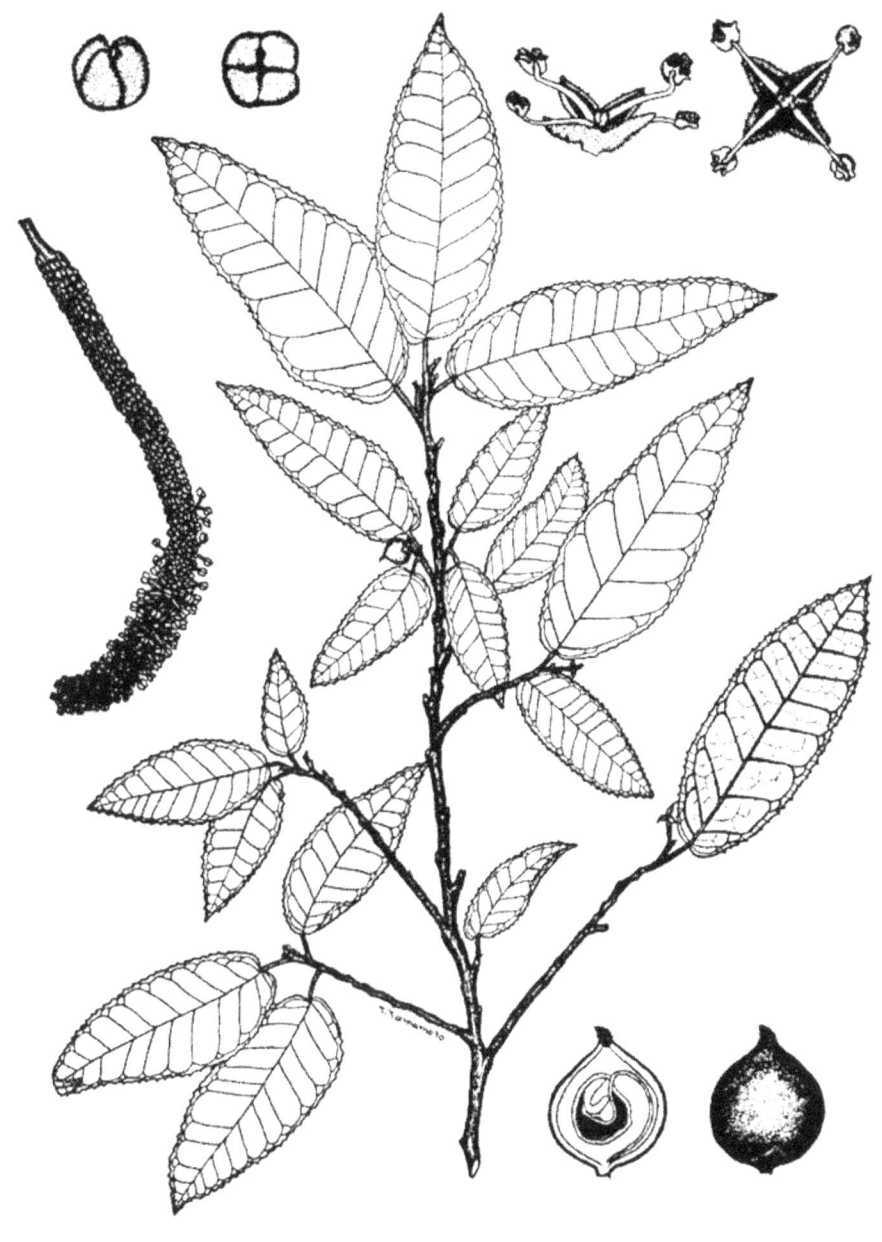

35. A'ia'I, Hawaiian false-mulberry — *Streblus pendulinus* (Endl.) F. Muell.
Twig with fruit, 1/2 X; female and male flowers (above), 5 X; male flower cluster (left), 1 X; fruits (lower right), 2 X (Degener).

stamens. Female flowers with urn-shaped 4-toothed finely hairy calyx and hairy pistil with ovary and long style.

Fruit a dry whitish ball about 1/2 in (13 mm) across. Individual fruits many, rounded, with enlarged dry calyx enclosing 1 seed (achene) 1/16 in (1.5 mm) long, elliptical and flattened, shiny.

Wood is dull reddish brown with pale whitish sapwood. Soft and fine-textured, it is easily worked.

The early Haweaiians prepared their tapa, kapa, or paper cloth from the bark of this native tree and from the introduced wauke or paper mulberry (*Broussonetia papyrifera* (L.) Vent.). This cloth served for clothing and bed covering. Rope and cord were made from the long strong fibers. The fruits were used in home remedies. The young leaves made a good tea that was used as a tonic.

This species of mamaki in the broad sense is distributed through the islands in moderately wet to wet forests at 200--6,000 ft (61--1,829 m) altitude. The 3 other Hawaiian species currently accepted are: *Pipturus forbesii* Krajina of east Maui, *P. kauaiensis* Heller of Kauai, and *P. ruber* Heller of Kauai.

37. Ōpuhe

Medium-sized native evergreen tree on the island of Hawaii or a shrub or small tree on the other islands, with slightly milky or watery sap, elliptical wavy toothed leaves, and very numerous tiny flowers along twigs partly back of leaves, male and female on different trees. To 35 ft (10.7 m) high, with straight trunk 1 ft (0.3 m) in diameter and long stout branches. Bark gray, smooth, very fibrous.

Leaves alternate, with leafstalks of 1 1/4--2 in (3--5 cm). Blades oblong or narrowly elliptical, 6--14 in (15--36 cm) long and 1 1/2--4 3/4 in (4--12 cm) wide, long-pointed at apex, blunt at base, finely wavy toothed in upper part, thin or thick and slightly fleshy, usually palmately 3-veined at base, with 12--15 parallel straight sunken veins on each side, beneath pale and often hairy along veins. Microscopic mineral growths (cystoliths) like crystals are present.

Flower clusters (cymes) at base of leaves or back of leaves, about 2--3 in (5--7.5 cm) in diameter, much forked regularly by 2. Flowers male and female on different plants (dioecious), very numerous, without corolla. Male flowers 8--20 almost stalkless in rounded balls, each about 1/8 in (3 mm) in diameter, composed of pale

Special areas--Wahiawa, Haleakala, Volcanoes, Kipuka Puaulu.
Champion--Height 28 ft (8.5 m), c.b.h. 2.8 ft (0.9 m), spread 33 ft (10.1 m). Hawaii Volcanoes National Park, Hawaii (1968).
Range--Kauai, Oahu, Molokai, Lanai, Maui, and Hawaii.
Other common name--waimea.
Botanical synonyms--*Pipturus brighamii* Skottsb., *P. gaudichaudianus* Wedd., *P. hawaiiensis* Lévl., *P. hel-* Skottsb., *P. oahuensis* Skottsb., *P. pachyphyllus* Skottsb., *P. pterocarpus* Skottsb., *P. rockii* Skottsb., *P. skottsbergii* Krajina.

Mamaki is the favorite food plant of the green caterpillar that becomes the beautiful reddish brown Kamehameha butterfly, according to Degener.

Another Hawaiian plant, 'ākōlea, *Boehmeria grandis* (Hook. & Arn.) Heller, is also reported in old accounts as mamake or mamaki. It is a shrub with reddish leaves arranged like *Pipturus* but opposite, smooth, and not hairy, and inflorescences that are long and dangling. Both species were used by the Hawaiians for similar purposes.

Urera glabra (Hook. & Arn.) Wedd.

reddish to whitish 4--5 lobed calyx and 5 stamens. Female flowers with 3--4-toothed calyx bordered by a cup and pistil with ovary and yellow stigma.

Fruits rounded, about 1/8 in (3 mm) in diameter, with enlarged fleshy orange yellow calyx enclosing 1 seed (achene), elliptical and rough, with yellow stigma.

The wood is soft and lightweight. The fibrous bark was used by the Hawaiians for fish nets and at times for their tapa cloth.

Widespread in moist forests through the islands, as a tree at 500--5,500 ft (152--1,676 m) altitude on the island of Hawaii and as a shrub or small tree on the other islands.

Special areas--Volcanoes, Kipuka Puaula, Wahiawa.
Range--Hawaiian Islands only.
Other common names--hōpue, hona.
Botanical synonym--*Urea sandwicensis* Wedd.

Treated here as a single variable species, though also separated as 2. Another, *Urera kaalae* Wawra, Kaala urera, is a small tree of the Waianae Range of Oahu, with palmately veined heart-shaped leaves, possibly or almost extinct.

PROTEA FAMILY (PROTEACEAE*)

38. Kahili-flower

Grevillea banksii R. Br.*

Small introduced ornamental tree, partly deciduous, with pinnate leaves of 3--11 very narrow leaflets, and showy dark red flowers. To 20 ft (6 m) tall, with short trunk to 1 ft (0.3 m) in diameter, and irregular branching

36. Mamaki *Pipturus albidus* (Hook. & Arn.) Gray
Flowering twig, 1/2 X (Degener).

37. Ōpuhe *Urera glabra* (Hook. & Arn.) Wedd.
Twig with leaves and female flowers, and fruit clusters (lower right), 1 X.

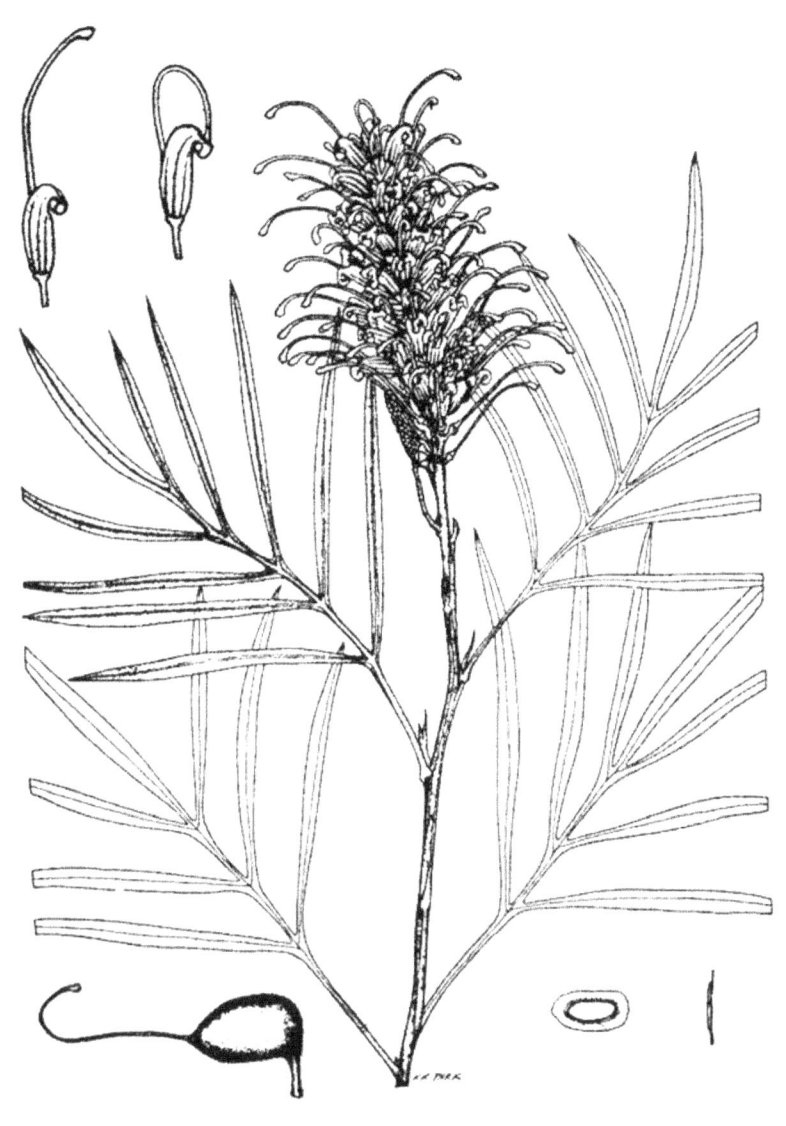

38. Kahili-flower   *Grevillea banksii* R. Br.*
Flowering twig, 1/2 X; flowers (upper left), fruit (lower left), and seeds (lower right), 1 X (Degener).

crown. Outer bark dark brown, inner bark reddish, fibrous, bitter. Twig, stout, covered with whitish pressed hairs.

Leaves alternate, pinnate, 4--8 in (10--20 cm) long, with slightly winged axis and deeply divided into 3--11 very narrow stalkless leaflets or segments 2 1/2--4 in (6--10 cm) long and 1/8--1/4 in (3--6 mm) wide, ending in tiny sharp point, thick and leathery, rolled under at edges, above shiny green and becoming nearly hairless, beneath with dense mat of pressed whitish hairs.

Flower clusters (racemes) terminal and lateral, 2--4 in (5--10 cm) long, with straight stout hairy persistent axis. Flowers many, paired, short-stalked, showy, irregular, composed of finely hairy red calyx about 3/4 in (2 cm) long including narrow tube split on 1 side and 4 curved narrow lobes, hairy on outside, no corolla, 4 stalkless stamens inserted near ends of lobes, and pistil with hairy ovary, slender threadlike curved style to 1 1/2 in (4 cm) long, and enlarged yellow stigma.

Fruit podlike (follicle), about 5/8 in (15 mm) long, flattened, gray hairy, splitting open on 1 edge, with long curved persistent style. Seeds 2, about 3/8 in (1 cm) long, elliptical, flat, blackish, bordered by narrow brown wing.

Wood light brown, hard, with large prominent rays that produce an oak-like figure. The wood resembles that of No. 40, silk-oak (*G. robusta*), but has not been used in Hawaii because the trees are small.

It is reported that some persons develop a skin rash from stiff hairs on the flower cluster when handling the plant. In Australia, it is toxic to horses.

This tree is planted as an ornamental in Hawaii for its showy flowers. It may be seen as an escape growing along the roadsides near Anahola, Kauai. There is no record of this tree having been planted in the forest reserves by the Division of Forestry. A form with creamy white flowers is known from Upper Manoa Valley, Honolulu. It is a mutant of unknown origin.

Range--Native of Australia.

Other common names--haʻikū, ʻokapua, ʻulaʻula, Banks grevillea, silk-oak.

Botanical synonym--*Stylurus banksii* (R. Br.) Deg.

The Hawaiian name haʻikū is from Haiku, Maui, where the species was first introduced. The scientific name honors Joseph Banks (1743--1820), British botanist and patron of sciences.

39. Silk-oak

Grevillea robusta A. Cunn.*

Silk-oak, a handsome Australian tree planted for shade, ornament, and reforestation, also widely naturalized, is recognized by the pinnate and deeply lobed fernlike leaves, dark green above, silky with whitish or ash-colored hairs beneath, by the showy yellowish flowers clustered on 1 side of slender stiff axes, and by black curved podlike fruits with slender stalks and styles.

Medium-sized to large deciduous tree 40--70 ft (12--21 m) high and 1--3 ft (0.3--0.9 m) in trunk diameter with straight axis and many branches, reported to reach 100 ft (30 m) or more where native. Bark light gray, rough and thick with many deep furrows, on branches gray and smooth. Inner bark with brownish layer and whitish within, fibrous, slightly bitter. Twigs stout, angled, with fine gray pressed hairs.

Leaves alternate pinnate, fernlike, 6--12 in (15--30 cm) long, almost bipinnate, with 11-21 side axes (pinnae) 1 1/2--3 1/2 in (4--9 cm) long, deeply divided into narrow long-pointed lobes 1/4--1/2 in (6--13 mm) wide, with borders turned under, shiny and hairless above, slightly thickened.

Flower clusters (racemes) 3--7 in (7.5--18 cm) long, unbranched, arise mostly from trunk, along twigs back of leaves, and at leaf bases. Flowers numerous with long slender stalks 3/8--3/4 in (1--2 cm) long, crowded on 1 side of axis. They are composed of 4 narrow yellowish sepals almost 1/2 in (13 mm) long, curved downward; no petals; 4 stalkless stamens inserted on sepals and opposite them; and pistil with stalk, ovary, long slender curved style, and enlarged stigma.

Fruits podlike (follicles) are broad, slightly flattened, leathery, with long slender curved style, splitting open on 1 side, remaining attached. Seeds 1 or 2, 3/8--1/2 in (10--13 mm) long, elliptical, flattened, winged, brown. Flowering from April to autumn, but predominantly April-May.

The pale pinkish brown wood is attractive because of the prominent rays, resembling oak, as the common name suggests. This was the former "lacewood" of world trade until it became scarce in Australia. It is of moderate density (sp. gr. 0.57), has low radial but high tangential shrinkage in drying and is one of the best woods in all-around machinability ever tested by the Forest Products Laboratory. Sawdust causes dermatitis to fair-skinned people. The wood is not resistant to decay or termites, and sapwood is highly susceptible to ambrosia beetle attack while in the log. Suitable for face veneer as well as paper pulp. It has very long fibers for a hardwood. In Hawaii, the wood has been utilized for furniture, cabinetmaking, paneling, interiors, and in one entire house including the framing.

The trees are propagated readily from the great quantities of seeds, grow rapidly, and are drought resistant. Along the road to Kokee State Park, Kauai, trees are unhealthy with ragged crowns. Elsewhere, trees have full crowns.

Introduced into Hawaii about 1880 from Australia, it is now seen as a shade tree and street tree through the islands from sea level to 4,000 ft (1,219 m). This species is the second most commonly planted tree in Hawaii

39. Silk-oak *Grevillea robusta* A. Cunn.*
Flowering twig, 1/2 X; flowers (lower left) and fruit and seeds (lower right), 2/3 X (P. R. v. 2).

after No. 107, robusta eucalyptus, *Eucalyptus robusta*. It is hardy in dry soils, quick growing and pest-free. Although it has been planted mostly in drier areas, it does best where planted in moist sites with 60--80 in (1,524--2,032 mm) rainfall. The trees scatter their winged seeds and have become undesirable weeds in some pastures and rangelands of Hawaii, requiring eradication. There are extensive naturalized stands on Kapapala Ranch in Ka'u and Huehue Ranch in North Kona. Hawaii has 3 million board feet of sawtimber and Oahu 2.5 million, mostly in the Honouliuli Forest Reserve.

In some countries, the trees have served as coffee shade. Some were planted for this purpose in Kona in the early part of the century. This species is classed as a honey plant.

Planted for ornament and shade in central and southern Florida, where it is persistent but not naturalized. There it is recommended as a fast-growing flowering tree for well-drained and sandy soils, being both cold hardy and drought resistant. Formerly planted in Puerto Rico, but now not used because of susceptibility to scale insects. Introduced also in southern Arizona and California. Northward in temperate climates, as in continental United States, the fernlike plants are grown indoors in pots.

Special areas--Waimea Arboretum, Kalopa, Tantalus.

Champion--Height 90 ft (27.4 m) c.b.h. 10.6 ft (3.2 m), spread 47 ft (14.3 m). State Forestry Arboretum, Hilo, Hawaii (1968).

Range--Native of Australia but widely introduced and naturalized in tropical and subtropical regions of the world.

Other common names--silver-oak, 'oka kilika, ha'iku ke'oke'o (Hawaii); roble de seda, roble australiano (Puerto Rico); grevilea (Spanish).

Botanical synonym--*Stylurus robusta* (A. Cunn.) Deg.

Grevillea was dedicated to Charles Francis Greville (1749--1809), British horticulturist.

SANDALWOOD FAMILY (SANTALACEAE)

40. 'Iliahi-a-lo'e, coast sandalwood

Santalum ellipticum Gaud.

Four species of sandalwood, genus *Santalum*, are native in Hawaii and formerly produced the Hawaiian sandalwood of commerce. Another has been introduced for forest planting. This species is the most widely distributed through the islands and is an example of the greenish to yellowish flowered sandalwoods.

A shrub or small tree to 50 ft (15 m) tall and 1 ft (0.3 m) or more in trunk diameter, evergreen, with rounded much branched crown. Bark dark gray, rough, deeply furrowed into scales and rectangular plates, thick. Inner bark red and brown streaked, bitter. Twigs green, hairless, becoming brown.

Leaves alternate, hairless, with yellow green leafstalk of 1/4--3/4 in (0.6--2 cm). Blades narrowly elliptical to nearly round, 1--2 1/2 in (2.5--6 cm) long and 3/4--1 1/2 in (2--4 cm) wide, blunt to rounded at apex, long- or short-pointed at base, not toothed on edges, thick and brittle, leathery or thin in a variety, flat or slightly curved upward, upper surface slightly shiny green with few fine side veins, lower surface dull light green.

Flower clusters (cymes) terminal and lateral, branched, 1--2 in (2.5--5 cm) long. Flowers fragrant, many, short-stalked, about 1/4 in (6 mm) long, composed of greenish yellow bell-shaped tube (hypanthium) with 4-lobed spreading calyx, 4 short stamens attached at throat of tube and opposite lobes; and pistil with 1-celled ovary half inferior, short style, and 2--4-lobed stigma.

Fruit (drupe) elliptical of nearly round, about 1/2 in (13 mm) long, with ring scar at apex, changing color from green to red and black with blue bloom when mature.

Sandalwood is well known for the fragrance of its heartwood. The Hawaiians called it lā'au 'ala (fragrant wood). They made a perfume from the powdered heartwood and added it to their bark cloth. In the Orient, sandalwood served in ornamental carving and cabinetwork and as a repellent against insects and was burned for incense. The heartwood is yellow brown, heavy and hard, and very fine-textured. Sapwood is pale brown.

Sandalwood of this and other species is widely scattered through the islands, mostly in dry forests and in lava fields, to about 8,000 ft (2,438 m) altitude. *Santalum ellipticum* is best known as a shrub near sea level though it is found up to 2,000 ft (610 m), rarely to 7,000 ft (2,134 m) altitude.

Champion (reported as this species but probably *S. paniculatum* Hook. & Arn.)--Height 65 ft (19.8 m), c.b.h. 7.7 ft (2.3 m), spread 48 ft (14.6 m). Honomolino, S. Kona, Hawaii (1968).

Range--Hawaiian Islands only.

Sandalwoods differ from most plants in being partial parasites. Their roots become attached to those of nearby plants and obtain some food by robbing these hosts. Thus, establishment of seedlings is uncertain, unless root contact is made with a suitable host.

The history of the sandalwood industry in Hawaii has been told in various references (Degener 1930, p.

40. 'Iliahl-a-lo'e, coast sandalwood   *Santalum ellipticum* Gaud.
Flowering twig, 1/2 X (Degener).

142--148). Before Capt. James Cook arrived in Hawaii in 1778, white sandalwood, *Santalum album* L., from the East Indies supplied the Orient with this valuable, fragrant cabinetwood. In 1791, Capt. John Kendrick, fur trader from Boston, began the sandalwood industry. Price by weight was $125 and up per ton. Export of this important timber brought money, mainly to the chiefs. These leaders were extravagant, purchasing several ships and going into debt. Over many years, they forced the men to labor in the forests, cutting and transporting the wood to harbors on the coasts. Maximum cutting was from about 1810 to 1820. Total sales reached 3 or 4 million dollars. The exports ended before 1845 when the forests became exhausted. There was no planting or forest management. Apparently, the parasitic habit may have limited natural regeneration of the wild trees. However, the species of sandalwoods persisted and are not threatened with extinction. Now, they are widespread through the islands, though not in commercial volumes or sizes.

Currently, India is the primary source of sandalwood and sandal oil, which still are important in commerce. Australia also produces some oil. Although much is said of Hawaii's sandalwood, many other Pacific Islands such as Fiji and the New Hebrides had extensive stands that also were heavily cut during the early 1800's.

41. 'Iliahi, Freycinet sandalwood (color plate, p. 31)

Santalum freycinetianum Gaud.

This species with its varieties will serve as an example of the red-flowered sandalwoods. Reported as a large tree to 82 ft (25 m) high and 3 ft (0.9 m) in trunk diameter in the natural forest, evergreen with slender drooping branches. Bark of small trunks gray, smoothish.

Leaves alternate, hairless, with leafstalk of 3/8--3/4 in (1--2 cm), often reddish. Blades narrowly elliptic of lance-shaped, 2--4 in (5--10 cm) long and 3/4--1 1/4 in (2--4 cm) wide, short-pointed at both ends, slightly turned under at edges, thin, more or less folded along midvein, somewhat curved, upper surface shiny dark green, lower surface dull and paler.

Flower clusters (cymes) terminal and lateral, branched. Flowers 3--9 or more, nearly stalkless, about 1/4 in (6 mm) long yellowish to white, turning red in age, composed of narrow bell-shaped tube (hypanthium) with 4-lobed calyx, 4 short stamens attached near base of tube and opposite lobes; and pistil with 1-celled ovary partly inferior, threadlike style, and stigma mostly 3-lobed.

Fruit (drupe) elliptical, 5/16--5/8 in (8--15 mm) long, with ring scar at apex, purplish black with a bloom, within greenish and juicy.

Fairly common in dry forests of both Waianae and Koolau Ranges, Oahu. More common in the Waianae Range which is drier, especially at 800--2,000 ft (244--610 m).

Special areas--Aiea, Waimea Arboretum, Kamehameha School.

Range--Oahu, Kauai, Lanai, and Maui.

This species was greatly reduced as a forest tree during the early part of the last century because of the export of the fragrant wood. Its name commemorates Henri Louis Claude de Saulces de Freycinet (1779--1840), leader of a French world expedition in 1817--20. The generic name is derived from the Greek name for sandalwood.

Besides the typical variety on Oahu, 2 varieties formerly treated as species, are distinguished: Kauai sandalwood, *Santalum freycinetianum* var. *pyrularium* (Gray) Stemm., on Kauai, and Lanai sandalwood, *S. freycinetianum* var. *lanaiense* Rock, on Lanai and Maui, listed as endangered.

The 2 other species of sandalwood native to Hawaii are:

- Hawaii sandalwood, *Santalum paniculatum* Hook. & Arn., grows on the Island of Hawaii in dry forests and lava fields at 1,500--6,500 ft (457--1,981 m), sometimes to 8,000 ft (2,438 m). It has greenish flowers and may be seen at Hawaii Volcanoes National Park near Volcano House.
- Haleakala sandalwood, *Santalum haleakalae* Hillebr., grows on Maui at 6,000--8,800 ft (1,829--2,682 m). Large clusters of flowers that become deep red as they age make this the most beautiful species. It is easily seen near Hosmer Grove in Haleakala National Park.

BUCKWHEAT FAMILY (POLYGONACEAE)

42. Seagrape

Coccoloba uvifera (L.) L.*

Seagrape, a small tree planted along sandy beaches as an ornamental or windbreak, is easily identified by the rounded or kidney-shaped thick and leathery leaves, which are slightly broader than long, often reddish when young or very old, and by the drooping grapelike clusters of crowded purple edible fruits about 3/4 in (2 cm) long.

A small introduced evergreen tree 20--30 ft (6--9 m) high and 1 ft (0.3 m) in trunk diameter, with widely spreading rounded crown of few coarse branches, often branching near base. Bark smoothish gray, thin, peeling

41. 'Iliahi, Freycinet sandalwood *Santalum freycinetianum* Gaud.
Flowering twig, 1/2 X; flowers and fruits (below), 2 X (Degener).

off in small flakes on large trunks, which become mottled whitish, light gray, and light brown. Inner bark light brown and bitter. Twigs stout spreading, green and minutely hairy when young, becoming gray, with leaf sheaths and ring scars at nodes.

Leaves alternate, with short leafstalks of 1/4--1/2 in (6--13 mm) and at base a reddish brown membranous sheath (ocrea) 1/4--3/8 in (6--10 mm) high around twig. Blades often turned on edge vertically, 3--6 in (7.5--15 cm) long and 4--8 in (10--20 cm) wide, rounded at apex and heart-shaped at base, with margins slightly curved under, hairless or nearly so, upper surface green or blue green, lower surface paler, and midvein and larger veins often reddish.

Flower clusters (narrow racemes) terminal and lateral, 4--9 in (10--23 cm) long have numerous small whitish or greenish white fragrant flowers 3/16 in (5 mm) across on short stalks of 1/16--1/8 in (1.5--3 mm), male and female on different trees (dioecious). Male flowers have greenish white basal tube (hypanthium) 1/16 in (1.5 mm) long and broad bearing 5 spreading rounded white calyx lobes more than 1/16 in (1.5 mm) long; 8 stamens united at base; and rudimentary pistil. Female flowers have small stamens and larger pistil with 1-celled ovary and 3 styles.

Fruits elliptical or egg-shaped, with thin fleshy covering (hypanthium) and with calyx at apex, sour or sweetish, enclosing 1 elliptical seed (achene) 3/8 in (10 mm) long. Flowering and fruiting through the year.

Sapwood light brown and heartwood reddish brown. Wood hard, moderately heavy (sp. gr. 0.7), and very susceptible to attack by dry-wood termites. It takes a fine polish but is little used where native except for posts, furniture, and cabinetwork. Straight pieces would be suited for wood turning. Because of the similarity of names, it is sometimes confused with cocobolo (*Dalbergia retusa*), an important commercial timber.

The bark contains tannin, and the astringent roots and bark have been used in medicines elsewhere. West Indian or Jamaican kino, an astringent red sap exuding or extracted from cut bark, formerly was in commerce for tanning and dyeing.

Jelly and a winelike beverage can be prepared from fruits, which also are eaten raw. Bunches of fruits in conelike packets formed by rolling the leaves have been sold on the streets in tropical America, the native home. Early Spanish colonists sometimes used the fresh thick leaves as a substitute for paper, scratching messages with a pin or other sharp point.

This is one of the first woody species to become established on sandy shores where native, being more hardy in these exposed places and more tolerant of salt than most trees. For these reasons it is often planted as an ornamental or windbreak along coasts. Since propagation is from cuttings, female plants should be selected for fruits. Frequently grown in southern Florida, it can be shaped as a hedge for landscaping. It is a good honey plant.

In Hawaii, the species is planted as an ornamental and windbreak along sandy beaches, it escapes and becomes naturalized locally. It may be seen along most shorelines. A total of 955 trees have been planted in the forest reserves by the Division of Forestry, mostly in the Honouliuli Forest Reserve. It is doubtful that they became established naturally so far from the shoreline.

Special area--Waimea Arboretum.

Range--Shores of central and southern Florida, Bermuda, and from Bahamas through West Indies including Puerto Rico and Virgin Islands. Also, Atlantic Coast from northern Mexico south to Venezuela and Guianas.

Other common names--uva de playa, uva de mar, uvero (Puerto Rico, Spanish); grape (Virgin Islands).

AMARANTH FAMILY (AMARANTHACEAE)

43. Pāpala *(color plate, p. 32)*

Charpentiera obovata Gaud.

Evergreen native shrub or small tree with small to large oblong hairless leaves and distinctive showy but tiny stalkless flowers very numerous along wiry branches of very large drooping clusters. Shrub or small tree 15--30 ft (4.6--9.1 m) high and 4 in (0.1 m) to more than 2 ft (0.6 m) in trunk diameter, often becoming enlarged into buttresses at base. Bark gray or light brown, smooth. Inner bark orange brown streaked, bitter. Twigs slender, green to brownish green, hairless, with half round slightly raised leaf scars and end bud of tiny leaves.

Leaves alternate, hairless, with flattened stalk 3/4--2 in (2--5 cm) long. Blade oblong or elliptical, 2--5 in (5--13 cm) long and 1--2 in (2.5--5 cm) wide, blunt at both ends and extending slightly down stalk, not toothed, thin or slightly thickened, upper surface dull green, lower surface shiny light green with fine parallel nearly straight side veins slightly raised.

Flower clusters (panicles) single at leaf bases, very slender and drooping, 4--16 in (10--40 cm) long, with several wiry branches to 8 in (20 cm), purplish or reddish, flowers many, stalkless, less than 1/8 in (3 mm) long, elliptical, composed of 3 tiny scales, 5 purplish sepals, and pistil with elliptical ovary and 2 spreading styles.

Fruits dry, elliptical (utricle), brown, about 1/16 in (2 mm) long, covered by 3 scales, with 2 spreading styles, membranous. Seed 1, elliptical, tiny, shiny black.

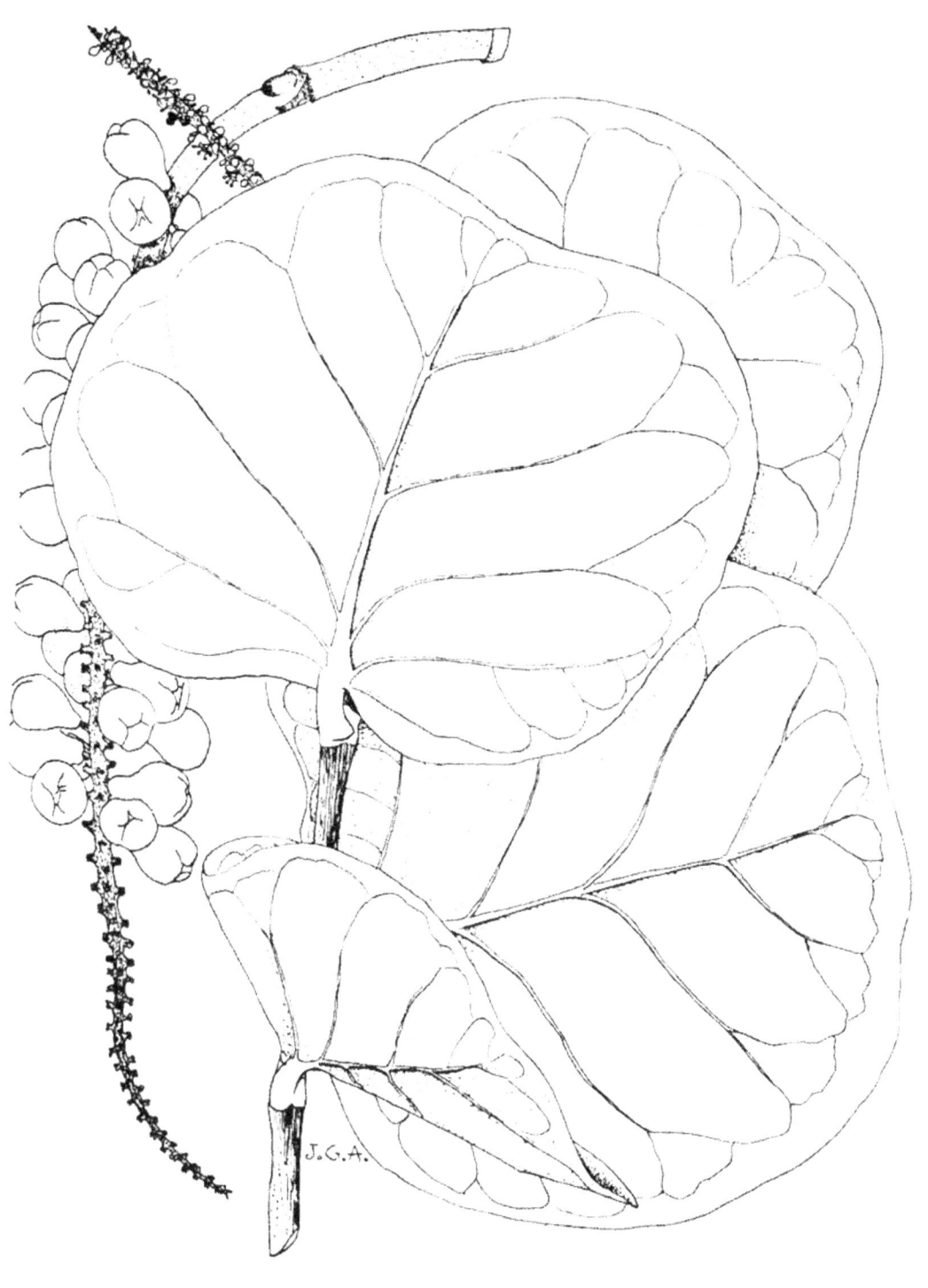

42. Seagrape *Coccoloba uvifera* (L.) L.*
Fruits (left), twig with male flowers, 2/3 X (P. R. v. 1).

Wood very soft and fibrous, composed of many rings, exceedingly lightweight when dry, and "will burn like paper." Heartwood is grayish yellow and sapwood lighter colored. Tends to "honeycomb" in air drying.

Hawaiians formerly used the wood for fireworks. These fireworks displays from pāpala sticks were well described a century ago, by Isabella Sinclair (1885) in her book of paintings, "Indigenous Flowers of the Hawaiians," perhaps the first on Hawaii's wildflowers. (See color plates p. 31-42.) Her observation, quoted by Rock (1913), merits repetition:

> On the northwest side of Kauai, the coast is extremely precipitous, the cliffs rising abruptly from the sea to a height of 1,000--2,000 ft (305--610 m), and from these giddy heights the ingenious and beautiful pyrotechnic displays take place.
>
> On dark moonless nights upon certain points of these awful precipices, where a stone would drop sheer into the sea, the operator takes his stand with a supply of pāpala sticks, and, lighting one, launches it into space. The buoyancy of the wood causes it to float in midair, rising or falling according to the force of the wind, sometimes darting far seaward, and again drifting towards the land. Firebrand follows firebrand, until, to the spectators (who enjoy the scene in canoes upon the ocean hundreds of feet below), the heavens appear ablaze with great shooting stars, rising and falling, crossing and recrossing each other, in the most weird manner. So the display continues until the firebrands are consumed, or a lull in the wind permits them to descend slowly and gracefully to the sea.

This species occurs in wet and dry forests through the islands at 600--5,700 ft (183--1,737 m). Of largest size in dry areas.

Special areas--Waimea Arboretum, Volcanoes, Kipuka Puaulu.

Range--Hawaiian Islands only.

This genus has 5 species (1 shrubby) in Hawaii and another in the Austral Islands (Sohmer 1972). It honors Jean G. F. de Charpentier (1786--1855), German-born Swiss geologist, conchologist, and botanist. The family has very few tree species, but many herbs, including amaranth garden weeds.

FOUR-O'CLOCK FAMILY (NYCTAGINACEAE)

44. Pāpala kēpau

Pisonia brunoniana Endl.

Small native evergreen tree of dry forests, with paired elliptical leaves, very large open flower clusters with many small flowers on long slender widely forking stalks, and narrow sticky fruits. To about 18 ft (5.5 m) and 4 in (10 cm) in trunk diameter. Bark light gray or light brown, smoothish; inner bark whitish, slightly bitter. Twigs green to brown. Young twigs and leaves with tiny pressed brown hairs.

Leaves opposite, with leafstalks of 1/2--1 in (13--25 mm). Blades elliptical or oblong, 2 1/2--5 1/2 in (6--14 cm) long and 1 1/4--3 1/4 in (3--8 cm) wide, blunt at apex, short-pointed at base, not toothed on edges, thin, hairless, shiny green above, dull light green beneath.

Flower clusters (panicles) terminal, very large and open or loose, 6--12 in (15--30 cm) long and broad. Flowers many, small, on many long slender widely forking stalks, 3/8 in (10 mm) long, composed of greenish tubular or narrowly funnel-shaped finely hairy calyx with 5 short, pinkish lobes, 8-12 stamens attached inside tube and extending beyond, and pistil with narrow ovary, long style, and dot stigma.

Fruit (anthocarp) narrowly cylindrical, consisting of enlarged calyx 3/4--1 3/8 in (20--35 mm) long and 1/8 in (3 mm) in diameter, widest at middle, spreading at apex, 5-ridged, sticky and exuding glue, enclosing the narrow 1-seeded fruit (achene) with style at apex.

Wood whitish yellow, very lightweight, very soft, and brittle. It "honeycombs" in air drying and is not used.

Widespread in dry forests and at edges of lava fields, mostly at 2,000--4,000 ft (610--1,219 m) altitude.

Special areas--Wahiawa, Volcanoes.

Champion--Height 50 ft (15.2 m), c.b.h. 6.3 ft (1.9 m), spread 31 ft (9.4 m). Hoomau Ranch, Honomalino, Hawaii (1968).

Range--Oahu, Lanai, Maui, Hawaii only.

Botanical synonyms--*Heimerliodendron brunoianum* (Endl.) Skottsb., *Pisonia inermis* G. Forst., not Jacq.

The Hawaiian common name pāpala kēpau is from kēpau, the name for tar, pitch, etc. A sticky liquid or glue exudes from the fruits and catches insects. Hawaiians formerly used the viscous fruits as birdlime to catch small birds.

45. Āulu

Pisonia sandwicensis Hillebr.

Small to medium-sized native evergreen tree of dry forests, recognized from a distance by the dark green color of the large oblong leathery leaves. To about 50 ft (15 m) high with up to 3 trunks 1--2 ft (0.3--0.6 m) in diameter. Bark dark gray, smoothish to finely fissured.

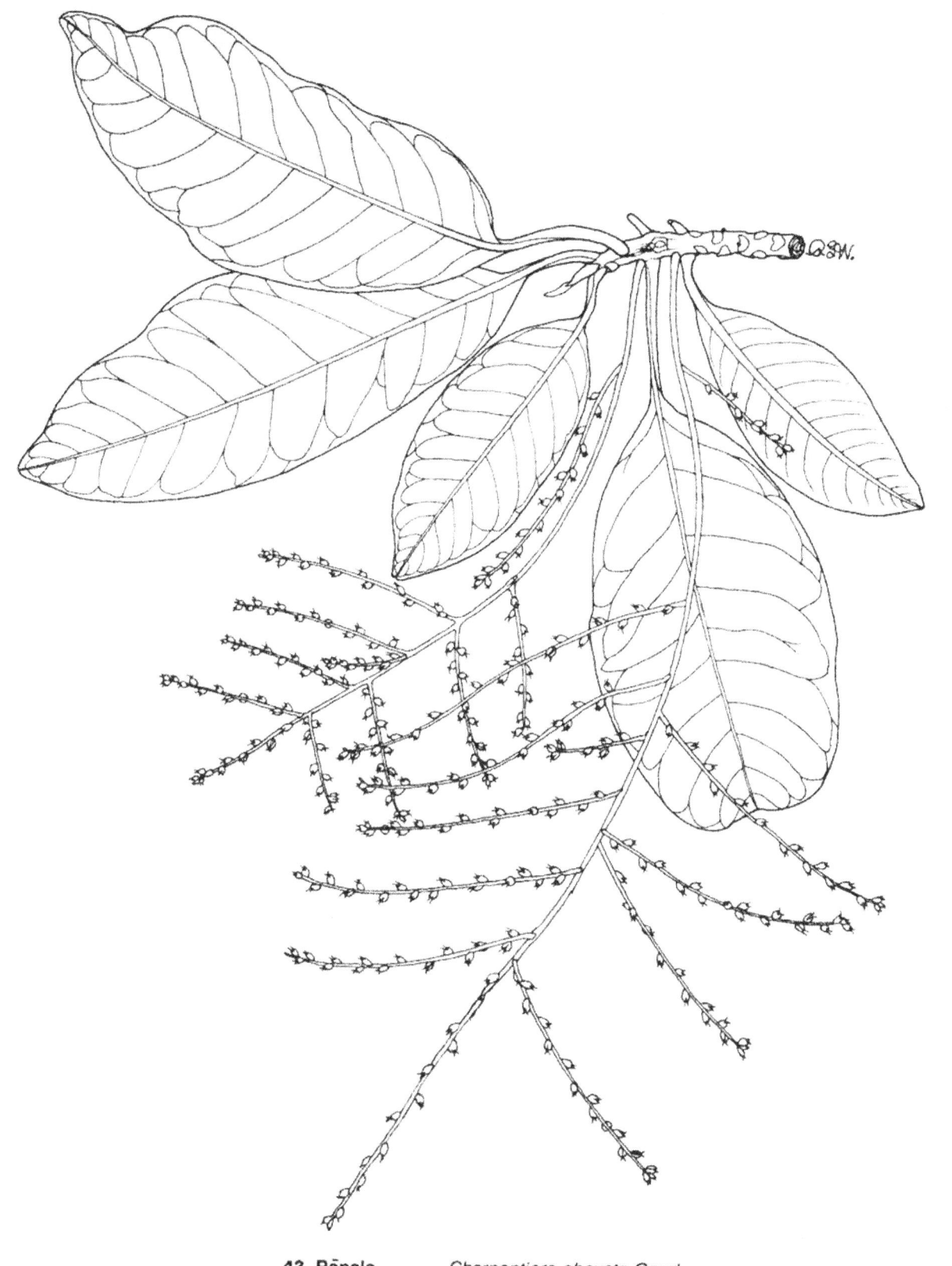

43. Pāpala *Charpentiera obovata* Gaud.
Flowering twig, 1 X.

44. Pāpala kēpau   *Pisonia brunoniana* Endl.
Twig with flowers, fruits (lower right), 1 X (Maiden).

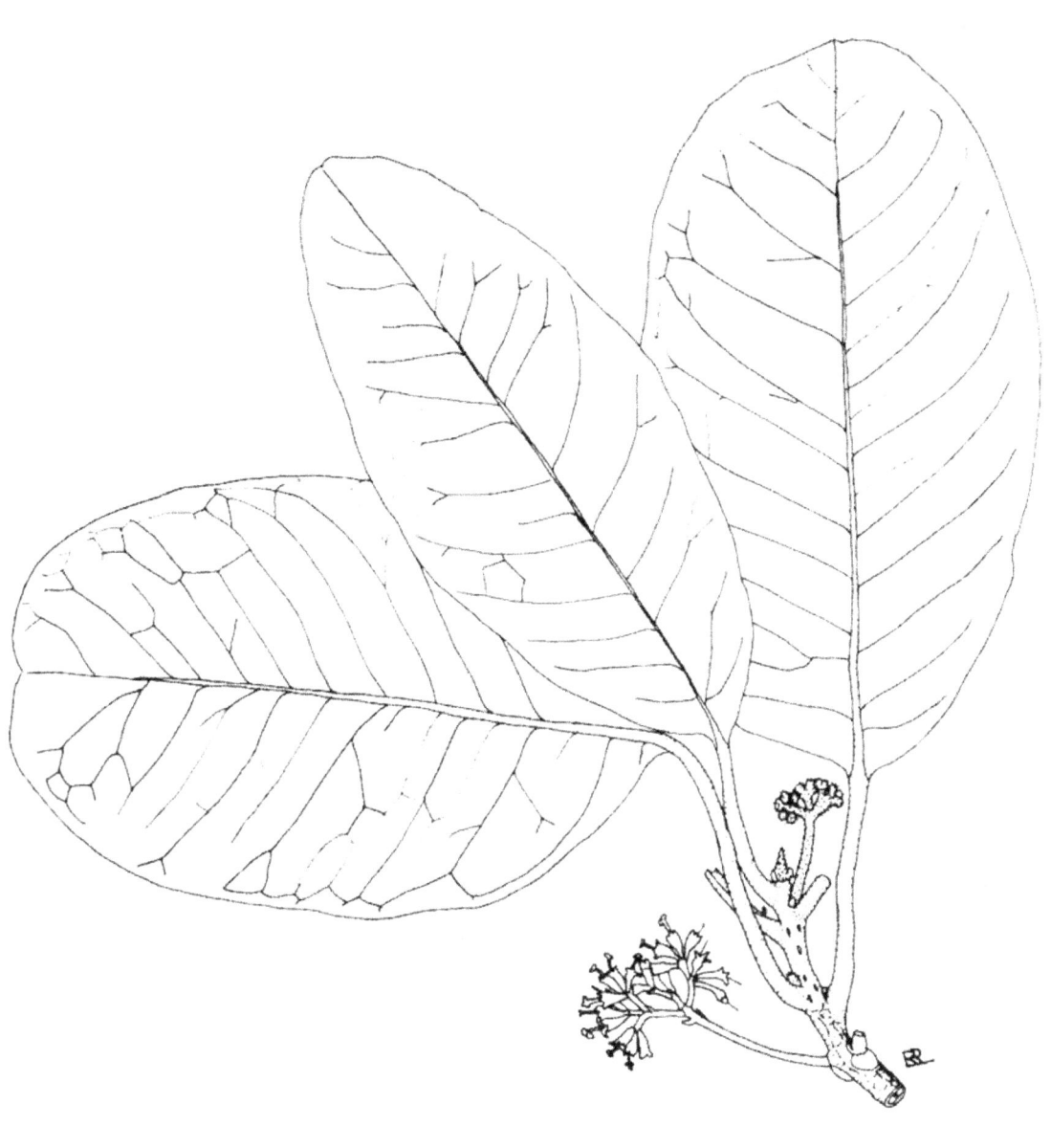

45. Āulu *Pisonia sandwicensis* Hillebr.
Twig with female flowers, 2/3 X.

Inner bark whitish, slightly bitter. Twigs light gray, with large raised half-round leaf-scars.

Leaves alternate, hairless, with slender leafstalks of 1 1/4--2 in (3--5 cm). Blades large, oblong, 4--12 in (10--30 cm) long and 2 1/4--6 in (6--15 cm) wide, leathery, blunt, rounded, or slightly notched at apex, broad and rounded at base, not toothed on edges, upper surface shiny dark green with inconspicuous side veins, paler beneath.

Flower clusters rounded on stalks of 1 1/4--5 1/4 in (3--6 cm) at leaf bases. Male and female flowers many, on different trees (dioecious), with 2--3 scales or bracts at base and narrow greenish tubular calyx about 1/4 in (6 mm) long, finely hairy, fragrant. Male flowers many, stalkless in rounded head about 2 in (5 cm) in diameter, consisting of deeply 5--6-lobed tubular calyx with 5 short lobes, about 20 minute sterile stamens inside tube, and pistil with narrow ovary, slender style, and enlarged fringed stigma.

Fruit (anthocarp) cylindrical, about 1 1/2 in (4 cm) long and very narrow, widest below middle, composed of enlarged calyx with lobes at apex and many faint lines and enclosing the narrow 1-seeded fruit (achene) with style at apex.

Wood whitish, soft, very lightweight and porous, brash, not used. Tends to separate or "honeycomb" when dried. Highly susceptible to fungal stain. Branches brittle and easily broken.

Widespread in dry forests through the islands, especially at 2,000--2,500 ft (610--762 m) altitude.

Special areas--Kokee, Waimea Arboretum.
Range--Hawaiian Islands only.
Other common name--pāpala kēpau.
Botanical synonym--*Rockia sandwicensis* (Hillebr.) Heimerl.

LAUREL FAMILY (LAURACEAE)

46. Camphor-tree
Cinnamomum camphora (L.) J. S. Presl*

Medium-sized to large introduced ornamental evergreen tree with dense rounded crown of 3-veined shiny dark green leaves and distinctive odor of camphor in crushed foliage. To 80 ft (24 m) high and 3 ft (0.9 m) in trunk diameter. Bark gray, smoothish, becoming thick, rough, and furrowed. Inner bark pinkish, spicy bitter. Twigs slender, greenish, hairless. End buds enlarged, elliptical, pointed, 1/4 in (6 mm) long, brownish, composed of many rounded overlapping scales which form rings of scars on twigs upon shedding.

Leaves alternate or paired (opposite), hairless, with slender leafstalks 1/2--1 1/4 in (1.3--3 cm) long. Blades elliptical, 2 1/2--4 in (6--10 cm) long and 1--2 1/4 in (2.5--6 cm) wide, long-pointed at apex and short-pointed or rounded at base, not toothed on edges, slightly thickened and leathery, with 3 main veins including 2 long curved side veins from near base of midvein and swollen glands in angles, pinkish and showy when young, shiny green above and dull light green beneath.

Flower clusters (panicles) on slender stalks at leaf bases, 1 1/2--3 in (4--7.5 cm) long, branched. Flowers several, yellowish, small, 1/8 in (3 mm) long and broad, composed of 6-lobed calyx, 9 stamens, and pistil with rounded ovary and short style.

Fruit (berry) 3/8 in (1 cm) in diameter, green to black, with short greenish cuplike base and enlarged stalk, the thin flesh with spicy taste of camphor. Seed 1, nearly 1/4 in (6 mm) in diameter, dark brown.

The wood is yellowish brown with darker streaks, lightweight (sp. gr. 0.45), soft, fine-textured, strongly scented, and takes a good polish. Elsewhere, it has served in cabinetwork, especially chests, because the odor is an insect repellent. A few trees have been cut in Hawaii and worked into chests and closet lining. Camphor gum and oil, used in medicine and industry, are prepared by steam distillation of leaf clippings and wood from plantations.

Planted as an ornamental and shade tree in Hawaii; elsewhere as windbreaks and hedges. On Oahu, Kauai, Lanai, and Maui, a total of 3,600 trees are recorded as having been planted in the Forest Reserves. In several wet forest areas, notably in Nuuanu Valley, dense thickets of this tree form an understory beneath Eucalyptus stands. It attains large size when grown as a plantation tree in the forest. There is a stand at about 1,100 ft (335 m) elevation along Tantalus Drive, Oahu, that attests to this.

Grown in subtropical regions of southern continental United States from Florida to southern Texas along the Gulf, and in California, it has escaped and is recorded as naturalized. Uncommon in Puerto Rico and Virgin Islands.

Special areas--Foster, Tantalus.
Champion--Height 83 ft (25.3 m), c.b.h. 22.7 ft (6.9 m), spread 100 ft (30.5 m). Ulupalakua, Maui (1968).
Range--Native of tropical Asia from eastern China to Vietnam, Taiwan, and Japan, and widely planted in tropical and subtropical regions.
Other common names--Japanese camphor-tree; alcanfor (Spanish).

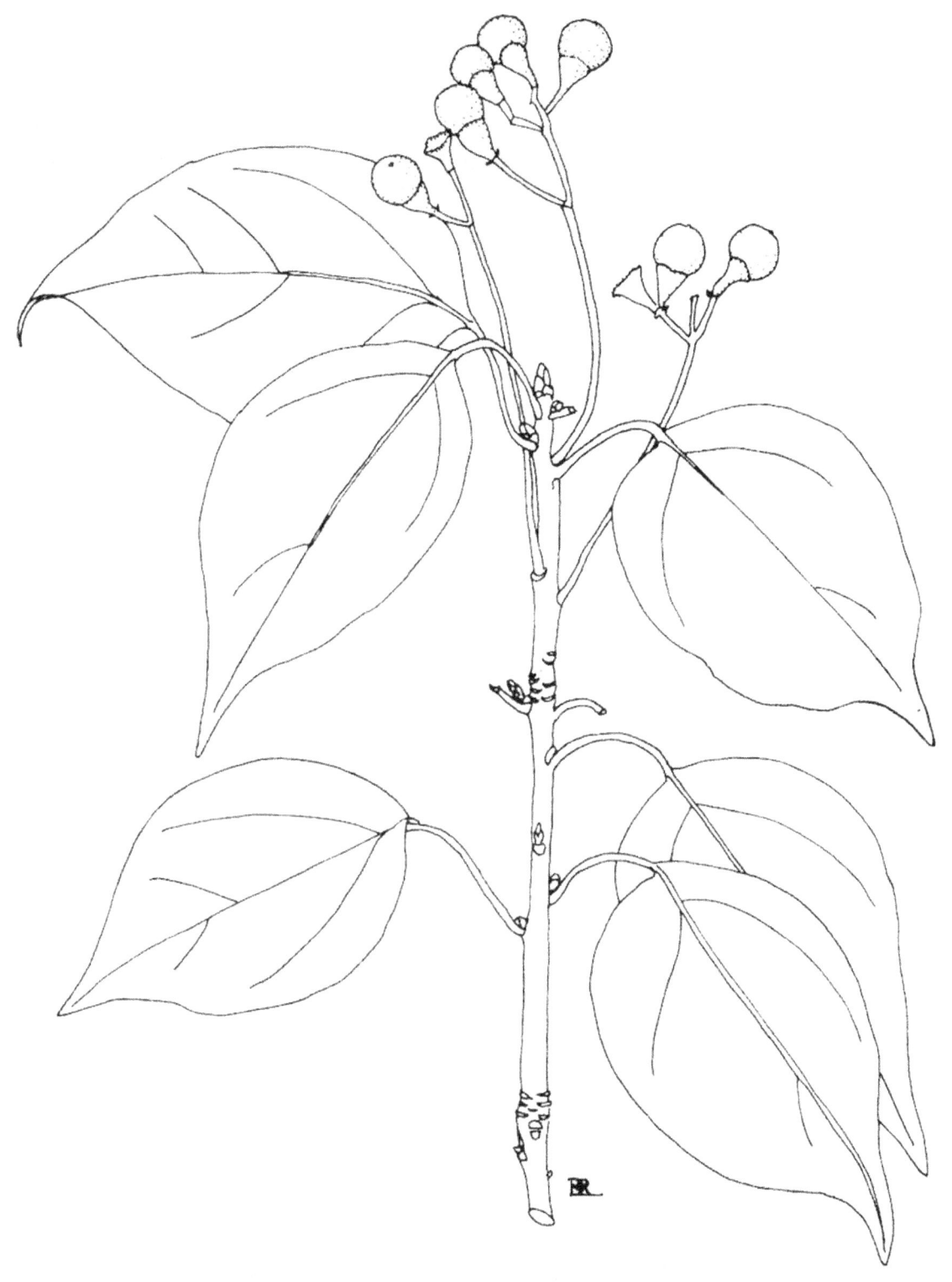

46. Camphor-tree *Cinnamomum camphora* (L.) J. S. Presl*
Twig with fruits, 1 X.

47. Avocado

Persea americana Mill.*

Avocado, the well-known fruit tree planted also in forests and sometimes growing as if wild, is known by its shiny yellow green pear-shaped or nearly round fruits about 4--5 in (10--13 cm) long and 3--4 in (7.5--10 cm) in diameter, with oily green and yellow edible flesh somewhat like butter and with very large egg-shaped seed.

Small to medium-sized deciduous tree commonly 15--30 ft (4.6--9 m) in height and 1 1/2 ft (0.5 m) in trunk diameter, to 60 ft (18.3 m) and 2 ft (0.6 m), with straight axis and symmetrical narrow or rounded crown. Old trees frequently lean. Bark brown or gray, slightly rough and fissured. Inner bark orange brown, slightly spicy and gritty to the taste. Twigs green, angular, and finely hairy, becoming brown.

Leaves alternate, crowded near ends of twigs, with yellow green leafstalks 1/2--1 1/4 in (1.3--3 cm) long. Blades slightly aromatic when crushed, elliptical, 3 1/2--7 in (9--18 cm) long and 2--3 1/2 in (5--9 cm) broad, long- or short-pointed at apex and short-pointed at base, without teeth on edges, slightly thickened, upper surface green to dark green, slightly shiny, hairless or nearly so, and lower surface dull gray green, finely hairy on veins.

Flower clusters (panicles) many, branched near ends of twigs and shorter than leaves. Flowers many on short hairy stalks, greenish yellow, about 3/8 in (1 cm) broad, composed of 6 widely spreading greenish yellow narrow hairy sepals about 3/16 in (5 mm) long, 9 greenish yellow stamens more than 1/8 in (3 mm) long and 3 smaller sterile stamens (staminodes), and whitish green pistil with 1-celled 1-ovuled ovary and slender style. Flowering when trees are leafless or nearly so.

Fruits (berries) borne singly, heavy, hanging down and bending twigs, with leathery skin and thick edible soft flesh somewhat like butter. Seed egg-shaped, about 2--2 1/4 in (5--6 cm) long and to 2 in (5 cm) in diameter, with thin brown skin, bitter.

Sapwood wide, cream-colored and heartwood light reddish brown. Wood moderately soft, lightweight (sp. gr. 0.45), easy to work, not durable, susceptible to attack by dry-wood termites. Seldom used because the tree is valued for its fruit.

The nutritious fruits are eaten raw as a vegetable or salad and are available in the stores of Hawaii virtually year-round from local sources and California. Hogs, other domestic animals, and wild animals are fond of the fruits. Commercial oils that can be used as a substitute for olive oil or as an oil for the hair have been extracted from the pulp, which has an oil content of about 5--25 percent. The seeds yield a reddish brown dye for marking clothing. Some parts of the plant, such as leaves, seeds, fruit rind, and bark, have been employed in folk medicines. The fragrant flowers are attractive to bees and make this tree a honey plant. Many races, varying in size, shape, color, and quality of fruit and time of ripening, are in cultivation in Hawaii. Propagation is from seed or, for superior varieties, by budding.

Planted as a fruit tree in moist lowlands of Hawaii to about 1,600 ft (488 m) altitude, it is also persistent about houses and naturalizes locally in lower forest and along roadsides. More than 57,000 trees have been planted in the Forest Reserves of all islands by the Division of Forestry and Wildlife, so it really is now a "forest" tree as well as an orchard tree. There are large stands in the Honuliuli and Nanakuli Forest Reserves on Oahu, and many trees may be seen along Tantalus Drive, growing in a forest situation.

Champion--Height 65 ft (19.8 m), c.b.h. 13.8 ft (4.2 m), spread 40 ft (12.2 m). Hawaiian Agricultural Co., Pahala, Hawaii (1968).

Range--Native of Mexico, Guatemala, and Honduras. Widely cultivated as a fruit tree and naturalized in tropical and subtropical regions. Grown in commercial orchards in southern Florida and southern California and naturalized locally in the former. Also, Puerto Rico and Virgin Islands.

Other common names--alligator-pear; aguacate (Spanish); pear, apricot (Virgin Islands); alageta (Guam); bata (Palau).

Introduced into Hawaii in the early part of the 19th century, according to Degener, probably by Don Marín. Not common till after several later attempts some years later, but by 1910 one of the most common trees in lowland gardens. Many improved varieties and hybrids are grown, maturing at different times. The main flowering season is from January to April and fruiting period from June to August. However, in Hawaii, there are so many varieties that some are in fruit throughout the year. Harvests are heavier in alternate years.

SAXIFRAGE FAMILY (SAXIFRAGACEAE)

48. Kanawao (color plate, p. 33)

Broussaisia arguta Gaud.

Kanawao is one of the most common shrubs in the understory of Hawaii's rain forests, sometimes a small tree. This handsome plant is recognized by the large narrowly elliptical or obovate leaves with finely toothed edges and many long curved side veins, paired or 3 at

47. Avocado   *Persea americana* Mill.*
Flowering twig, fruit (right), 2/3 X (P. R. v. 1).

a node and by the large clusters of many small round bluish or dark red berries with narrow ring at tip.

An evergreen shrub of 5--10 ft (1.5--3 m) or small tree to 20 ft (6 m). Bark gray brown, smoothish, slightly fissured. Inner bark pinkish, astringent. Twigs stout, slightly succulent, hairy when young, with large raised half-round leaf-scars.

Leaves 2 or 3 at a node (opposite or whorled), with stout fleshy leafstalk 3/4--2 in (2--5 cm) long, grooved above and enlarged at base. Blades large, narrowly elliptical or obovate, 4--10 in (10--25 cm) long and 1 1/2--3 1/2 in (4--9 cm) wide, long- or short-pointed at both ends, often widest beyond middle, slightly thick and leathery, with finely toothed edges, upper surface shiny dark green and hairless with veins often sunken, lower surface light green with raised veins finely hairy.

Flower clusters (corymbs) terminal and erect, broad, flattened or rounded, 2--4 1/2 in (5--11 cm) long and broad, with many short-stalked small flowers, male and female on different plants (dioecious). Male flowers about 1/2 in (13 mm) long and broad, composed of short calyx with 5 pointed lobes, corolla of 5 spreading petals almost 3/8 in (10 mm) long, white or tinged with blue or pink, 10 spreading stamens nearly 1/2 in (13 mm) long, and small nonfunctioning pistil. Female flowers about 3/8 in (10 mm) long, composed of basal cup (hypanthium) bearing 5-lobed calyx 1/8 in (3 mm) long, 5 petals 1/16 in (1.5 mm) long, and pistil with inferior 5-celled ovary, short style, and 5 rounded stigmas.

Fruit (berry) round, 3/8 in (10 mm) in diameter, bluish or dark red, fleshy. Seeds many, elliptical, minute.

Common and widespread in wet forests at low and middle altitudes of 1,000--6,000 ft (305--1,829 m) through the Hawaiian Islands.

Special areas--Haleakala, Volcanoes.
Range--Hawaii only.
Other common names--pu'aha'nui, kanawau, kupuwao, pi'ohi'a, akiahala, nawao.
Botanical synonyms--*Broussiasia pellucida* Gaud., *B. arguta* var. *pellucida* (Gaud.) Fosberg.

The genus *Broussaisia* with only 1 species is confined to Hawaii. The name honors Francois Joseph Victor Broussais (1772--1838), French medical doctor and physiologist. This species is also the only native Hawaiian example of its family, though several others have been introduced as ornamentals. Two varieties differing in leaf arrangement and flower color originally named as separate species have been distinguished.

PITTOSPORUM FAMILY (PITTOSPORACEAE)

49. Hō'awa

Pittosporum confertiflorum Gray

This variable species is an example of its genus, which has about 10 Hawaiian species known as hō'awa, also 2 naturalized species. Small evergreen shrubs or trees with large narrow leathery leaves on stout twigs, many small whitish flowers crowded at base of leaves, and large rounded or 4-angled deeply wrinkled fruits that split open into 2 parts.

To 30 ft (9 m) in height and 8 in (0.2 m) in trunk diameter, with open crown of few stout, stiff erect branches. Bark gray, smooth to fissured. Inner bark orange or light yellow within green outer layer, bitter. Twigs stout, gray, with pressed brown hairs when young, smoothish, with clustered large half-round leaf-scars and long portions without leaf-scars.

Leaves alternate, many crowded near end of erect twigs, with stout light yellow leafstalks of 3/8--2 in (1--5 cm). Blades obovate to oblong, mostly 2 1/4--4 in (6--10 cm) long and 1--1 1/2 in (2.5--4 cm) wide, the largest to 8 in (20 cm) by 4 in (10 cm), thick, stiff, blunt at apex, widest beyond middle and gradually narrowed toward base, curved under at edges. Upper surface dull green, densely gray hairy when young, becoming nearly hairless, with sunken light yellow midvein and network of prominently sunken veins; lower surface densely brown hairy, with raised veins.

Flower clusters (corymbose racemes) mostly terminal at leaf bases, about 1 in (2.5 cm) long. Flowers many, fragrant, crowded on short brown hairy stalks, perfect or male and female, about 3/8 in (1 cm) long, composed of cup-shaped 5-lobed brown hairy calyx about 1/4 in (6 mm) long; white corolla with cylindrical tube about 3/8 in (10 mm) long with 5 spreading lobes 1/4 in (6 mm) long. Male flowers have 5 alternate stamens attached at base of tube and extending beyond. Female flowers have 5 minute nonfunctioning stamens and narrow pistil with hairy slightly 2-lobed ovary, 2-celled and containing many ovules, and slender style.

Fruits (seed capsules) usually 1 (sometimes 2--3), rounded or 4-angled, 3/4--1 1/2 in (2--4 cm) long, brown, with point at apex, hard and thick-walled, the surface finely hairy, rough, deeply wrinkled, splitting into 2 parts, 1 celled, inner wall orange, resinous or mucilaginous within. Seeds many, elliptical, flat, more than 1/4 in (6 mm) long, shiny black.

The most widespread and common species of this genus in Hawaii, occurring from dryland forests to moist forests at 600--7,200 ft (183--2,194 m).

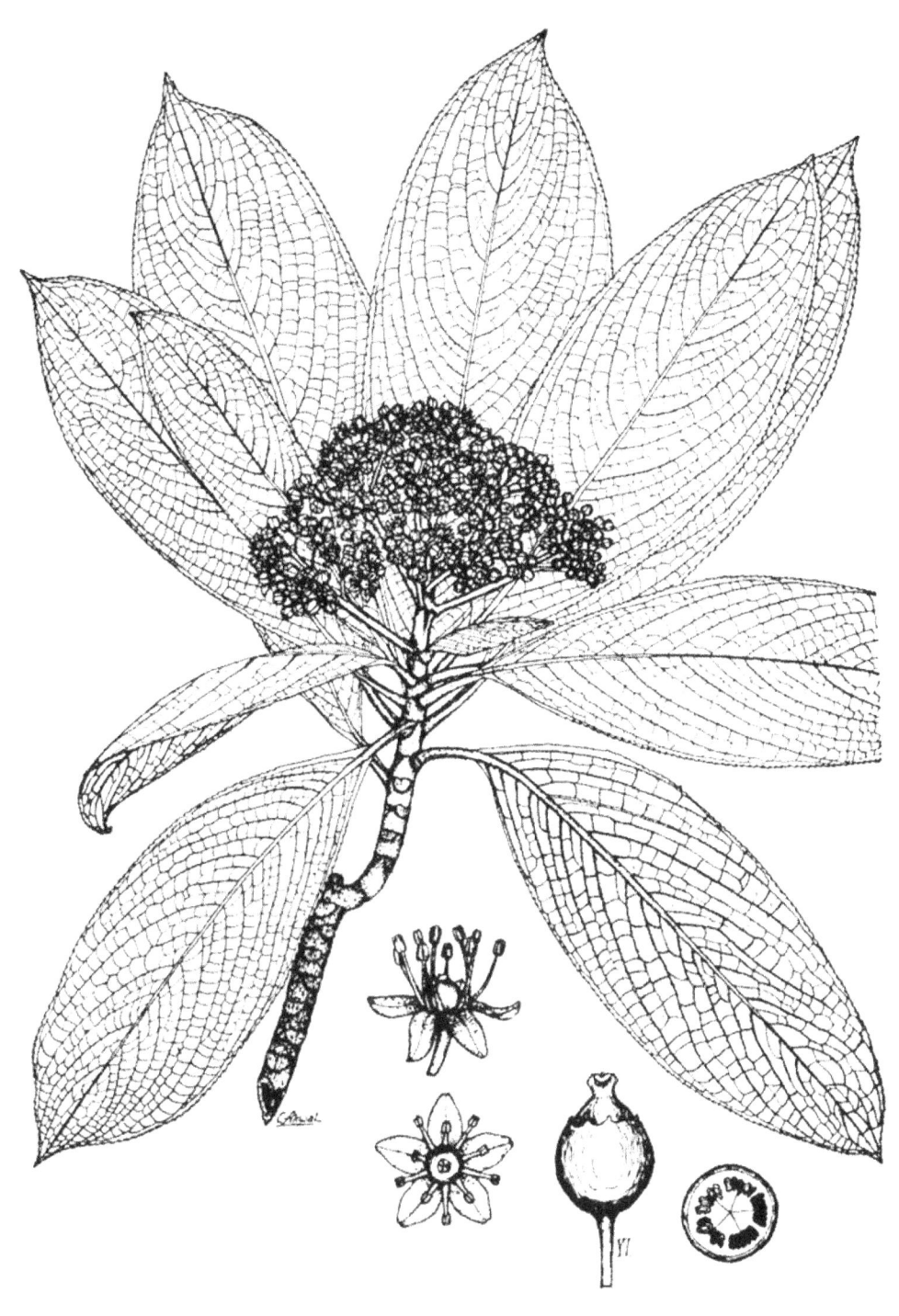

48. Kanawao *Broussaisia arguta* Gaud.
Twig with male flowers, 1/2 X (Degener).

Special areas--Haleakala, Volcanoes.
Range--Oahu, Lanai, Maui, and Hawaii only.
Other common name--hā'awa.
Botanical synonyms--*Pittosporum cauliflorum* Mann, *P. cladanthum* Sherff, *P. confertiflorum* Gray, *P. halophiloides* Sherff, *P. halophilum* Rock, *P. lanaiense* St. John.

Rock (1913) observed that the trees in this genus are very variable and that it is difficult to render the exact limitation of each species. He found capsules of 3 different "species" on a single twig on the island of Lanai, where the genus is exceedingly well represented. As there are as many different forms as trees, one would be naming individual trees. Insect pollination is a factor, he concluded.

One native and a few introduced species are planted as ornamentals. In Hawaii, a home remedy was obtained from the pulp of the pounded fruits. Plants of another native species, *Pittosporum hosmeri* Rock, can be seen on the grounds of the Bishop Museum.

LEGUME FAMILY (LEGUMINOSAE)

50. Formosa koa

Acacia confusa Merr.*

Handsome spreading evergreen tree, introduced for ornament and forest planting, with compact rounded crown, curved or sickle-shaped "leaves" like koa, flowers in small light yellow balls, and narrow flat dark brown pods not smaller between seeds. A tree 20--50 ft (6--15 m) tall, with 1 or more forking trunks to 1 ft (0.3 m) in diameter and widely spreading branches. Bark gray, smooth; inner bark whitish, fibrous, slightly bitter. Twigs slender, brown, hairless. Mimosa subfamily (Mimosoideae)

Leaves alternate, modified as sickle-shaped flattened leafstalks (phyllodes) like koa, narrowly lance-shaped, curved, 2 1/2--3 1/2 in (6--9 cm) long and 3/16--3/8 in (5--10 mm) wide, slightly thickened, hairless, gradually narrowed to both ends, with tiny curved point at apex, with several fine parallel veins from base, dull green.

Flower clusters of light yellow balls (heads) 3/8 in (1 cm) in diameter, 1 or 2 on slender stalks about 1/2 in (13 mm) long at leaf base. Flowers slightly fragrant, tiny, numerous, stalkless in balls, nearly 1/4 in (6 mm) long, consisting of 5 sepals, no petals, many spreading threadlike separate stamens ending in dot anther, and narrow pistil with threadlike style.

Fruits (pods) narrow flat, 2--4 in (5--10 cm) long, 5/16--3/8 in (8--10 mm) wide, dark brown, short-pointed at apex, narrowed into stalk at base, not smaller between seeds, splitting open. Seeds 4--8, beanlike, 3/16 in (5 mm) long, elliptical, slightly flattened, dark brown, slightly shiny.

The heartwood is pale brown, not sharply demarcated from the paler sapwood. The wood is moderately heavy and hard and should be suitable for use in furniture manufacture. In Hawaii, the wood has not been used, probably because the tree is almost always too small and the stem too crooked to be suitable for sawing into lumber.

More than 295,000 trees have been planted by the Division of Forestry in the Forest Reserves of all islands since introduction from Taiwan about 1925. The tree is quite hardy in adverse conditions and has been used extensively for revegetating eroded sites in Hawaii, also in Taiwan and the Northern Mariana Islands. It grows well on both wet and dry sites from sea level to 2,000 ft (610 m). Trees can be seen at Ualakaa State Park (Round Top), Oahu, or on the mall near Hamilton Library, University of Hawaii.

Special areas--Waimea Arboretum, Tantalus.
Champion--Height 65 ft (19.8 m), c.b.h. 5.9 ft (1.8 m), spread 88 ft (26.8 m). Kohala Forest Reserve, Muliwai, Hawaii (1968).
Range--Native of Taiwan and Philippines.
Other common names--small Philippine acacia; yanangi (Palau).

The native species No. 51, koa (*Acacia koa* Gray), is a taller tree with larger, more curved "leaves" and longer pods.

51. Koa

Acacia koa Gray

Koa, the largest native tree and second most common, is well known. This large to very large evergreen tree becomes 100 ft (30 m) tall and 5 ft (1.5 m) or more in trunk diameter but may be only half that size. Trunk straight and tall or becoming crooked and branched; crown spreading, rounded to irregular, dark green. Bark light gray, smooth on small trunks, becoming very rough, thick, and deeply furrowed, scaly and shaggy. Twigs brown, becoming hairless. Mimosa subfamily (Mimosoideae).

Leaves alternate, mostly modified as sickle-shaped leafstalks or petioles (phyllodes), narrowly lance-shaped, curved, 4--6 in (10--15 cm) long and 1/4--1 in (6--25 mm) wide, slightly thickened and leathery, hairless, very long-pointed at both ends, with dotlike gland near base, several fine parallel veins from base, dull

49. Hō'awa *Pittosporum confertiflorum* Gray
Twig with fruit (above) and twig with flowers (below), 2/3 X.

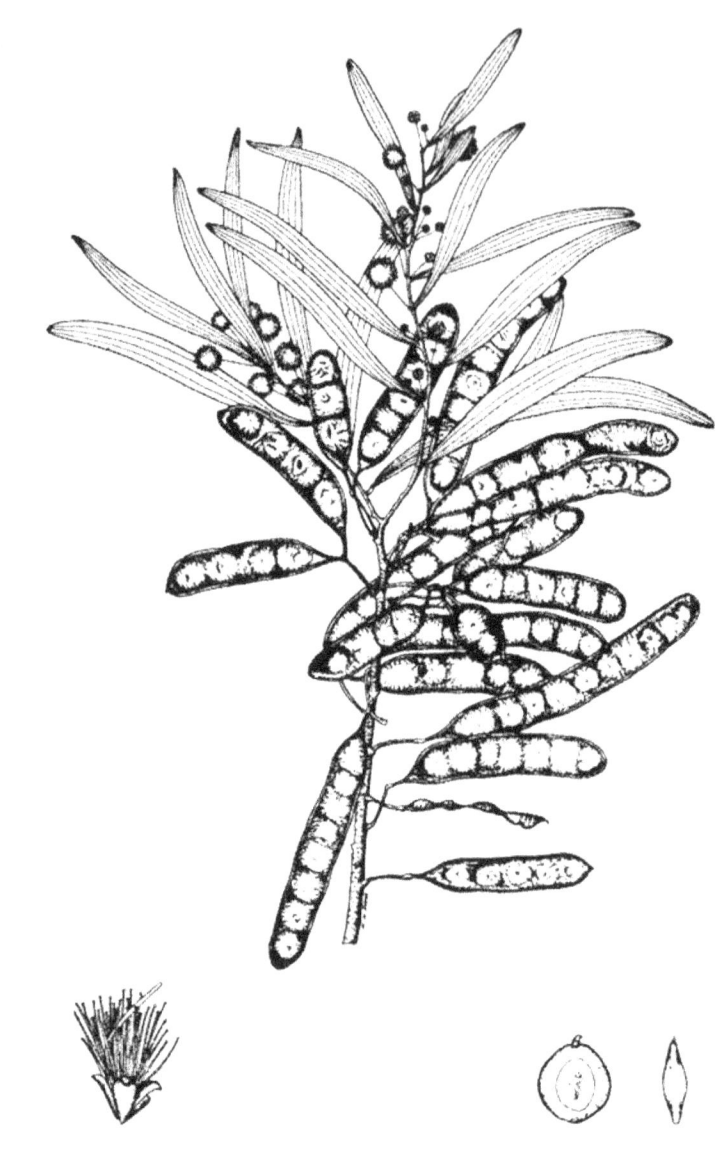

50. Formosa koa   *Acacia confusa* Merr.*
Twig with phyllodes, flowers, and fruits, 1/2 X; flower (lower left), 3 X; seeds (lower right), 2 X (Degener).

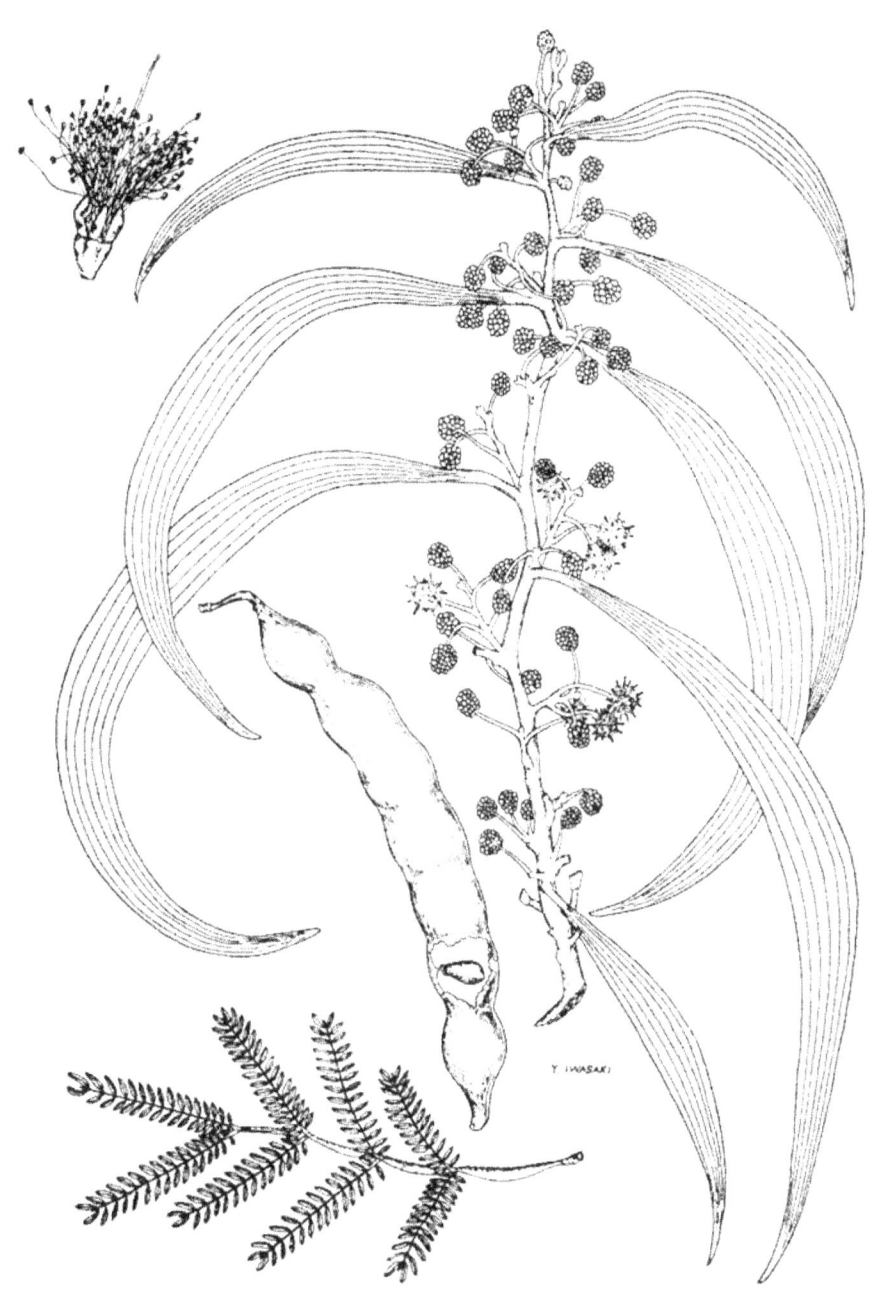

51. Koa   *Acacia koa* Gray
Single flower (upper left), 5 X; twig with flowers, phyllodes, and pod, broken to show enclosed seed, 1/2 X; compound leaf (below), 1/4 X (Degener).

green to dark green. True (juvenile) leaves on seedlings and young twigs twice compound or divided (bipinnate), 6--7 in (15--18 cm) long, with 5--7 pairs of axes, each with 24--30 paired stalkless oblong leaflets 1/4 in (6 mm) long.

Flower clusters of light yellow balls (heads) 3/8 in (1 cm) in diameter, 1 or few on slender stalks about 1/2 in (13 mm) long at leaf base. Flowers tiny, numerous, stalkless in balls, nearly 1/4 in (6 mm) long, consisting of cup-shaped calyx, 5 narrow petals slightly united at base, many spreading threadlike separate stamens ending in dot anther, and narrow pistil with threadlike style. Flowering mostly in late winter and early spring.

Fruits (pods) broad, flat, 3--6 in (7.5--15 cm) long, 5/8--1 in (1.5--2.5 cm) wide, brown, mostly not splitting open. Seeds several, beanlike, 5/16 in (8 mm) long, oblong, flattened, straight, dark brown or blackish, slightly shiny.

Koa is an excellent cabinet wood of reddish brown color that is often highly figured. It is moderately heavy wood (sp. gr. 0.55) identical in weight and strength properties with black walnut (Skolmen 1968). This stable wood works and seasons well and takes a high polish. It is not resistant to decay and is quite susceptible to drywood termites.

The wood is used for furniture, cabinetwork, carved bowls and turnery, gunstocks, and veneer; formerly for construction and surfboards. Many large offices as well as homes in Hawaii have paneling and furniture of koa. It is Hawaii's best known wood. Native Hawaiians had many uses, such as house timbers, carved dugout canoes, and paddles. Koa canoes are prized for competitive paddling by outrigger canoe clubs today and logs suitable for making them are extremely scarce and costly. The bark served in tanning.

Koa is widely distributed in both dry and rain forests at 600--7,000 ft (183--2,134 m) altitude. Koa forests are an important habitat for rare birds. This species may be seen on all the larger islands, near Kokee on Kauai, along Likelike and Nuuanu Pali Highways on Oahu, the Hāna Road near Keanae, Maui, and Hawaii Volcanoes National Park on Hawaii. There are 2.5 million cu ft (0.7 million cu m) of sawtimber in Hawaii. Only Hawaii and Kauai have tall straight-stemmed koa suitable for lumber. The trees on other islands have short crooked stems. Trees on the Island of Hawaii growing in closed forest at 4,000--6,000 ft (1,219--1,829 m) are by far the largest.

Special areas--Waimea, Arboretum, Tantalus, Haleakala, City, Volcanoes, Kipuka Puaulu.

Former champion--Height 140 ft (42.7 m), c.b.h. above bulge, 37.3 ft (11.4 m), spread 148 ft (45.1 m). District of Ka'u, Hawaii (1969). This giant was probably the largest native tree in Hawaii, the tallest as well as greatest in trunk circumference. Unfortunately, it split in two from the groundline in the late 1970's and is now quite small. A new champion has not been selected.

Range--Known only from Hawaiian Islands.

A few botanical varieties have been named from different islands, distinguished by shape of the "leaves."

Kauai koa (*Acacia kauaiensis* Hillebr.) is a closely related species endemic to Kauai and common at Kokee. This large tree is easily recognized by the flower clusters of many light yellow balls (heads) at ends of twigs).

Koaia (*Acacia koaia* Hillebr., another closely related species, is endemic to dry areas of Molokai, Maui, and Hawaii and now rare and considered endangered. It is distinguished by seeds that lie parallel rather than perpendicular to the pod axis. Wood heavier, harder, and more finely textured than koa.

52. Black-wattle acacia

Acacia mearnsii De Wild.*

Small handsome evergreen introduced tree with thin spreading crown of finely divided leaves, flowers in small light yellow balls, and narrow flattened beadlike pods. To 40 ft (12 m) tall and 10 in (25 cm) in trunk diameter. Bark brown to gray, smooth to finely fissured. Inner bark light yellow with brown steaks, bitter. Twigs brownish, finely angled, with tiny hairs. Mimosa subfamily (Mimosoideae).

Leaves alternate, finely divided, twice pinnate (bipinnate), 3--6 in (7.5--15 cm) long, with tiny hairs and slender angled axis. Side axes 4-15 pairs, very slender, with dotlike gland at base of each pair. Leaflets very numerous, 30--80 crowded featherlike on each axis, not paired, stalkless, very narrow (linear), 3/16--3/8 in (5--10 mm) long, gray green to dark green.

Flower clusters of light yellow balls (heads) 1/4 in (6 mm) in diameter, several on short stalks at leaf bases and shorter than leaves. Flowers fragrant, tiny, numerous stalkless in balls, 1/8 in (3 mm) long, with tiny 5-toothed calyx and corolla and many spreading threadlike separate stamens ending in dot anther.

Fruits (pods) several clustered, 2--4 in (5--10 cm) long, 1/4 in (6 mm) wide, flattened, gray brown to blackish, finely hairy, slightly narrowed between seeds, splitting open. Seeds several, beanlike, 1/8 in (3 mm) long, elliptical, dull black.

The wood is light brown with reddish streaks resembling pale koa. It is heavy (sp. gr. 0.55), hard, and relatively difficult to work. It is not resistant to decay. Elsewhere, the wood is used in turnery and for fuel, and the bark for tannin. It is not used in Hawaii.

Considered a noxious weed in most places in Hawaii, particularly in Kula, Maui, where it is very prolific. Propagated from seeds and easily established. Short-lived, but new sprouts are produced by roots. The Division of Forestry has planted 65,000 trees in the Forest

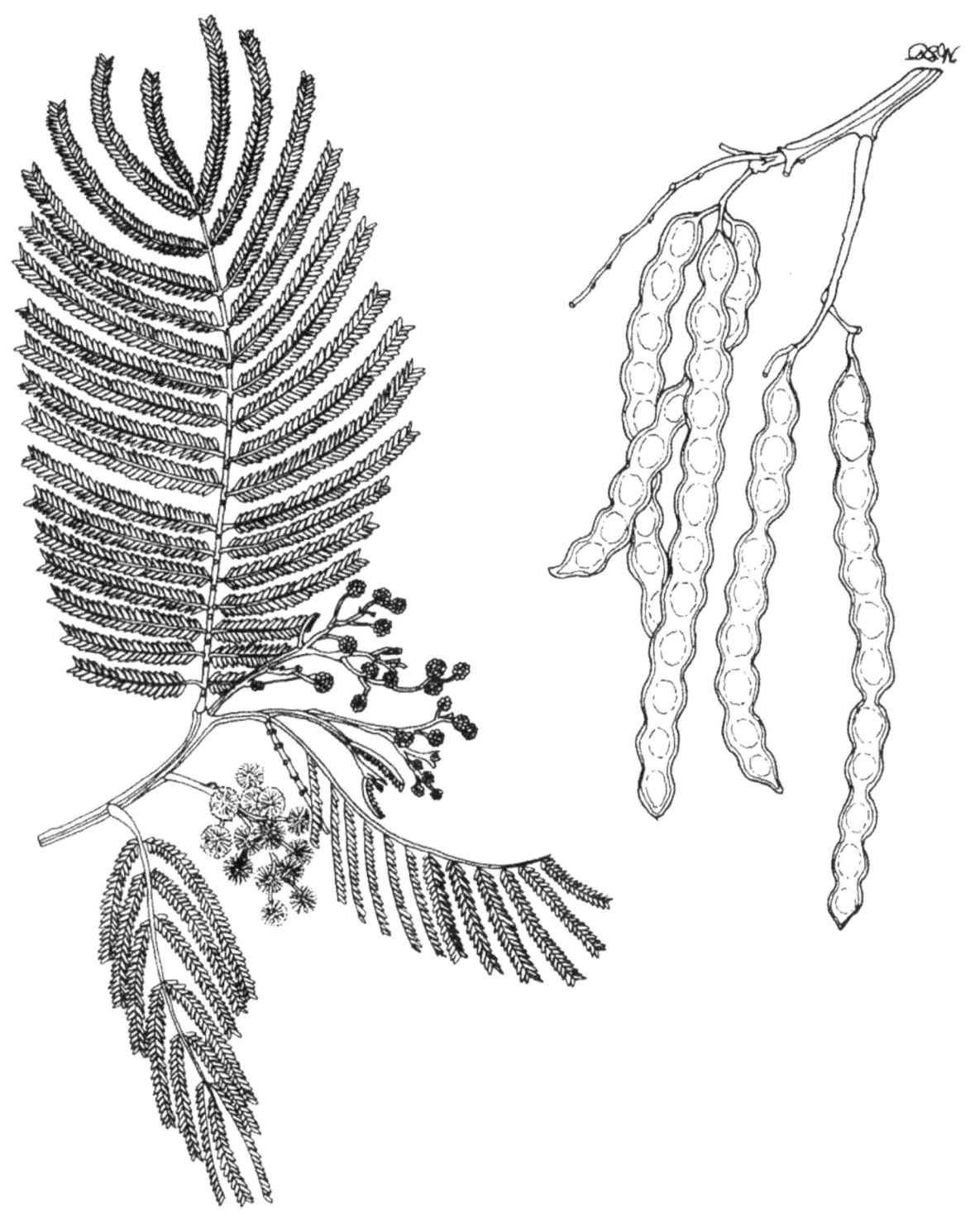

52. Black-wattle acacia *Acacia mearnsii* De Wild.*
Flowering twig (left), fruits (right), 1 X.

Reserves, primarily at Mokuleia, Oahu and Kula, Maui. The trees at the Kula Forest Reserve were mostly planted in 1938 and 1939 and may be the progenitors of the escapes. Also at Kamuela, Hawaii. Introduced as an ornamental in southern California and other subtropical regions.

53. Blackwood acacia

Blackwood acacia, or Australian blackwood, is an evergreen tree introduced in forestry tests and as an ornamental. It differs from species No. 51, the native koa, *Acacia koa*, in that its "leaves" are shorter and less curved, and its pods are narrow and curved. Mimosa subfamily (Mimosoideae).

Medium-sized tree 40 ft (12 m) high and 1 ft (0.3 m) in trunk diameter. In southeastern Australia, where it is native, this acacia is a large tree to 80--110 ft (24--30 m) in height and 2 1/2--4 ft (0.8--1.2 m) in diameter. With straight axis, erect branches, and regular narrow crown of dense foliage. Bark gray, rough, finely furrowed. Twigs slender, slightly angled, with tiny pressed hairs, brownish.

Leaves alternate, modified as sickle-shaped flattened leafstalks or petioles (phyllodes) narrowly sickle-shaped or lance-shaped, slightly curved, 2 1/2--4 1/2 in (6--12 cm) long and 3/8--3/4 in (1--2 cm) wide, thickened and leathery, hairless, long-pointed at both ends, with dotlike gland near almost stalkless base, with 3--6 fine parallel veins from base, dull green. Leaves of young plants (juvenile) are compound (bipinnate), to 5 in (13 cm) long, with 4--10 pairs of branches, each with 15--20 pairs of oblong thin dull green leaflets 3/8--1/2 in (10--13 mm) by 1/8 in (3 mm).

Flower clusters (racemes) of 3--5 short-stalked light yellow balls (heads) 3/8--1/2 in (10--13 mm) in diameter at leaf base. Flowers tiny, 30--50 stalkless in ball, consisting of cup-shaped 5-toothed calyx, corolla to 5 narrow petals united in lower part, many spreading threadlike separate stamens 1/4--5/16 in (6--8 mm) long, ending in dot anther, and narrow pistil with hairy 1-celled ovary and threadlike style.

54. Molucca albizia

Large deciduous introduced tree with tall trunk and thin very broad spreading crown of large twice compound leaves. To 100 ft (30 m) tall, with long trunk to 4 ft (1.2 m) in diameter, not enlarged at base. Bark light gray, smooth with corky warts, showy. Inner bark pink, astringent and slightly bitter. Twigs stout, light gray. Mimosa subfamily (Mimosoideae).

Leaves alternate, bipinnate, 9--12 in (23--30 cm) long, with tiny pressed hairs and slender angled axis bearing gland above base. Side axes 10--12 pairs 2--4 in (5--10 cm) long. Leaflets paired, 30--40 on an axis, small,

Special areas--Kokee, Kula, Tantalus.
Range--Native of Australia.
Other common names--green-wattle acacia, black-wattle, green-wattle.

This species was formerly called *Acacia decurrens* (Wendl.) Willd.

Acacia melanoxylon Br.*

Fruits (pods) oblong, narrow, 3--5 in (7.5--13 cm) long and less than 3/8 in (1 cm) broad, flat with thick borders, reddish brown, curved and twisted. Seeds 6--10, beanlike elliptical, 3/16 in (5 mm) long, shiny black, hanging and encircled by red ringlike stalk.

The wood is golden brown to dark brown, sometimes tinged or streaked with red, with a beautiful figure very similar to koa. It is hard, moderately heavy, moderately strong and durable, works easily, turns well, and takes a high polish. The wood has been compared with black walnut, as has koa.

Known as blackwood or "hickory," this species is one of the most ornamental Australian timbers, used principally for cabinetwork, paneling, also veneer, split staves, and furniture. Wood of 5 trees grown at Kokee, Kauai, was tested in 1966 and found to be similar in density and appearance to koa, but with a somewhat higher shrinkage in drying.

Propagated from seed, also by root sprouts. Relatively slow growing. Elsewhere planted as a street tree and ornamental, though not recommended for Hawaii for these purposes because of root suckering problem. Introduced also in California where it grows very well as a street tree. A total of 17,000 trees was planted in the Forest Reserves before 1960. Since 1960 several large plantings have been made in the Waiakea Forest Reserve on Hawaii, but the trees there are slow growing.

Special area--Kokee.
Champion--Height 78 ft (23.8 m), c.b.h. 8.2 ft (2.5 m), spread 57 ft (17.4 m). Kokee State Park, Kauai (1968).

Range--Native of southeastern Australia.
Other common names--Australia blackwood, blackwood, black acacia.

Albizia falcataria (L.) Fosberg*

stalkless, oblong, 1/4--1/2 in (6--13 mm) long and 1/8--3/16 in (3--5 mm) wide, short-pointed at apex, unequal-sided and blunt at base, not toothed, thin, above dull green, beneath paler.

Flower clusters (panicles) large, lateral, branched, 8--10 in (20--25 cm) long. Flowers many, clustered, stalkless, 1/2 in (13 mm) long, whitish, composed of light green bell-shaped 5-toothed calyx 1/8 in (3 mm) long, greenish white corolla 1/4 in (6 mm) long with 5 narrow pointed lobes, very many theadlike spreading stamens

53. Blackwood acacia *Acacia melanoxylon* Br.*
Twig with phyllodes and flowers, twig with phyllode and compound leaf (left),
and fruit (lower right), 1 X (Maiden).

more than 1/2 in (13 mm) long, and slender pistil with narrow ovary and long threadlike style.

Fruits (pods) narrow, flat, 4--5 in (10--13 cm) long, 3/4 in (2 cm) wide, green, turning brown, thin-walled, splitting open. Seeds 15--20, beanlike, 1/4 in (6 mm) long, oblong, flattened, dull dark brown.

The heartwood is pale yellow brown with a pink tinge. Sapwood is white. Wood lightweight (sp. gr. 0.33), coarse-textured, and essentially unfigured. It has strength equivalent to ponderosa pine and machines well except for occasional fuzzy grain caused by tension wood. The lumber seasons well but is subject to staining unless dried rapidly. Dust from machining operations is very irritating to mucous membranes, both when the wood is green and when it is dry. The wood is nonresistant to decay or termites.

Over 1 million board feet of timber have been cut on the island of Hawaii. Most of it was made into corestock veneer, for which it is highly suited. It has also been used in lumber form for lightweight pallets, boxes, and shelving. It is also suitable for internal furniture parts. Tests in the Philippines indicate it is a good pulpwood.

Plants become established naturally in abandoned sugarcane fields as well as in the forest wherever there are seed trees. The tree reproduces prolifically in areas below 1,000 ft (305 m) elevation with 80--150 in (2,032--3,810 mm) annual rainfall. The lightweight pods are blown by winds, and seeds are abundant. This species has very rapid growth, as much as 15 ft (4.5 m) a year, but has a short life. It does well on poor, heavy clay, moist soils. The tree is a nitrogen fixer. Fallen leaves improve the soil.

Introduced to Hawaii in 1917 by Joseph Rock as an ornamental and for reforestation, it is now naturalized. The largest stand is at the Moloaa Forest Reserve on Kauai where there are more than 5 million board feet of sawtimber. It may be seen near Lyon Arboretum in Manoa Valley and at Foster Botanic Garden on Oahu, at the Lava Tree State Park in Puna, Hawaii, and along roadsides and in gardens through the State. A stand in Palolo Valley, Oahu, had trees 27 in (69 cm) in diameter at only 13 years of age.

Champion--Height 110 ft (33.6 m), c.b.h. 29.8 ft (9.1 m), spread 167 ft (50.9 m). Hilo, Hawaii (1968).

Range--Native to Molucca.

Other common name--sau.

Botanical synonyms--*Albizia moluccana* Miq., *A. falcata* (L.) Backer (in part), *Paraserianthes falcataria* (L.) Nielson.

55. Siamese cassia

Cassia siamea Lam.*

This introduced tree planted in lowlands, especially along roadsides, has large showy clusters of numerous bright yellow flowers in late summer and many long narrow flat dark brown pods that remain attached and become unattractive. Medium-sized evergreen tree 60 ft (18 m) high with straight trunk and axis 1 ft (0.3 m) in diameter and erect crown. Bark gray or light brown, smoothish, becoming slightly fissured. Inner bark light brown, gritty and tasteless. Twigs greenish and minutely hairy when young, turning brown. Cassia subfamily (Caesalpinioideae).

Leaves alternate, even pinnate, 9--13 in (23--33 cm) long, with slender grooved green and reddish tinged, finely hairy axis. Leaflets 12--22 on short stalks of 1/8 in (3 mm), oblong, of uniform size, 1 1/4--3 in (3--7.5 cm) long and 1/2--7/8 in (13--22 mm) broad, rounded at both ends, with tiny bristle tip, not toothed, thin, upper surface slightly shiny green and almost hairless, and lower surface gray green with sparse tiny hairs.

Flower clusters (panicles) large erect terminal, 8--12 in (20--30 cm) or more in length and 5 in (13 cm) broad. Flowers almost regular, on straight yellow green finely hairy stalks of 1--1 1/4 in (2.5--3 cm). Calyx is composed of 5 concave pointed greenish yellow, finely hairy sepals 5/16 in (8 mm) long; corolla of 5 short-stalked spreading nearly equal bright yellow petals 5/8--3/4 in (15--19 mm) long; 7 stamens of different lengths and 3 smaller sterile stamens; and pistil with pale green, minutely hairy 1-celled ovary and curved style. Flowering from July to October.

Pods, so numerous that they sometimes give an untidy appearance to the tree, are 6--10 in (15--25 cm) long, about 1/2 in (13 mm) broad, and 1/16 in (1.5 mm) thick, stiff and often slightly curved, splitting up sides into 2 parts. Seeds many, beanlike, elliptical, 5/16 in (8 mm) long, shiny dark brown.

The sapwood is light brown and the heartwood dark brown. The wood is heavy (sp. gr. 0.75) and hard. It has a beautiful figure on flat-sawn faces imparted by prominent parenchyma tissue that is reminiscent of a pheasant's tail, causing the wood frequently to be called "pheasant wood." The wood, which is very susceptible to attack by dry-wood termites, is used for small turnery and carvings and is very popular. Elsewhere, it is employed for posts, fuel, construction, furniture, and similar purposes. Tannin has been extracted from the bark.

Only about 1,500 trees have been planted in the Forest Reserves, almost all on Oahu. It is not commonly planted now as an ornamental because of the messy appearance of its pods. However, many old trees are still seen along roads and in old gardens. A prime example is a labeled tree near the Ewa-makai gate of Iolani Palace grounds.

The species has been widely planted in Puerto Rico and the Virgin Islands in recent years for ornament, shade, and windbreaks. The trees form good windbreaks because they retain a deep closed crown. They

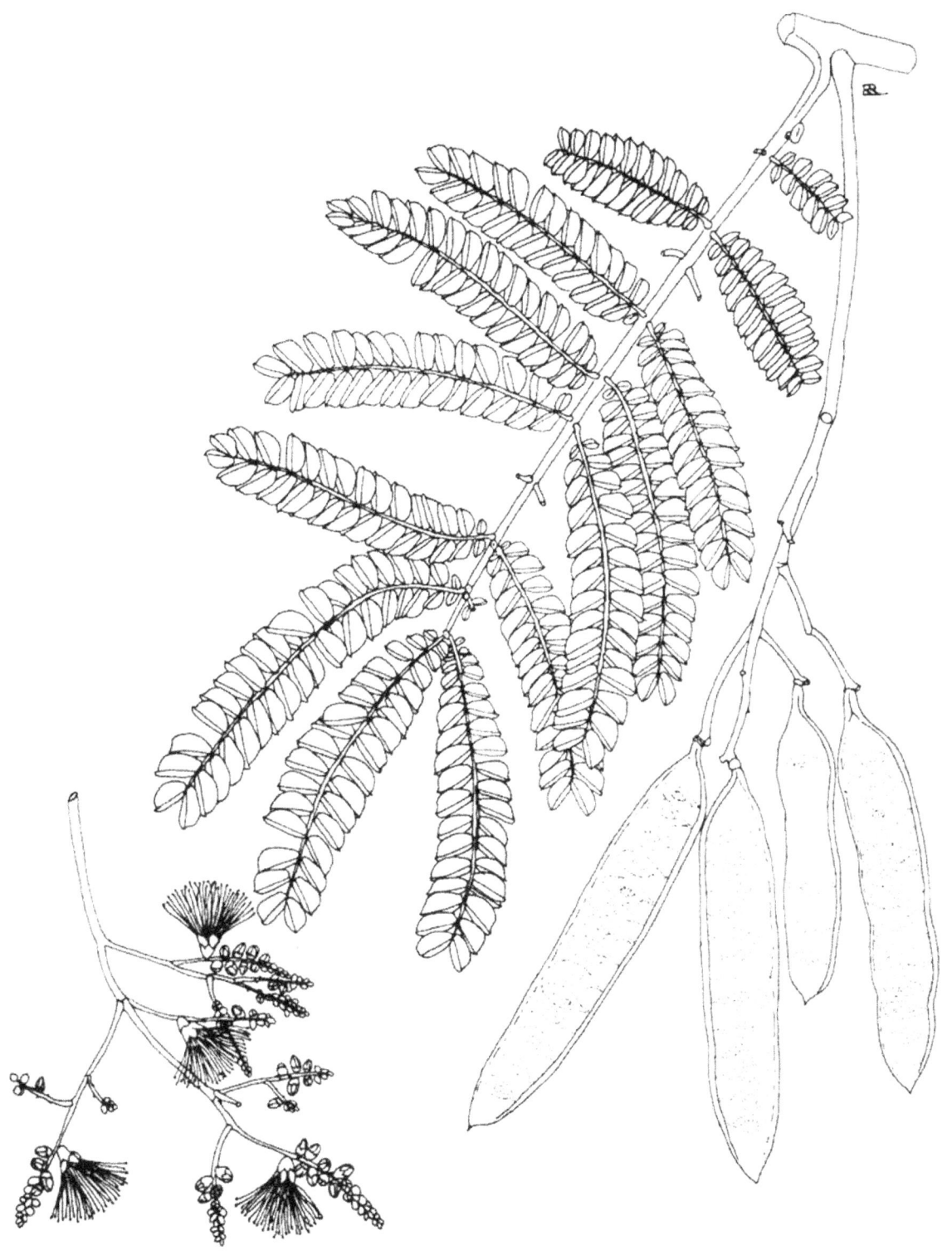

54. Molucca albizia *Albizia falcataria* (L.) Fosberg*
Twig with fruits (above), flowers (lower left), 2/3 X.

are propagated by seeds, grow very rapidly in full sunlight, and are suitable for fuel within a few years. However, they are very susceptible to attack by scale insects there.

The seeds, pods, and foliage are toxic to hogs and cause death quickly after being eaten. As hogs relish the poisonous leaves, farmers in Puerto Rico have suffered losses. Trees blown over or broken by storms increase the danger. Thus, swine and perhaps other livestock should be kept away from these trees.

Planted in lowlands, especially along roadsides in Hawaii and escaping. Introduced about 1865.

56. Kolomona, scrambled-eggs

Introduced ornamental evergreen shrub or small tree, escaping at low altitudes, with abundant bright yellow 5-petaled flowers throughout the year. To 20 ft (6 m) tall, with gray brown bark, smoothish, becoming slightly fissured. Inner bark light yellow beneath green outermost layer, fibrous, bitter. Twigs finely hairy. Cassia subfamily (Caesalpinioideae).

Leaves alternate, pinnately compound, 4--5 in (10--13 cm) long, with slender finely hairy axis enlarged at base and grooved above. Leaflets mostly 12--20, paired, almost stalkless with gland at base of 3 lowest pairs, elliptical, 1/2--1 3/4 in (1.3--4.5 cm) long and 1/4--5/8 in (6--15 mm) wide, thin, rounded or slightly notched at apex, rounded at base, not toothed, dull green above, paler beneath with fine pressed hairs.

Flower clusters (corymbs) terminal and lateral, about 3 in (7.5 cm) long, branched. Flowers several on slender hairy stalks 1/2--1 in (13--25 mm) long, 1 1/2--2 in (4--5 cm) across, composed of calyx of 5 unequal rounded greenish yellow sepals, 5 spreading elliptical slightly unequal petals to 7/8 in (22 mm) long, short-stalked at 3-veined base, 10 short unequal sta-

57. Wiliwili *(color plate, p. 34)*

Small deciduous native tree, originally one for the most common in dry regions, characterized by short spines, leaves with 3 broadly triangular leaflets, and showy orange, yellow, salmon, greenish or whitish flowers when leafless. Pea subfamily (Faboideae).

Tree 15--30 ft (4.5--0 m) tall, with short stout crooked or gnarled trunk 1--3 ft (0.3--0.9 m) in diameter, stiff spreading branches, and broad thin crown becoming wider than high. Bark smoothish, light to reddish brown, with scattered stout gray or black spines to 3/8 in (1 cm) long, becoming slightly fissured, thin. Inner bark light yellow beneath green outer layer, gritty and slightly bitter. Twigs nearly horizontal, stout, green and with yel-

Special area--Iolani.
Range--Native of East Indies, Malaysia, India, and Sri Lanka, spread by cultivation. First described from Siam (now Thailand), as the common and scientific names indicate. Widely planted in West Indies and naturalized locally. Planted also in southern Florida, Central America, and northern South America.
Other common names--kassod-tree, kolomona; casia de Siam (Puerto Rico).
Botanical synonym--*Sciacassia siamea* (Lam.) Britton.

This species is known for its showy flowers in late summer after most legumes cease blooming. However, other species are more attractive generally.

Cassia surattensis Burm. f.*

mens, and slender pistil with curved hairy ovary, short style, and dot stigma.

Fruits (pods) 2--4 in (5--10 cm) long, 1/2 in (13 mm) wide, flat, thin, dull dark brown, almost hairless, with long narrow point at end, splitting open on both edges. Seeds 10--20, beanlike, oblong, flattened, shiny dark brown, 3/16 in (5 mm) long.

Wood whitish, hard. It is reported that the bark is medicinal.

This planted ornamental has escaped to roadsides, pastures, and waste places at lower altitudes and is quite common; it is often used as a hedge plant. Only 224 trees have been planted in the Forest Reserves, all on Maui in the Koolau Forest Reserve.

Special areas--Waimea Arboretum, Volcanoes (coastal forests).
Range--Native from India to Australia and Polynesia.
Other common names--kalamona, glossy-shower senna.
Botanical synonyms--*Cassia glauca* Lam., *Psilorhegma glauca* (Lam.) Deg. & I. Deg., *Senna surattensis* (Burm. f.) Irwin & Barneby.

Erythrina sandwicensis Deg.

lowish hairs when young, with scattered blackish prickles or spines.

Leaves alternate, compound, 5--12 in (13--30 cm) long, with long slender leafstalk 3 1/2--10 in (9--25 cm) long. Leaflets 3, short-stalked, 2 paired and 1 largest at end, broadly triangular, 1 1/2--4 in (4--10 cm) long and 2 1/2--6 in (6--15 cm) wide, slightly broader than long, short-pointed at apex and almost straight at base, thin, becoming nearly hairless above, beneath yellow hairy with raised veins, with 2 dotlike glands at base of each leaflet and 1 or 2 glands at base of leafstalk.

Flower clusters (racemes) near ends of twigs, at end of yellow hairy stalk of 3 in (7.5 cm) or less. Flowers many, crowded in mass 3--6 in (7.5--15 cm) long, short-

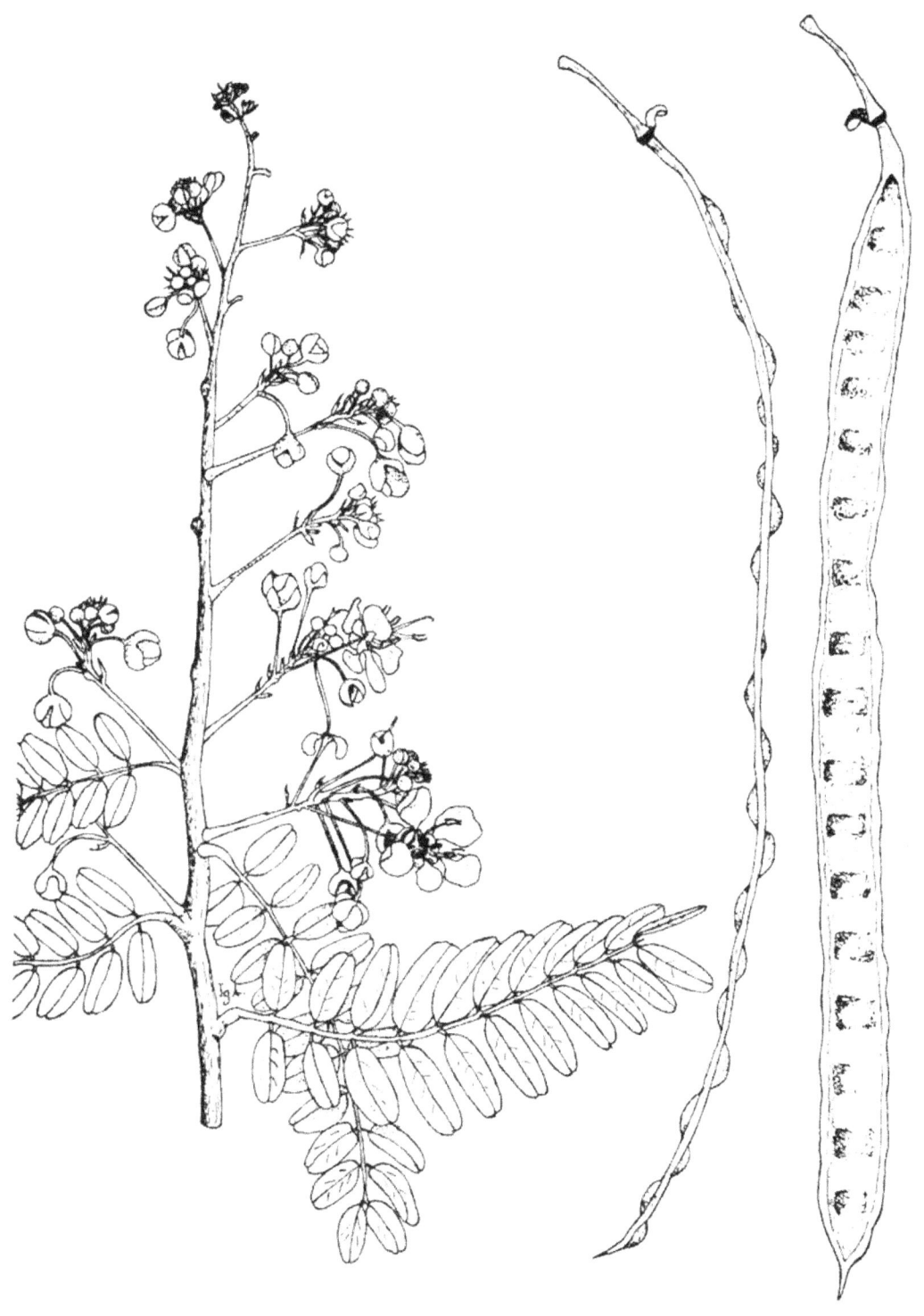

55. Siamese cassia *Cassia siamea* Lam.*
Flowering twig (left), fruits (right), 2/3 X (P. R. v. 1).

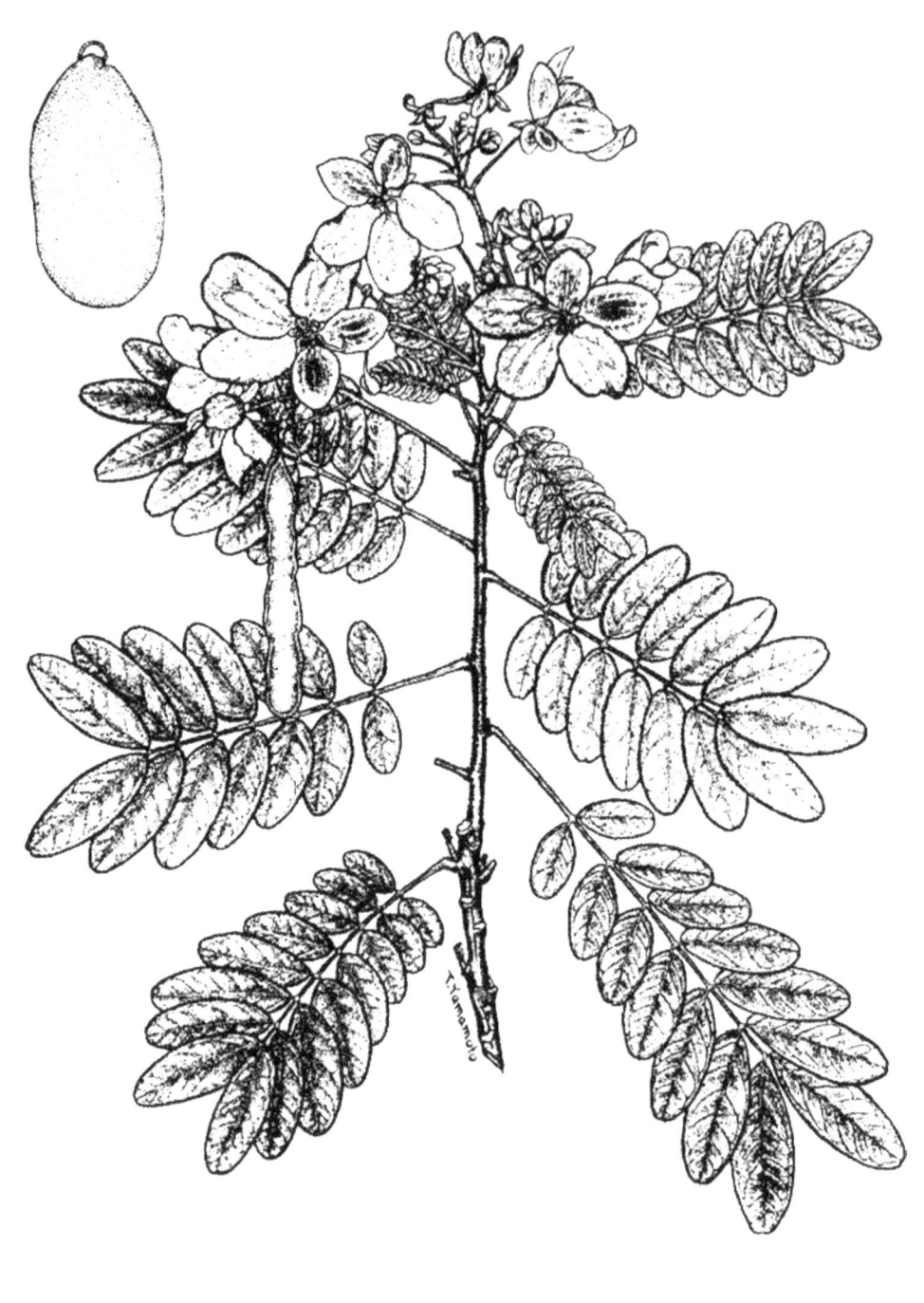

56. Kolomona, scrambled-eggs *Cassia surattensis* Burm. f.*
Twig with flowers and fruits, 1/2 X; seed (upper left), 5 X (Degener).

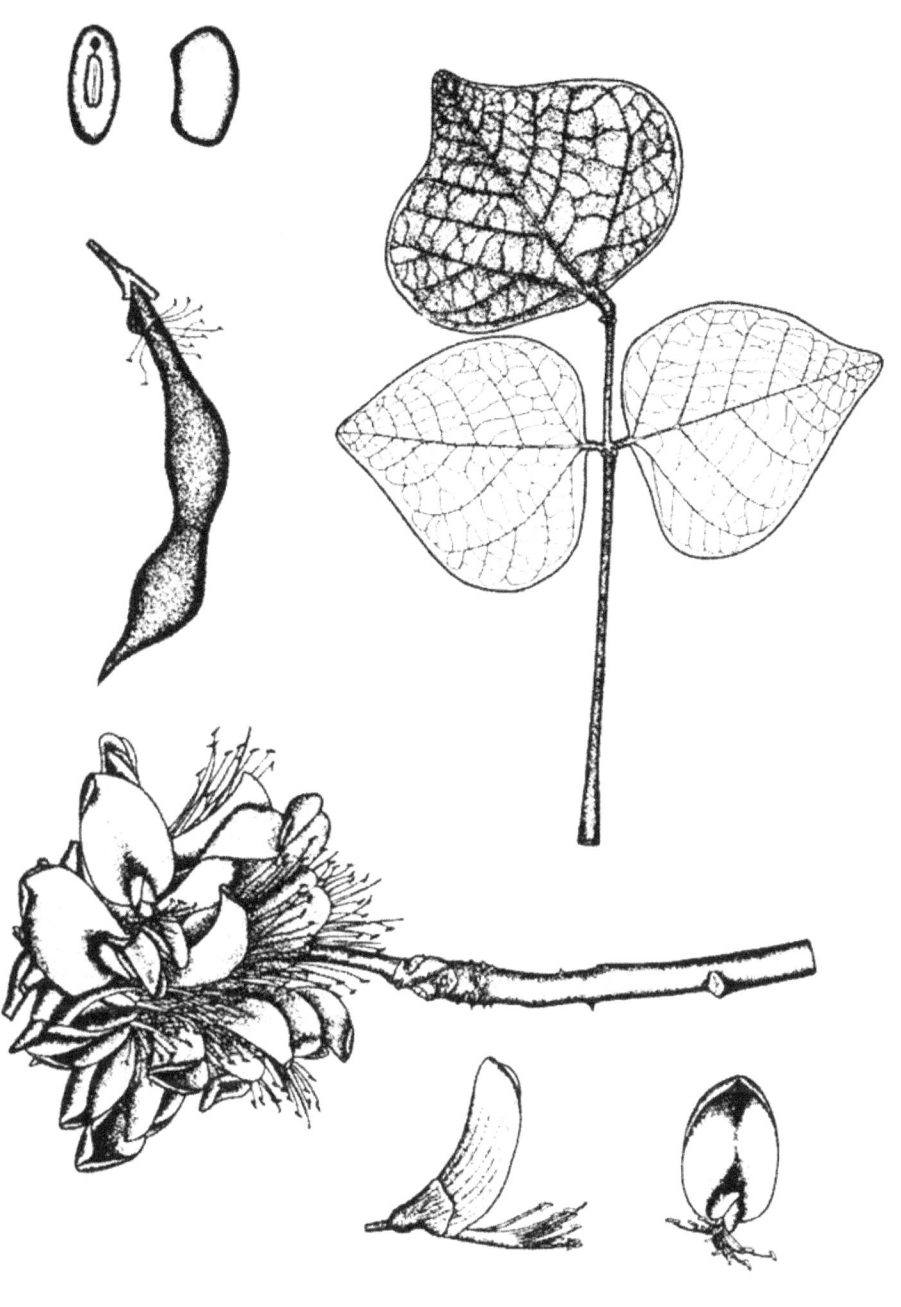

57. Wiliwili   *Erythrina sandwicensis* Deg.
Leaf, fruit, and flowers, 1/2 X; seeds (upper left), 1 X (Degener).

stalked, composed of cuplike densely yellow hairy calyx 5/8 in (15 mm) long, curved and split open on 1 side; corolla orange, yellow, salmon, greenish, or whitish with 1 very large elliptical curved petal (standard) about 1 1/2 in (4 cm) long and 4 small petals (wings and keel) about 1/2 in (13 mm) long; 10 orange or yellow curved stamens about 1 1/2 in (3 cm) long, 9 united in lower half and 1 separate; and pistil with stalked narrow densely hairy ovary and slender curved or straight style.

Fruits (pods) about 4 in (10 cm) long and 1/2 in (13 mm) broad, flattened, long-pointed at both ends, slightly narrowed between seeds, blackish, hard-walled, splitting open. Seeds 1-5, beanlike, elliptical, 1/2--5/8 in (13--15 mm) long, shiny orange red.

The wood is pale yellow brown, soft, very lightweight and coarse-textured. Hawaiians used it for fishnet floats, outriggers of canoes, and surf boards.

The flowers, mostly orange but varying in color on different trees, are borne from early spring to July, while the trees are leafless.

Hawaiians strung into leis the bright red seeds, which probably are poisonous. Seeds of several other members of the genus contain alkaloids similar to curare. However, a large number of seeds would have to be cooked and eaten to be harmful. Captain James Cook was given leis made of wiliwili seeds and the worn bases of conus shells ("puka" shells) when he visited the islands in 1778.

Wiliwili was one of the most common native trees in the dry forests at low altitudes of 500--2,000 ft (152--610 m) on the lee side of the Hawaiian Islands. It is characteristic of the barren, rough aa lava flows. At present, it has been largely replaced by No. 62, kiawe (*Prosopis pallida*), but may still be seen in the dry gullies on the lee side of all islands. It is particularly common along the road from Ulupalakua to Kaupo on Maui and frequent in the dry forest near Puuwaawaa on Hawaii. There are several trees still surviving on the goat-ravaged island of Kahoolawe. The Division of Forestry has planted more than 6,000 wiliwili trees in Forest Reserves, 4,000 of them on Molokai on the land of Palaau.

Wiliwili should be planted more often, since it thrives where other trees cannot survive. It is easily propagated by seeds and cuttings.

Special areas--Foster, Koko, Wahiawa, Waimea Arboretum, Volcanoes.

Champion--Height 55 ft (16.8 m), c.b.h. 12.5 ft (3.8 m), spread 57 ft 917.4 m). Puuwaawaa Ranch, North Kona, Hawaii (1968).

Range--Known only from Hawaiian Islands. A closely related species in Tahiti.

Other common names--Hawaiian erythrina, Hawaiian coraltree.

Botanical synonym--*Erythrina monosperma* Gaud., not Lam.

Trees of many species of *Erythrina* from around the world can be seen at Waimea Arboretum.

58. India coralbean

Erythrina variegata L.*

Introduced deciduous ornamental tree with prickles, leaves with 3 broadly triangular leaflets, and showy red flowers when leafless, related to the native wiliwili. Medium-sized tree to 50 ft (15 m) or more in height. Bark gray, smooth, thin, with many small sharp black or brownish prickles. Twigs nearly horizontal, stout, green and finely hairy when young. Pea subfamily (Faboideae).

Leaves alternate, compound, 8--12 in (20--30 cm) long, with long slender leafstalk 4--5 in (10--15 cm) long. Leaflets 3, short-stalked, 2 paired and 1 largest at end, broadly triangular, 2 1/2--6 in (6--15 cm) long and broad, slightly broader than long, short-pointed at apex and almost straight at base, thin, shiny green above, becoming hairless or finely hairy beneath, with 2 dotlike glands at base of each leaflet.

Flower clusters (racemes) 6 in (15 cm) or more in length on stout unbranched stalks 3--4 in (7.5--10 cm) long, spreading and curved downward. Flowers many, crowded, short-stalked, spreading horizontally around axis and falling promptly, showy, more than 2 in (5 cm) long, composed of narrow calyx 1--1 1/4 in (2.5--3 cm) long, finely hairy and 3--5 toothed; corolla deep bright red or scarlet, with 1 very large curved petal (standard) 2--2 1/4 in (5--6 cm) long and 1--1 1/4 in (2.5--3 cm) wide and 4 small petals (wings and keel) about 3/4 in (2 cm) long; 10 brilliant red stamens 2 1/4 in (6 cm) long, 9 united in lower half and 1 separate; and pistil with stalked narrow hairy ovary and curved red style. In Hawaii flowers are borne on leafless twigs after the leaves fall in January and February.

Fruits (pods) 6--12 in (15--30 cm) long and 1 in (2.5 cm) wide, flattened, slightly narrowed between seeds, dark brown or black, hairless, splitting open late. Seeds 6--10, beanlike, elliptical, 5/8--3/4 in (15--20 mm) long, dark reddish brown.

The wood is described as very lightweight and soft and is used elsewhere for boxes and fuel. It is reported that the bark and leaves have served in home remedies and that the bark has been employed for dyeing and tanning and for its fiber.

A rapidly growing ornamental, it is cultivated for the showy red flowers in late winter when leafless. The dark red seeds have been used for leis. These seeds, like those of related species, probably are poisonous.

Introduced primarily in lowlands of Hawaii. The Division of Forestry has planted about 900 trees in the Forest Reserves. An example may be seen along Round Top Drive above Manoa Valley in Honolulu. It is strictly

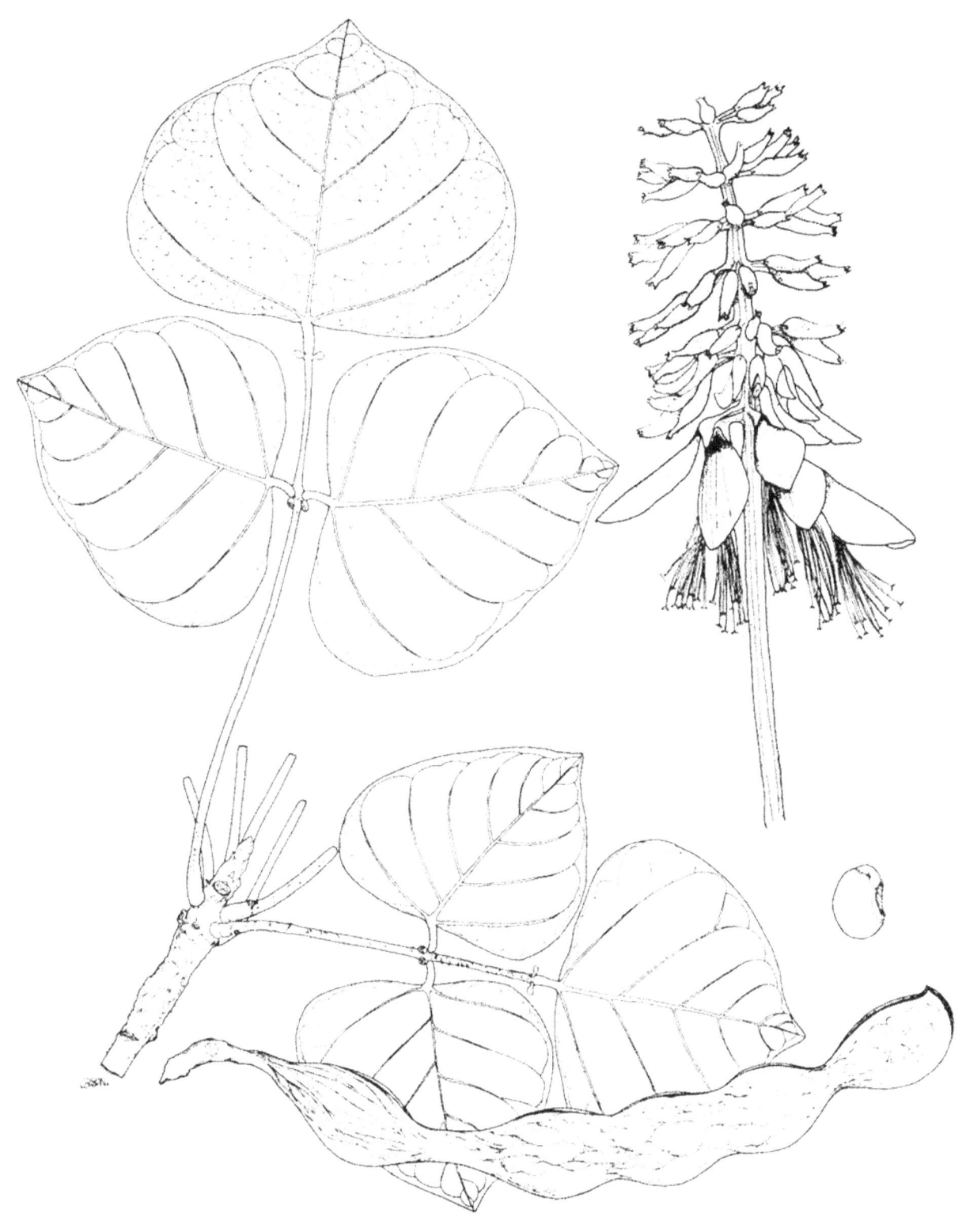

58. India coralbean   *Erythrina variegata* L.*
Leafy twig, flower cluster (upper right), pod (below), and seed (lower right), 2/3 X.

an ornamental that has been planted in a few parklike settings in the forests.

Range--Native from India to southern Polynesia.

Other common names--tigers-claw, variegated coralbean, variegated coraltree, Indian coraltree, Indian wiliwili; gabgab (Guam); gaugau (N. Marianas); roro (Palau); par (Yap); par (Truk); pahr (Pohnpei); gatae (Am. Samoa).

Botanical synonym--*Erythrina indica* Lam.

The common name tigers-claw is said to refer to the flower buds. One form has variegated leaves, with yellowish midveins and leafstalks.

59. Koa haole, leucaena

Leucaena leucocephala (Lam.) de Wit*

Koa haole (foreign koa), or leucaena, is a vigorous shrub or small tree of dry lowlands throughout the Hawaiian Islands, also of larger size on moderately wet sites. This naturalized deciduous species is characterized by twice-pinnate leaves with numerous small gray green leaflets, many flowers in whitish round balls 3/4--1 in (2--2.5 cm) across the spreading threadlike stamens, and many clustered dark brown flat pods. Mimosa subfamily (Mimosoideae).

A rapidly growing small tree 20--30 ft (6--9 m) tall and 4 in (10 cm) in trunk diameter. Bark light gray to brownish gray, smooth with many dots or warts (lenticels). Inner bark light green or light brown and slightly bitter. Twigs gray green and finely hairy, becoming brownish gray.

Leaves alternate, twice-pinnate (bipinnate), 4--8 in (10--20 cm) long, with 3--10 pairs of lateral axes (pinnae), the axes gray green and finely hairy, with swelling at base. Leaflets 10--20 pairs on each lateral axis, stalkless, narrowly oblong or lance-shaped, 5/16--5/8 in (8--15 mm) long and less than 1/8 in (3 mm) wide, short-pointed at apex and unequal or oblique at short-pointed base, thin, gray green and nearly hairless, slightly paler beneath, folding upward together at night.

Flower heads are whitish round balls 3/8--1/2 in (10--13 mm) across on stalks of 3/4--1 1/4 in (2--4 cm) in terminal clusters (racemelike) at ends or sides of twigs. Flowers many, narrow, stalkless. Each individual flower 5/16 in (8 mm) or more in length has a tubular greenish white hairy 5-toothed calyx more than 1/16 in (1.5 mm) long, 5 narrow greenish white hairy petals nearly 3/16 in (5 mm) long, 10 threadlike white stamens about 5/16 in (8 mm) long, and slender stalked pistil nearly 1/4 in (6 mm) long with narrow green hairy ovary and white style.

Pods many oblong, 4--6 in (10--15 cm) long and 5/8--3/4 in (15--19 mm) wide, flat and thin, with raised border, dark brown, short-pointed at apex, narrowed into stalk at base, and minutely hairy. Many hang down in a cluster from end of stalk, splitting open on both edges. Seeds many in a central row, beanlike, oblong, flattened, pointed, shiny brown, 5/16 in (8 mm) long. Flowering and fruiting nearly through the year.

Wood hard and heavy (sp. gr. 0.7); sapwood light yellow and heartwood yellow brown to dark brown. It makes excellent firewood and charcoal and has potential as a source for pulp and paper, roundwood, and construction material.

The seeds, after being softened in boiling water, are strung into necklaces, leis, table mats, purses, and curiosities for tourists.

The plants contain a poisonous alkaloid called mimosine, which can cause the loss of long hair in humans, horses, and some other animals, and sickness in ruminant animals. Mimosine is easily leached from leaves by soaking in water, and cooking will also remove it.

The trees are easily propagated from seeds or cuttings and coppice well. However, because of the hard seed coat, the seeds should be treated or scarified first. In some countries this species has been used for coffee shade, cacao shade, and hedges. Being hardy it can be planted in pastures, to be followed afterwards by timber trees. In some areas the trees are planted and managed for fuel or charcoal on a short rotation of 3 or 4 years between cuttings. Young plants have been harvested also as a green manure for tea and coffee plantations. In the Far East this legume is grown to rebuild the soil and as a forage crop.

Abundant as a weed in dry lowlands of Hawaii, often forming dense thickets in lowlands and lower mountain slopes of 2,500 ft (762 m) altitude. According to Degener, this species was unknown in Hawaii in 1864, but reported as "frequent" 20 years later. It is reported that seeds have been broadcast from airplanes.

Special areas--Waimea, Koko, City.

Range--Native apparently in southeastern Mexico but the distribution has greatly extended by introduction beyond. Now widely naturalized through New and Old World tropics. Naturalized in Hawaii, Mariana Islands, southern Texas, southern Florida, and Puerto Rico and Virgin Islands. Planted also in California.

Other common names--false koa, lili-koa, ekoa, ipilipil, wild tamarind; zarcilla (Puerto Rico); tantan (Virgin Islands); tangan-tangan (Guam); taln tangan (N. Marianas); telentund (Palau); ganitnityuwan tangantan (Yap); tangan-tangan (Marshalls); lopā-samoa (Am. Samoa).

Botanical synonym--*Leucaena glauca* Benth.

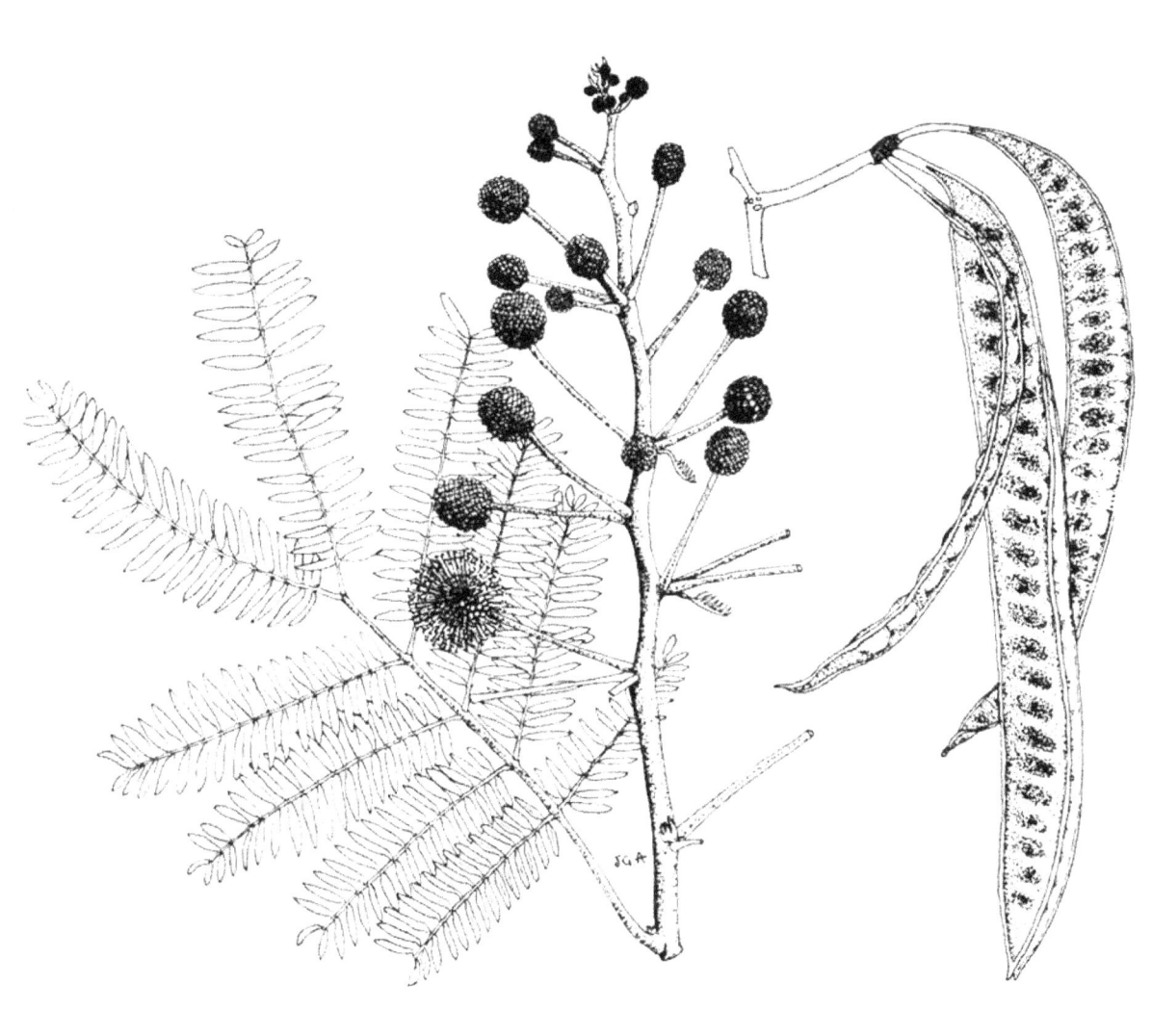

59. Koa-haole, leucaena *Leucaena leucocephala* (Lam.) de Wit.*
Flowering twig, fruits (right), 2/3 X (P. R. v. 1).

60. 'Opiuma
Pithecellobium dulce (Roxb.) Benth.*

This tree, introduced about 1870 for shade in dry lowlands, has become naturalized. It is identified by paired sharp spines usually present at base of leaf, twice pinnate leaves with 4 oblong leaflets, small creamy white flowers in balls of 3/8 in (1 cm), and curved or coiled pink to brown pods with several shiny black seeds mostly covered by whitish pulp. Mimosa subfamily (Mimosoideae).

Small to medium-sized tree to 60 ft (18 m) high and 2 ft (0.6 m) in trunk diametr, the short trunk and branches often crooked, with broad spreading crown and slender drooping twigs. Nearly evergreen but shedding old leaves as new pinkish or reddish foliage appears. Bark light gray, smoothish, becoming slightly rough and furrowed. Inner bark thick, light brown, and bitter or astringent. Twigs slender and drooping, greenish and lightly hairy when young, becoming gray, covered with many small whitish dots (lenticels).

Leaves alternate, bipinnate, with pair of slender, sharp spines (stipules) 1/16--5/8 in (1.5--15 mm) long usually present at base, very slender green leafstalk of 1/4--1 1/2 in (6--38 mm) with tiny round gland near apex, and 2 lateral axes (pinnae) only 1/8--1/4 in (3--6 mm) long. Leaflets 4 in pairs, nearly stalkless, oblong or ovate, 1/2--2 in (13--50 mm) long and 3/16--5/8 in (5--15 mm) wide, rounded at apex, the oblique base rounded or short-pointed, not toothed on edges, thin or slightly thickened, hairy or hairless, dull pale green above, and light green beneath.

Flower clusters (heads) many, short-stalked in slender drooping terminal or lateral axes, each covered with whitish hairs and composed of 20--30 densely hairy flowers. Each flower has tubular hairy 5-toothed calyx about 1/16 in (1.5 mm) long; funnel-shaped tubular hairy 5-toothed corolla about 1/8 in (3 mm) long; about 50 spreading long threadlike stamens united into short tube at base; and pistil with hairy ovary and threadlike style.

Pods 4--5 in (10--13 cm) long, 3/8--5/8 in (1--1.5 cm) wide, slightly flattened, inconspicuously hairy. Seeds beanlike, elliptical, 3/8 in (1 cm) long, hanging down from open pod inside pulpy edible mass (aril) as much as 3/4 in (2 cm) long. With flowers in spring and fruits from April to June.

Sapwood is yellowish, and heartwood yellowish or reddish brown. Wood moderately hard, heavy, strong, and durable. It takes a high polish but is brittle and not easily worked.

Elsewhere, the wood is employed for general construction, boxes and crates, posts, and fuel. The bark has been harvested for its high tannin content. It also yields a yellow dye and is an ingredient in home remedies. A mucilage can be made by dissolving in water the transparent deep reddish brown gum that exudes from the trunk.

The thick whitish sweetish acid pulp around the seeds can be eaten or made into a fruit drink. Livestock browse the pods under the trees. The flat black seeds are strung into leis in Hawaii. The tree is also a honey plant.

This attractive species makes a good highway and street tree, especially in dry areas, growing rapidly and enduring drought, heat, and shade. It withstands close browsing and pruning and is suitable for fences and hedges. Formerly, it was a popular street tree in southern Florida. However, it was susceptible to hurricane damage and did not recover well.

In Hawaii, this species is planted and naturalized in pastures and waste places through the dry lowlands. It is of frequent occurrence along the highway near Haleiwa, Oahu, and in the scrub forest near Lahaina, Maui. One cultivated form has variegated green and white leaves. Degener (1930) reported that the false mynah bird eats the fleshy seed covering and spreads the seeds. According to Neal (1965), the Hawaiian name 'opiuma is from the resemblance of the seeds to the opium of commerce.

Champion--Height 66 ft (20.1 m), c.b.h. 24.9 ft (7.6 m), spread 107 ft (32.6 m) Napoopoo, Hawaii (1968).

Range--Mexico (Baja California, Sonora, and Chihuahua southward) through Central America to Colombia and Venezuela. Widely planted and naturalized in New and Old World tropics. Introduced in southern Florida and Puerto Rico and Virgin Islands.

Other common names--gaumuchil, Manila-tamarind, Madras-thorn; guamá americano (Puerto Rico); guamuche (Mexico, commerce); kamachili (Guam, N. Marianas); kamatsiri (Palau).

This species was named and described botanically in 1795 from Coromandel, India, where it had been introduced. The specific name, meaning sweet, refers to the edible seed pulp.

61. Monkeypod, 'ōhai
Pithecellobium saman (Jacq.) Benth.*

Monkeypod is a familiar beautiful shade tree with large trunk and very broad arched crown of dense foliage. It has bipinnate leaves with many paired diamond-shaped leaflets that fold together at night and on cloudy days. The flower heads are a mass of threadlike stamens pink in outer half and white in inner half, and the flattened brown or blackish pods with sweetish pulp do not open. Mimosa subfamily (Mimosoideae).

A large introduced tree attaining 50--75 ft (15--23 m) in height, with relatively short trunk 2--4 ft (0.6--1.2 m) in

60. 'Opiuma *Pithecellobium dulce* (Roxb.) Benth.*
Flowering twig, leafy twig, fruits (below, 1 X (P. R. v. 1).

diameter. Crown of long stout nearly horizontal branches broader than tall, becoming 80--100 ft (24--30 m) across. Bark gray, rough, thick, furrowed into long plates or corky ridges. Inner bark pink or light brown, bitter. Twigs stout greenish, minutely hairy. Nearly evergreen but almost leafless for a few weeks in spring after old leaves fall in February and March.

Leaves alternate, bipinnate, 10--16 in (25--40 cm) long. The axis and 2--6 pairs of branches (pinnae) are green and finely hairy with swelling at base of each and gland dot on axis where branches join. Each branch bears 6--16 paired stalkless leaflets with gland dot between each pair. Branches toward apex are longer, with more leaflets. Leaflet blades 3/4--1 1/2 in (2--4 cm) long and 3/8--3/4 in (1--2 cm) broad, the outer largest, blunt with minute point at apex, short-pointed at base, not toothed on edges, the sides unequal, slightly thickened, upper surface shiny green and lower surface paler and finely hairy.

Flower clusters (heads or umbels) several near end of twig, each on green hairy stalk of 2 1/2--4 in (6--10 cm), about 2 1/2 in (6 cm) across and 1 1/2 in (4 cm) high, composed of many short-stalked narrow tubular pinkish flowers tinged with green. The narrow green calyx is tubular, about 1/4 in (6 mm) long, 5-toothed, finely hairy; the narrow pink and greenish tinged corolla 3/8--1/2 in (10--13 mm) long, tubular, 5-lobed, and finely hairy; many stamens united in tube near base, with spreading very long threadlike filaments about 1 1/2 in (4 cm) long, ending in dotlike anthers, delicate and soon wilting and shriveling; and pistil consisting of 1-celled light green ovary 3/16 in (5 mm) long and threadlike pinkish style 1--1 1/4 in (2.5--3 cm) long. Flowering from April to August in Hawaii.

Pods narrowly oblong, 4--8 in (10--20 cm) long, 5/8--3/4 in (15--19 mm) wide, and 1/4 in (6 mm) thick, with raised border, straight or a little curved. Seeds several beanlike, oblong reddish brown, about 5/16 in (8 mm) long.

Sapwood thin and yellowish and heartwood dark chocolate brown when freshly cut, becoming attractive light to golden brown with darker streaks. The wood is moderately hard, lightweight (sp. gr. 0.52), of coarse texture, and fairly strong. It is resistant to very resistant to decay and resistant to dry-wood termites. It takes a beautiful finish but is often cross-grained and difficult to work. The wood shrinks very little in drying and consequently can be carved into bowls while green and dried later without serious degrade. Machining characteristics made on wood of low density in Puerto Rico were as follows: planing, mortising, sanding, and resistance to screw splitting were good; shaping and boring were fair; and turning was poor. When green, Hawaii-grown monkeypod turns very well. It is the wood around which the carved bowl trade of Hawaii was built beginning in 1946. The wood has been employed occasionally for furniture, interior trim, and flooring. It is a popular wood for large frame members in wooden boats. It is suitable also for boxes and crates, veneer, plywood, and paneling. In Central America, cross sections of thick trunks have served as wheels of ox carts.

Souvenirs made from the golden brown wood with darker streaks have been purchased in quantities by tourists in Hawaii. Beautifully polished bowls and plates of various sizes and shapes are in demand. However, at present, most of these bowls are made in the Philippines, Thailand, or Malaysia, where monkeypod is also plentiful. Seeds are used in leis.

Monkeypod was introduced to Hawaii in 1847 by Peter A. Brinsmade, then consul from the Kingdom of Hawaii at Mexico City, who brought in 2 seeds. One became a tree which was at Bishop and Hotel Streets in downtown Honolulu until 1899, when it was cut down to permit construction of a building. The other seed was planted at Koloa, Kauai, and produced a tree that was the parent of a large stand of monkeypod there (Anon. 1938).

The trees in Hawaii are valued mainly for shade and beauty. The nutritious pods are relished by cattle, hogs, and goats. Some people chew the pods for the sweetish flavor like licorice. The tree is a honey plant. In a few countries this species has been employed as shade in plantations of coffee and cacao, but less so now than formerly. Because of their enormous growth the trees compete heavily for water and soil nutrients, inhibiting the shrubs planted beneath. The trees are easily propagated from seed and cuttings and grow rapidly. Cattle disseminate the seeds in pastures.

Monkeypod is a favorite shade tree of streets and parks in lowlands of Hawaii and is planted and naturalized in pastures. Because of its large size it is less suited for planting around homes. Sometimes trees become top-heavy and dangerous along highways and near houses. The many surface roots may also be objectionable. The trees are also messy, dropping sticky flower parts and pods on cars parked beneath. Perhaps the species is better suited to dry localities, where the size is smaller.

Special areas--Keahua, Foster.

Champion--Height 104 ft (31.7 m), c.b.h. 30.8 ft (9.4 m), spread 140 ft (42.7 m). Kohala Forest Reserve, Muliwai, Hawaii (1968) A very large monkeypod in Moanalua Gardens, Oahu, has a crown that covers over 3/4 acre (0.3 ha).

Range--Native from Mexico (Yucatan Peninsula) and Guatemala to Peru, Bolivia, and Brazil. Widely planted and naturalized elsewhere in continental tropical America from Mexico southward, throughout the West Indies and in Old World tropics. Common in Puerto Rico and the Virgin Islands and grown also in southern Florida.

Other common names--raintree; samán (Puerto Rico, Spanish); licorice, giant tibet (Virgin Islands); gumorni spanis (Yap).

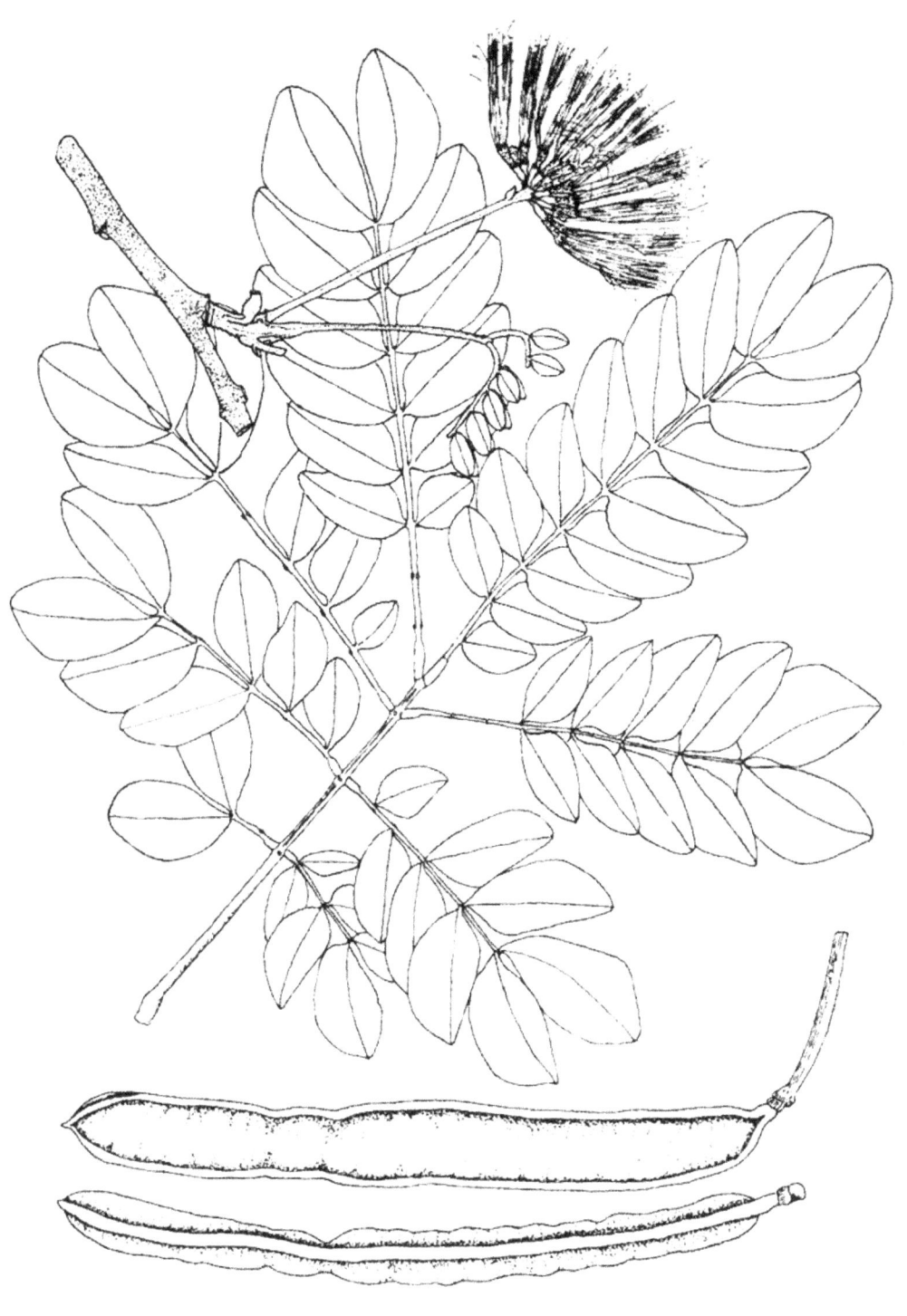

61. Monkey-pod, ʻōhai *Pithecellobium saman* (Jacq.) Benth.*
Flowers (above), leaf, fruits (below), 2/3 X (P. R. v. 1).

Botanical synonyms--*Albizia saman* (Jacq.) Polhill & Raven, *Samanea saman* (Jacq.) Merr.

The generic name *Pithecellobium*, from Greek, "ape's earring," refers to the fancied resemblance of the coiled pods of some species such as No. 60, 'opiuma, *Pithecellobium dulce*. Somehow, the less suitable common name "monkeypod" has been applied in Hawaii to this species with its flat pods!

The Spanish word "samán" and the specific name are from a South American aboriginal name. Several origins of the English word raintree and its French equivalent have been given. Early travelers reported that the tree mysteriously produced "rain" at night and would not sleep underneath it. Others observed the grass to be greener beneath the trees during droughts. Probably, the "rain" was the excreta of cicada insects inhabiting the trees. The Spanish name dormilón refers also to the movements of the leaflets, which close in the dark, suggesting sleep at night. In droughty areas, grass is often green under the large spreading crowns while grass in the open is brown, simply because it is exposed to the sun. Insects defoliate many trees each spring, but trees recover leaves rapidly.

Earpod-tree (*Enterolobium cyclocarpum* (Jacq.) (Griseb.*) is a similar giant deciduous tree introduced for shade and ornament in Hawaii, mainly in dry lowlands. This tree is recognized by: (1) short stout trunk with gray bark and thin spreading crown broader than high; (2) alternate twice-pinnate leaves with numerous paired oblong leaflets 3/8--1/2 in (10--13 mm) long; (3) flowers in whitish balls about 1 in (25 mm) across numerous threadlike white stamens; and (4) very distinctive blackish seed pod, flattened and curved in a circle 3 1/2--4 1/2 in (9--11 cm) in diameter, slightly resembling an ear. Planted as a street tree in Hawaii, growing rapidly but large for use around homes. A very large tree is in Foster Botanic Garden, Honolulu. Under the Spanish names guanacaste and genicero, the wood is used elsewhere for construction, carpentry, interiors, furniture, and veneer. Native from Mexico south through Central America to northern South America and introduced as a shade tree in other tropical regions.

62. Kiawe, algarroba

Kiawe or algarroba, an introduced deciduous usually spiny tree, is one of the most common trees and perhaps most useful of the dry lowlands of Hawaii. Recognized by the irregular short trunk and wide spreading thin crown of small twice compound leaves and by the beanlike narrow, slightly flattened, yellowish pods. Mimosa subfamily (Mimosoideae).

Small to medium-sized tree 30--60 ft (9--18 m) tall) with trunk 1 1/2 ft (0.5 m) or more in diameter, usually smaller, often angled and fluted, twisted, and crooked, and with widely forking branches. Bark gray brown, finely fissured. Outer bark brown, inner bark orange brown, fibrous, bitter. Twigs green, hairless, slightly zigzag, the nodes or joints at leaf bases often with 1--2 spreading spines to 1 in (2.5 cm) long.

Leaves alternate on long twigs or short spurs, dull light green and often finely hairy, bipinnate, 3 in (7.5 cm) or less in length, consisting of short axis less than 1 in (2.5 cm) long and 2--3 pairs of side axes 1--1 1/2 in (2.5--4 cm) long, with gland dot between each pair. Leaflets many (8--11 pairs), stalkless, narrowly oblong, about 1/4 in (6 mm) long and less than 1/8 in (3 mm) wide, rounded at apex, rounded and unequal-sided at base, thin.

Flower clusters (spikes) lateral, 3--4 in (7.5--10 cm) long and 5/8 in (1.5 cm) wide, unbranched, hanging down. Flowers very numerous, crowded, light yellow, about 1/4 in (6 mm) long, composed of cuplike green 5-toothed calyx, corolla of 5 narrow petals, 10 threadlike stamens, and narrow pistil with hairy ovary, curved threadlike style, and dot stigma. Flowering mainly in spring and summer.

Prosopis pallida (Humb. & Bonpl. ex Willd.) H.B.K.*

Fruits (pods) few hanging from slender stalks, beanlike, yellowish, narrow and slightly flattened, 3--8 in (7.5--20 cm) long, 3/8 in (1 cm) wide, and 3/16 in (5 mm) thick, long-pointed, not splitting open, with whitish slightly sweet pulp. Seeds 10--20, each within whitish 4-angled cover, beanlike, elliptical and slightly flattened, 1/4 in (6 mm) long, shiny light brown.

Wood is dark reddish brown, very heavy (sp. gr. 0.85), extremely hard, and has low shrinkage in drying. In Hawaii, it has been made into cement floats and mallets, as well as heavy rifle stocks in match shooting. Its most common uses are charcoal, fuelwood, and fence posts. The heartwood is very resistant to decay. Though attacked by marine borers, the timber has served for piling elsewhere. The bark reportedly contains tannin and yields a brownish gum suitable for varnish, glue, and medicine.

Kiawe is one of the most useful introduced trees of Hawaii, primarily because it occupies barren lands that are otherwise unproductive. The pods serve as valuable feed for livestock in rangelands and are harvested for this purpose. It is reported that a mature tree bears up to 200 pounds (91 kg) of pods annually. The foliage is eaten also. Flowers are an important nectar source for bees. Following the introduction of honeybees in 1857, Hawaii exported 200 tons (182 t) of kiawe honey a year. At present, most honey is produced on Niihau and Molokai.

This species has become established as the most common tree in the lowland dry zone of Hawaii from sea level to about 2,000 ft (610 m) altitude. It covers an estimated 90,000 acres (36,473 ha) of barren soils throughout the islands. It grows on coastal sand, old

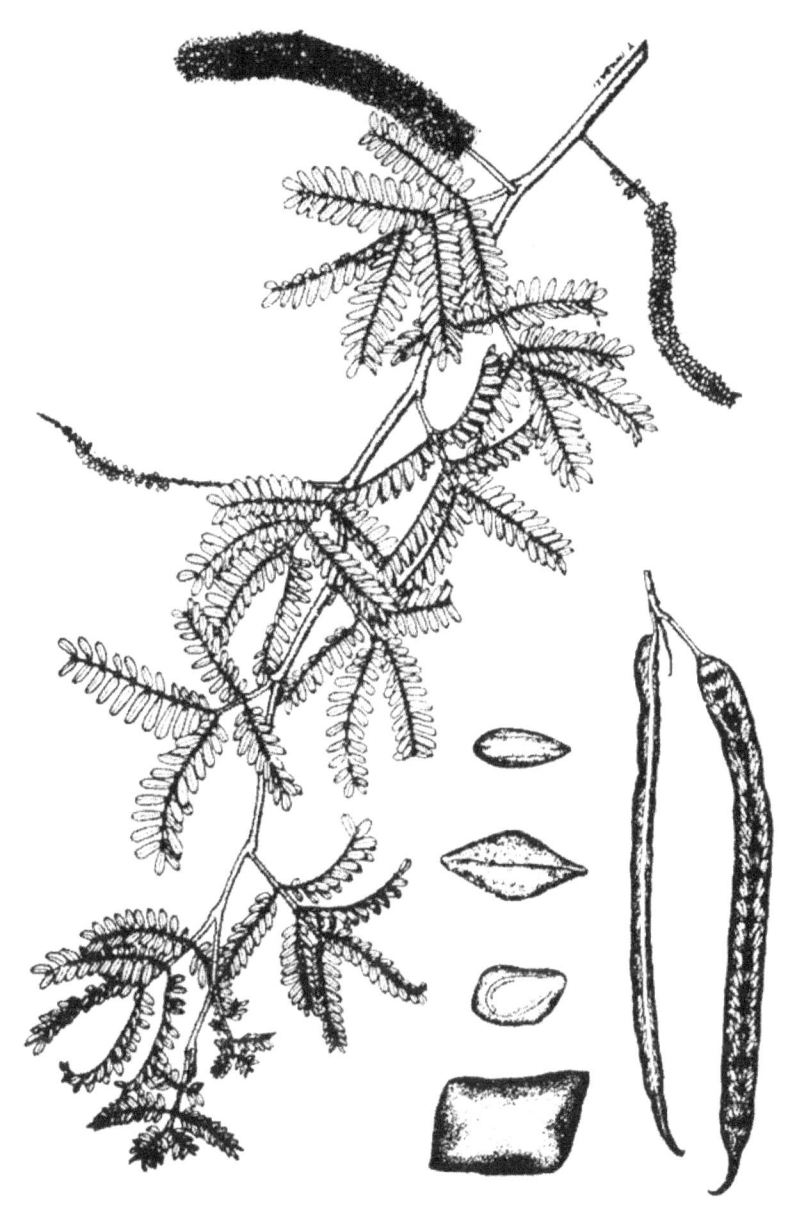

62. Klawe, algarroba *Prosopis pallida* (Humb. & Bonpl. ex Willd.) H.B.K.*
Twig with flowers, 1/2 X; seeds (lower right), 5 X; fruits (lower right), 1/2 X (Degener).

lava flows, and clay soils with rainfall as low as 10 in (250 mm) per year. It is not salt tolerant and is defoliated by salt spray from winter storms.

Though large trees are ornamental and grow rapidly, planting near buildings is not recommended. The trees have shallow root systems and may be uprooted during storms. Also, and more importantly, the large thorns on twigs that fall from the trees make walking barefoot beneath the tree very hazardous and painful. The thorns readily penetrate soft-soled shoes.

Introduction of this species has been traced to seed from a Peruvian tree growing in the royal garden at Paris (Judd 1916). The seed was planted by Father Bachelot in 1828 in the Catholic Mission ground in Honolulu. By 1840 it was common as a shade tree throughout Honolulu, which prior to the introduction of this tree was almost treeless. For many years it was identified as *Prosopis juliflora* (Sw.) DC., a related species with a broader distribution in tropical America, and as *P. chilensis* (Mol.) Stuntz, of Chile. Continued selfing over the years has resulted in distinct genotypes showing up frequently, for example, a spineless variety. Kiawe is present on all the islands. One of the oldest trees in Honolulu stands in front of Kawaiahao Church. It was a good size tree when photographed in 1855. A portion of the stump of the original tree is displayed at the Catholic Church on Fort Street, close to the site where it grew until 1919.

Special area--Koko.

Champion--Height 91 ft (27.7 m), c.b.h. 13.4 ft (4.1 m), spread 81 ft (24.7 m). Puako, Kawaihae, Hawaii (1968).

Range--Native of dry Pacific coastal region of Ecuador, Peru, and Chile. Introduced from Hawaii to Australia, South Africa, and other tropical areas. Naturalized also in Puerto Rico and Virgin Islands.

Other common names--mesquite; algarroba (Spanish); bayahonda (Puerto Rico).

63. Mamane

Sophora chrysophylla (Salisb.) Seem.

Evergreen native tree or shrub widespread but mainly in dry mountain forests, identified by compound (pinnate) leaves with mostly silvery gray leaflets, showy golden yellow beanlike flowers, and distinctive 4-winged pod. Small to medium-sized tree 20--40 ft (6--12 m) tall and to 2 ft (0.6 m) in diameter, or a shrub. Bark gray brown, smoothish, becoming furrowed into scaly ridges. Twigs silky hairy when young. Pea subfamily (Faboideae).

Leaves alternate, pinnate, 5--6 in (12.5--15 cm) long, blunt or slightly notched at apex, rounded at base, stalkless, thin, silky hairy when young, mostly silvery gray, hairy beneath or sometimes hairless.

Flower clusters (racemes) at ends and sides of twigs, unbranched, less than 2 in (5 cm) long. Flowers several on slender stalks, beanlike, 1/2--1 in (13--25 mm) long, golden yellow, consisting of cup-shaped hairy calyx less than 3/8 in (10 mm) long with short teeth, 5 yellow petals 1 in (25 mm) long, broad curved standard, 2 wings, and keel, 10 separate stamens, and pistil with hairy ovary and long slender style.

Fruits (pods) 4--6 in (10--15 cm) long and more than 1/4 in (6 mm) wide, with 4 long wings, deeply narrowed between the seeds, hard and not splitting open. Seeds 4--8, beanlike, elliptical, 5/16 in (8 mm) long, slightly flattened, yellow, very bitter.

The sapwood is pale brown and the heartwood yellowish brown with reddish streaks, coarse textured, heavy, very hard, and very durable in the ground. It has a spicy odor and distinct growth rings. It is used for fenceposts at high elevations near where it grows and was formerly used by the Hawaiians for tool handles. The Hawaiians sometimes cut the wood for posts and beams of their houses and for runners of their sleds for sliding down steep mountain courses paved with rocks.

Livestock, particularly sheep, browse the foliage and destroy seedlings. It has been demonstrated that animal damage has seriously depleted the mamane forests on Hawaii. On Mauna Kea, this tree is the primary food source for the endangered endemic bird palila. Other birds such as the 'i'iwi and 'apapane also feed on the tree.

This species is one of the most widely distributed trees in Hawaii. It is common mainly in dry mountain forests at 4,000--8,000 ft (1,219--2,438 m) altitude, ranging down almost to 100 ft (30 m) and up to 9,500 ft (2,896 m) as the timberline on the highest mountains of the island of Hawaii, Mauna Kea, Mauna Loa, and Hualalai. It reaches its best development as a tree on the higher slopes of Mauna Kea and Mauna Loa. Elsewhere, except for portions of Haleakala on Maui, it grows predominantly as a shrub.

Found on 5 of the 6 large islands of Hawaii, apparently extinct on Lanai. It is rare on Kauai, Oahu, and Molokai, but common at higher elevations on Maui and Hawaii.

Special areas--Kokee, Waimea Arboretum, Haleakala, Volcanoes, Kipuka Puaulu, Pohakuloa State Park on Hawaii.

Champion--Height 39 ft (11.9 m), c.b.h. 12.2 ft (3.7 m), spread 42 ft (12.8 m). Mauna Kea Forest Reserve, Humuula, Hawaii (1968).

Range--Known only from Hawaii. Two closely related species are native to New Zealand.

The plants are hardy, deep-rooted and endure heavy browsing. They vary greatly in size and shape, from low much-branched shrubs to medium-sized trees. The foliage differs from green and hairless at low altitudes to silvery and hairy near timberline. One variety

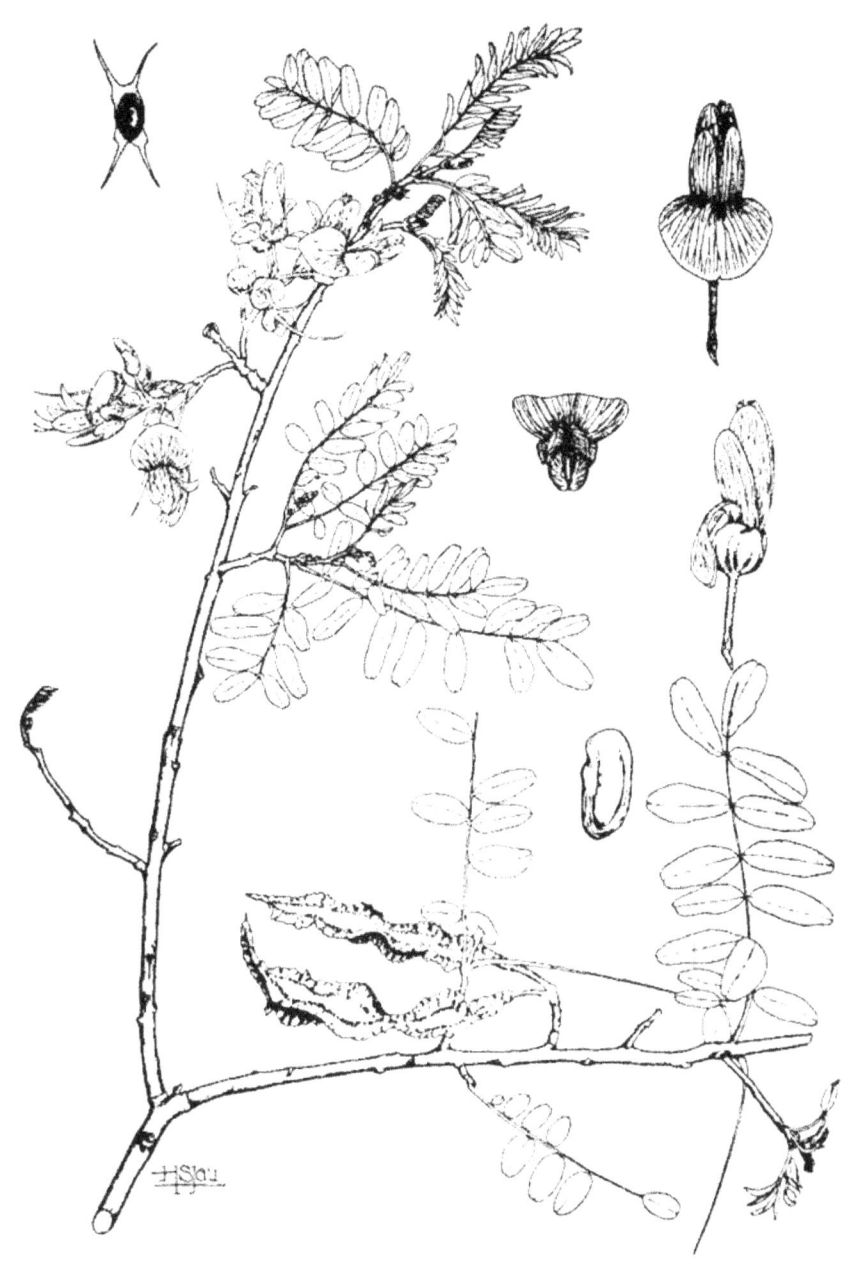

63. Mamane   *Sophora chrysophylla* (Salisb.) Seem.
Twig with flowers and fruits, 1/2 X; cross-section of fruit (upper left), 1 1/2 X; flowers (upper right), 1 X; seed (lower right), 3 X (Degener).

from Hawaii and Maui has simple leaves with large blades.

Other common names--mamani, mamano.

64. Tamarind

Tamarindus indica L.*

Tamarind is a handsome introduced shade tree of lowlands, mainly in dry areas, where it has escaped from cultivation. It has a dense crown of blue green feathery, pinnate leaves, showy pale yellow flowers tinged with red, giving a yellowish color to the tree, and thick brown pods with sour edible pulp. Cassia subfamily (Caesalpinioideae).

Medium-sized evergreen tree to 40 ft (12 m) high and 2 ft (0.6 m) in trunk diameter, with rounded crown of dense foliage. Bark gray or brown, rough, thick, much fissured. Inner bark brownish, gritty, and slightly bitter. Twigs green and minutely hairy when young, turning gray or brown.

Leaves alternate, even pinnate, 2--4 1/2 in (5--11 cm) long, hairless, with slender pale green axis. Leaflets 10--18 pairs, close together and almost stalkless, oblong, 3/8--7/8 in (10--22 mm) long and 1/8--1/4 in (3--6 mm) broad, rounded at both ends and unequal at base, not toothed, thin, blue green above and slightly paler beneath, folding together against axis at night.

Flower clusters (racemes) terminal and lateral, 1 1/2--6 in (4--15 cm) long, several on slender stalks from dark red buds, showy, about 1 in (25 mm) across, irregular shaped and delicate. Narrow pale green basal tube (hypanthium) is 3/16 in (5 mm) long; calyx of 4 pale yellow sepals 1/2 in (13 mm) long; corolla of 3 pale yellow petals with red veins, keeled and broader toward finely wavy apex, 2 outer 5/8 in (15 mm) long, central petal 3/8 in (10 mm) long, and 2 others reduced to minute scales; 3 greenish stamens 1/2 in (13 mm) long, united by filaments to middle, and 2 minute sterile stamens; and green beanlike pistil 5/8 in (15 mm) long with stalked 1-celled ovary and curved style.

Pods oblong, often curved, 1 1/2--4 1/2 in (4--11 cm) long, 3/4--1 in (2--2.5 cm) wide, and 3/8--5/8 in (1--1.5 cm) thick, slightly narrowed between seeds, brown, rough, heavy, with brittle outer shell and dark brown fibrous pulp, edible though very sour, not splitting open. Seeds usually 3--4, beanlike, flattened shiny brown, 5/8 in (15 mm) long.

Sapwood is light yellow and moderately soft, and the small heartwood dark purplish brown. The wood is described as very hard, heavy (sp. gr. 0.9), and takes a fine polish. It is strong and durable, although very susceptible to attack by dry-wood termites. It has been used occasionally in Hawaii for chopping blocks and rated as excellent.

The wood is used in other tropical areas chiefly for fuel and is reported to generate great heat. In other places where the species is sufficiently common, the wood is employed for construction, tool handles, furniture, and articles in wood turning but is considered very difficult to work. Gunpowder was formerly manufactured from its charcoal.

Elsewhere, this tree is planted around homes for its fruits. A refreshing beverage like lemonade, as well as candy and preserves, are prepared from the edible fruit pulp of the pods. The young tender sour fruits have been cooked for seasoning meats, and the young leaves and flowers are reportedly consumed as food. The ornamental flowers attract bees and are an important source of honey. However, the litter of the pods is objectionable in street planting. In India, the trees are planted on forest firebreaks because the ground underneath usually remains bare of other plants.

The fruit pulp is employed in home medicine and was formerly an official drug source of a laxative. It contains sugar as well as acetic, tartaric, and citric acids, and is antiscorbutic. Medical decoctions have been obtained from flowers, seeds, young leaves, and bark. A yellow dye can be made from the leaves.

In Hawaii, planted mainly for shade and along roadsides in dry areas, it persists and escapes from cultivation. Easily propagated from seed, the tree is not popular now as an ornamental. Introduced in 1797 by Don Marín, it is seen mostly in the older parts of towns where it was planted during the 1800's. Such a tree is growing on Mililani Street, adjacent to the Judiciary Building in Honolulu.

Range--Native of the Old World tropics and widely planted and naturalized in tropical and subtropical regions. It was introduced into the New World at a very early date. Cultivated and often naturalized throughout the West Indies and from Mexico to Brazil. Planted and naturalized in Puerto Rico and Virgin Islands. Introduced also in southern Florida and naturalized locally.

Special areas--Waimea Arboretum, Foster.

Champion--Height 62 ft (18.9 m), c.b.h. 21.7 ft (6.6 m), spread 47 ft (14.3 m). Pioneer Mill Co., Lahaina, Maui (1968).

Other common names--wi 'awa'awa; tamarindo (Spanish); kamalindo (Guam).

The scientific and common names are from Arabic *tamr hindi*, Indian dried date, through Spanish and Italian *tamarindo*.

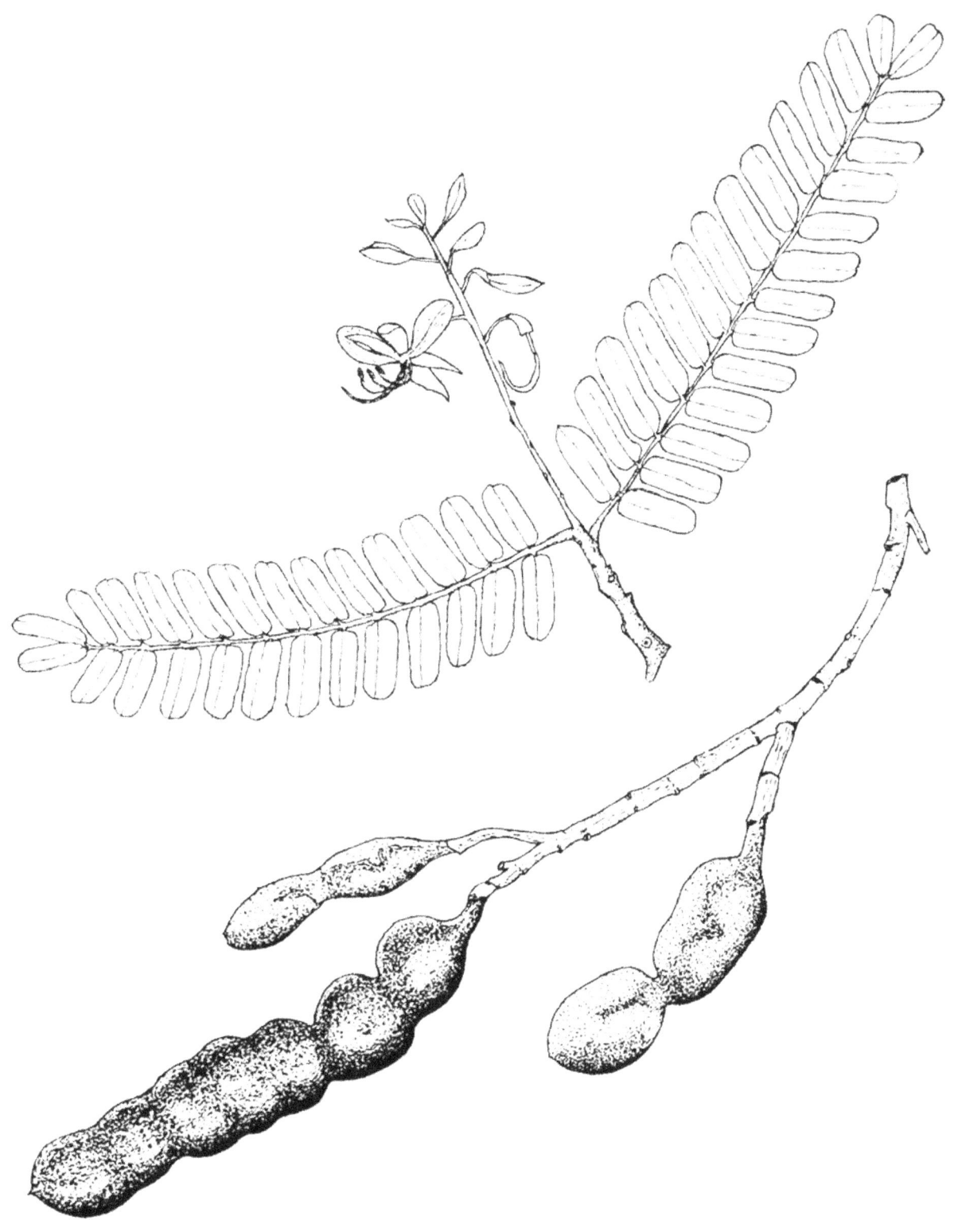

64. Tamarind *Tamarindus indica* L.*
Flowering twig (above), fruits (below), 1 X (P. R. v. 1).

RUE OR CITRUS FAMILY (RUTACEAE)

65. Queensland-maple *Flindersia brayleyana* F. Muell.*

Large evergreen tree introduced in 1935 for trial in forest plantations, characterized by large paired pinnate leaves with 8--12 paired leaflets and by large cylindrical seed capsules splitting open into 5 parts.

To 70 ft (21 m) high and 3 ft (0.9 m) in trunk diameter. Bark gray brown, smoothish, slightly fissured into warty ridges. Inner bark dark red near surface, beneath light pink streaked, with gritty bitter taste. Twigs stout, dark green, with tiny hairs and raised light brown dots and slightly triangular light brown leaf scars about 1/4 in (6 mm) wide.

Leaves paired (opposite), large, pinnate, 12--20 in (30--51 cm) long, with minutely hairy dark green round axis 10--14 in (25--36 cm) long. Leaflets 8--12 paired, with stalks of 3/8 in (1 cm), large, oblong to elliptical, 5--7 in (13--18 cm) long, 2--3 in (5--7.5 cm) wide, long-pointed at apex, short-pointed or rounded at base, edges slightly turned under, hairless, thin, with numerous tiny gland dots visible under lens.

Flower clusters (panicles) large. Flowers many, fragrant, small; 1/8 in (3 mm) long, with 5 white petals.

Fruits (seed capsules) few, clustered, hanging on long stalks, cylindrical, about 2 1/2 in (6 cm) long, 1 in (2.5 cm) in diameter, dark brown, becoming slightly 5-angled, 5-celled, and splitting from apex into 5 parts. Seeds several, flat and winged, 2 in by 3/8 in (5 cm by 1 cm), light brown, with thin body surrounded by membranous wing.

Sapwood is pink and heartwood a lustrous pale brown, often with interlocked and wavy grain giving a pronounced figure. Called silkwood because of its resemblance to silk. This lightweight wood (sp. gr. 0.45) machines well but lacks resistance to decay or insect attack. It is an excellent cabinetwood, one of Australia's finest, and has been used for everything from aircraft propellers to boat planking. In Hawaii, trees now growing are expected to be used primarily for veneer production.

This species was first planted in 2 locations on Oahu in 1935. In 1957, one of these stands was evaluated and found to be growing very rapidly and reproducing naturally. Tests of the wood of 2 trees indicated that the wood produced was the same as that of trees grown in Australia. On this basis, the Division of Forestry began more extensive planting of the species in the Waiakea Forest Reserve on Hawaii where there are now about 400 ac (162 ha) of trees 15 to 25 years old. This species has been introduced to all the large islands and has been found to grow best in wet forest conditions. Extremely tolerant of shade as seedlings, the trees can seed in on the forest floor and grow up through very heavy overstory shade.

Special area--Waiakea.
Range--Native to Queensland, Australia.
Other common name--silkwood.

This genus honors Matthew Flinders (1774--1814), British explorer.

66. Mokihana *Pelea anisata* Mann

This species known only from Kauai is easily recognized by its strong pleasant fragrance, like anise, in all parts, that persists for years. It is a small shrubby evergreen tree to 26 ft (8 m) high, with trunk to 10 in (25 cm). Bark gray, smooth, thin. Inner bark with green cap, light yellow, spicy, with strong taste and odor like anise or licorice. Twigs with minute pressed gray hairs when young, soon nearly hairless, with raised half-round leafscars. End bud narrow, composed of paired very young hairy leaves.

Leaves paired or opposite, with leafstalks 1/4--1 1/2 in (6--40 mm) long, with tiny hairs when young. Blades elliptical, thin, hairless, mostly rounded at both ends or slightly notched at apex and short-pointed at base, straight on edges, mostly 1 1/2--6 in (4--15 cm) long and 1--3 in (2.5--7.5 cm) wide, with several fine side veins nearly at right angle with midvein, light green above, paler beneath, with tiny gland dots visible under a lens.

Flower clusters (cymes) at leaf bases, usually less than 3/4 in (2 cm) long. Flowers of one or both sexes (polygamous), short-stalked, 1--3 (rarely 5), 1/4 in (6 mm) long. Male flowers composed of 4 blunt sepals less than 1/8 in (3 mm) long, 4 pointed petals slightly more than 1/4 in (6 mm) long, 4 stamens as long as petals and 4 half as long, and minute pistil. Female flowers similar, with 4 petals nearly 1/4 in (6 mm) long, 8 minute sterile stamens, and on a disk the pistil with broad 4-lobed ovary, short style, and 4-lobed stigma.

Fruits (capsules) cube-shaped, 3/8--1/2 in (10--13 mm) broad, light green, hairless, splitting into 4 parts. Seeds egg-shaped, shiny black, 1 or 2 in a part. Fruits borne abundantly in early fall, especially in September, as well as in other times of the year.

The seed capsules and wood retain their fragrance for many years and were favorite perfumes of the Hawaiians, often placed in their tapa cloth. The capsules, which are symbolic of the island of Kauai, were gathered

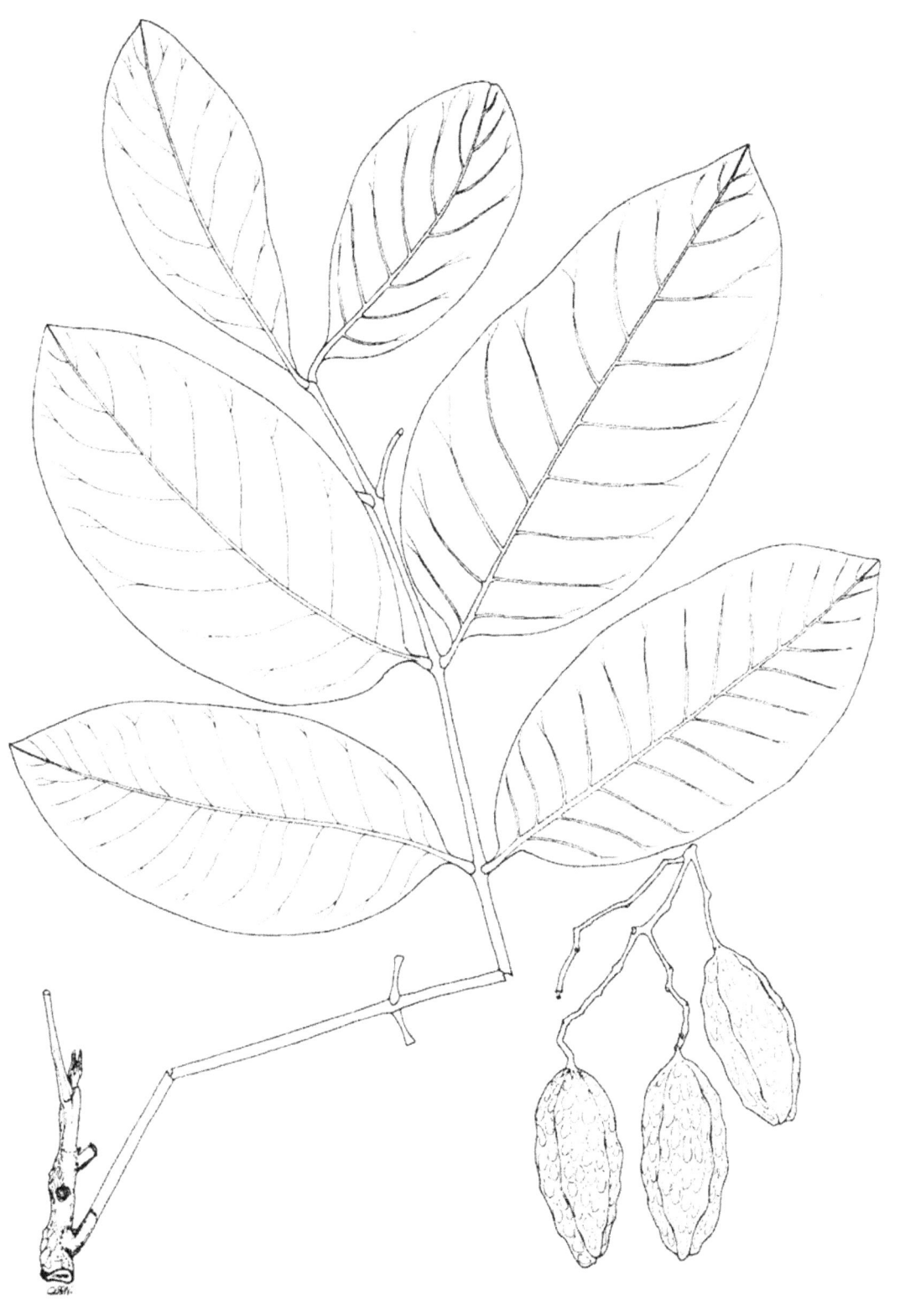

65. Queensland-maple *Flindersia brayleyana* F. Muell.*
Leaf and fruits (lower right), 1/2 X.

and strung into leis. However, *fresh and dry capsules can cause a painful irritation* on a perspiring neck, even burning through clothing. The skin will redden, blister, and peel. The chemical coumarin in the leaves causes the burning sensation, which can recur later when sensitized skin is moistened by perspiration. The leaves served also in home remedies.

The wood in this genus is described under No. 67, alani, *P. clusiifolia.*

Common at altitudes of 2,500--4,000 ft (762--1,219 m) in mountains on Kauai. Collecting capsules for leis has caused the plants to become less common. Also, in the Kokee region, the plants have been crowded out by the introduced Larsen berry or blackberry (*Rubus argutus* Link).

Special areas--Kokee, Wahiawa.
Range--Kauai only.

A few species on other islands have similar fragrance, but not so lasting. The fragrances of other species of *Pelea*, called alani, vary from citrus and anise to root beer scents. Hikers crush the leaves and put them in their pockets.

67. Alani, Clusia-leaf pelea

Pelea clusiifolia Gray

Shrubs and small trees known as alani are easily recognized. However, this variable genus, *Pelea*, is one of the largest of woody plants in Hawaii in number of species. The genus is characterized by leaves alternate, paired, or in whorls of 3 or 4, rarely 5 or 6, with gland dots visible under a lens; small flowers at leaf bases, mostly greenish, 4-parted; and seed capsules 4-lobed or splitting into 4 parts. One species will serve as an example, though a distinctive second is No. 66, mokihana, *Pelea anisata* Mann.

This species with several varieties is one of the most common and is widely distributed through the islands. It is characterized by short-stalked thick leathery elliptical leaves broadest beyond middle, attached mostly 4 at a node. A shrub or small tree to 30 ft (9 m) tall, hairless nearly throughout. Twigs gray, with ringed nodes, ending in narrow bud of very young hairy leaves.

Leaves whorled, usually with 4 at a node, rarely 3--6, with short leafstalk of 3/8--3/4 in (1--2 cm). Blades elliptical or obovate, mostly 2--4 1/2 in (5--11 cm) long and 1 1/4--2 in (3--5 cm) wide, stiff and brittle, rounded or slightly notched at apex, blunt or short-pointed at base, turned under at edges, above very shiny green, beneath shiny light green with purplish black midvein, many fine parallel side veins with network of smaller veins, and with tiny gland dots visible under a lens.

Flower clusters (cymes) on twigs mostly back of leaves, less than 3/4 in (2 cm) long. Flowers of one or both sexes (polygamous), short-stalked, few, nearly 1/4 in (6 mm) long. Male flowers consist of 4 blunt sepals less than 1/8 in (3 mm) long, 4 narrow petals nearly 1/4 in (6 mm) long, 8 stamens, and rudimentary pistil. Female flowers similar, with 8 minute sterile stamens and on a disk the pistil with 4-lobed ovary, short style, and 4-lobed stigma.

Fruits (capsules) clustered on twigs back of leaves, shallowly 4-lobed, 3/8--5/8 in (10--15 mm) in diameter, hard, wrinkled, greenish, the rounded lobes splitting on a line. Seeds 1--2 in each lobe, 1/8 in (3 mm) long, rounded, shiny black.

The wood of all species is yellowish white, fine-textured, tough, but rather soft. It was used by the Hawaiians for kapa beaters and for canoe trim and rigging.

This species with several varieties is widespread and common in wet forests through the islands.

Special areas--Kokee, Haleakala, Volcanoes.
Range--Through the 6 large islands of Hawaii only.

This genus was named in honor of the Hawaiian goddess of fire and volcanoes, Pélé, in 1854, by Asa Gray, U.S. botanist. It was based upon a specimen of this species collected on Oahu by the U.S. Exploring Expedition under Captain Charles Wilkes. Today, about 50 species are known (more than 80 have been named), several doubtfully distinct. All are from Hawaii except 2 in the Marquesas Islands. The monograph by Stone (1969) serves for identification of species.

68. Pilo kea, spatula-leaf platydesma

Platydesma spathulatum (Gray) Stone

This aromatic evergreen shrub or small tree has a strong odor of pepsin when crushed, paired often very large oblong to obovate leaves with gland dots visible under a lens, few large creamy white 4-parted flowers at leaf bases, and 4-lobed capsules. To 40 ft (12 m) high and 8 in (20 cm) in trunk diameter, usually much smaller, 10--20 ft (3--6.1 m), with spreading branches leafy toward end. Twigs stout, greenish when young, becoming pale gray and hairless, with large raised triangular leafscars.

Leaves opposite, with long slender leafstalk of 3/8--2 1/4 in (1--6 cm), enlarged at base, finely hairy when young. Blades oblong to obovate, 4--20 in (10--51 cm) long and 1--8 in (2.5--20 cm) wide, blunt at both ends, often turned under at edges, thick and leathery, with fine nearly straight side veins, above green to dark green and hairless, beneath paler and sometimes finely hairy on veins, with gland dots visible under a lens.

Flower clusters (cymes) at leaf bases, with mostly 3--5 short-stalked flowers at end of stalk of about 1 in (2.5 cm). Flowers about 5/8--3/4 in (15--19 mm) long, not

66. Mokihana *Pelea anisata* Mann

Leafy twig (left) and twig with flowers and fruits (right), 1/2 X; female and male flowers (below), 5 X; fruits (lower right), 1 X (Degener).

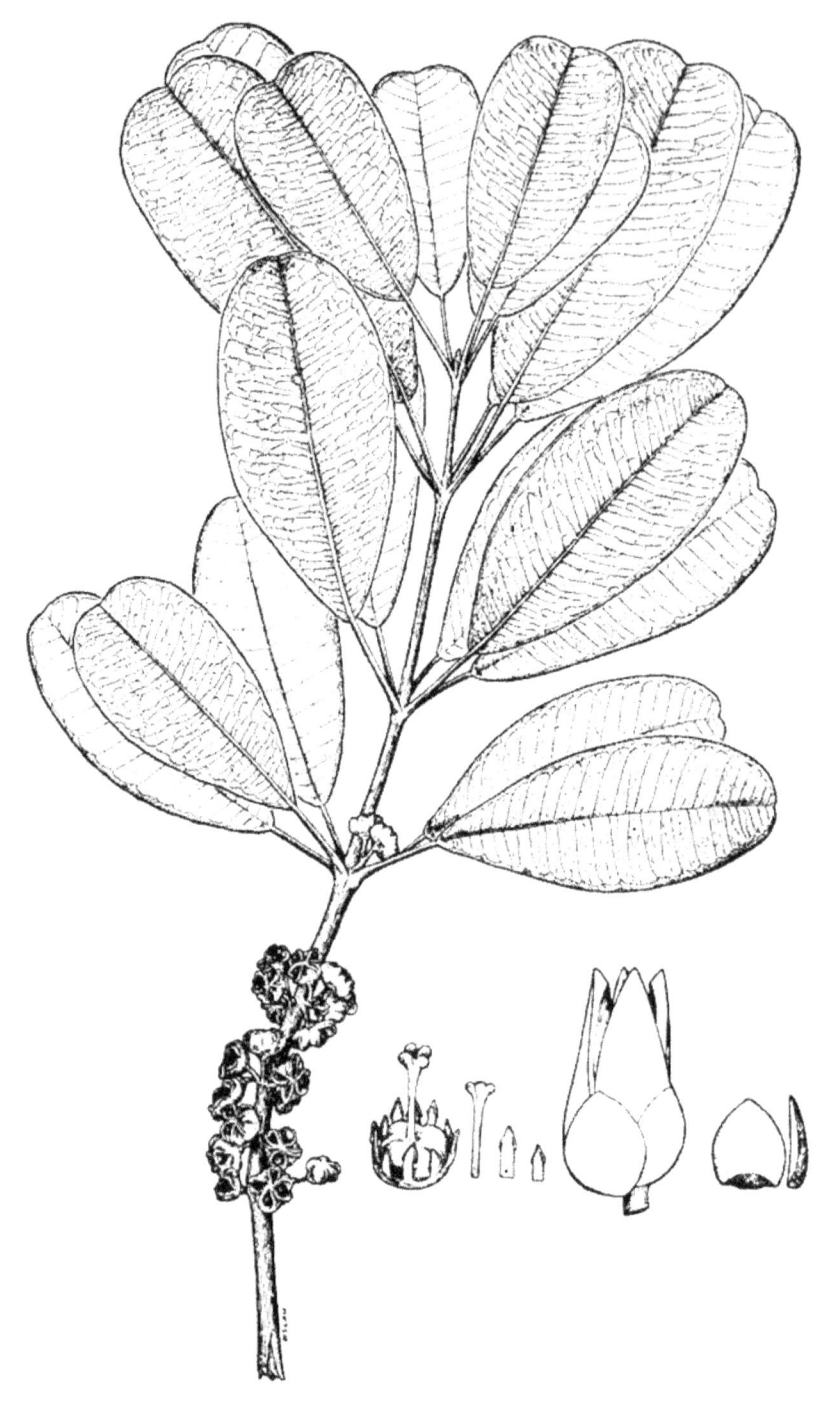

67. Alani, Clusia-leaf pelea *Pelea clusiifolia* Gray
Twig with fruits, 1/2 X; female flowers and parts (lower right), 3 X (Degener).

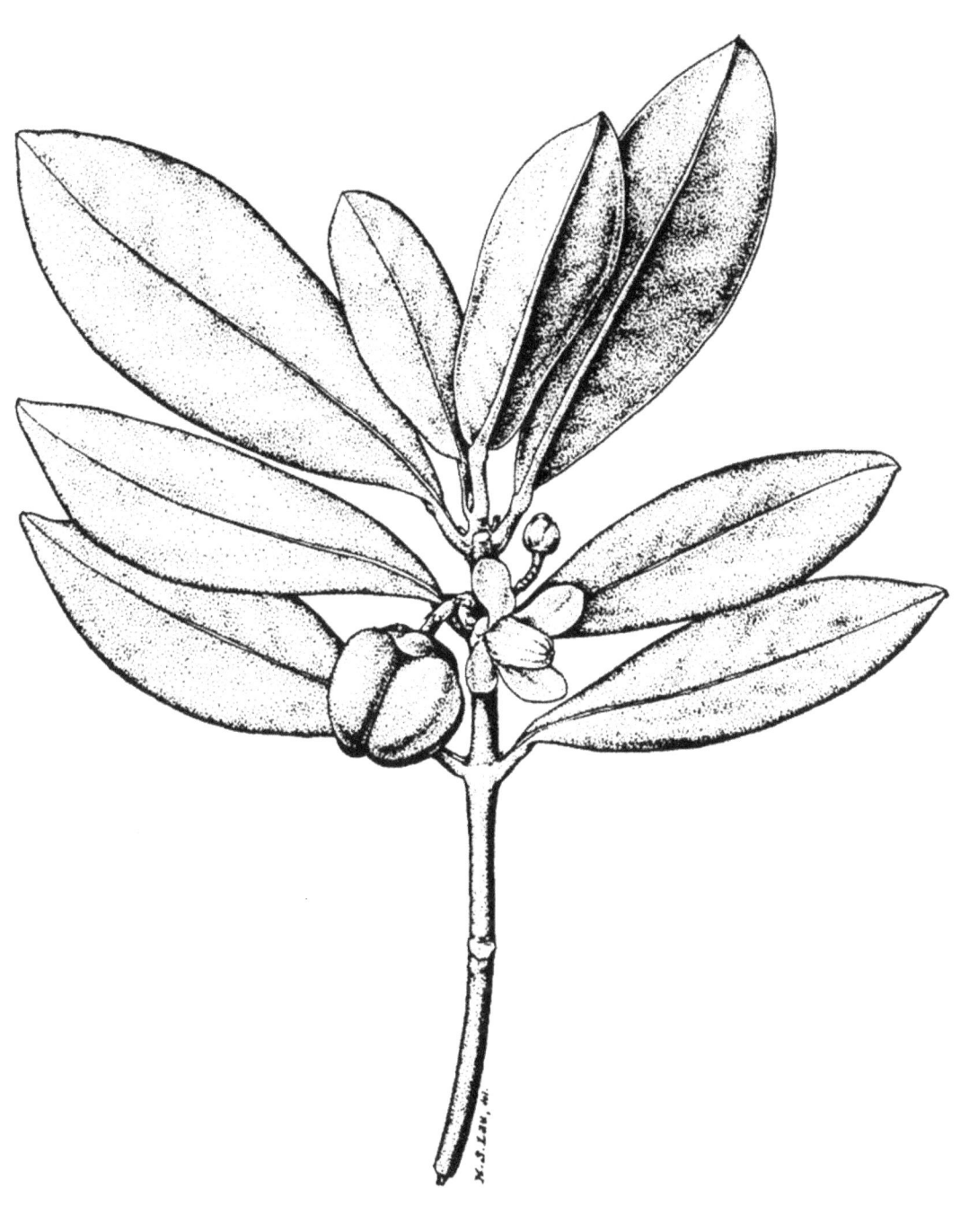

68. Pilo kea, spatula-leaf platydesma *Platydesma spathulatum* (Gray) Stone
Twig with flower and fruit, 3/4 X (Degener).

fragrant, composed of calyx of 4 overlapping rounded green sepals about 3/8 in (10 mm) long, 4 large spreading creamy white petals 5/8--3/4 in (15--19 mm) long, 8 white stamens united in tube of 1/2 in (13 mm), and pistil with 4-lobed 4-celled ovary, slender style, and 4-grooved stigma.

Fruits (capsules) rounded, 4-lobed, about 1 in (25 mm) in diameter, hairless, splitting into 4 parts, with calyx at base. Seeds 5--8, elliptical, rounded, shiny black.

69. A'e

A handsome small aromatic evergreen tree of wet forests of Oahu, characterized by leaves with 3 long-stalked ovate or rounded leaflets, small greenish flowers, and small podlike fruits that split open on 1 side. To 16 ft (5 m) in height. Bark smoothish, with yellow dots, thin. Twigs stout, rough, hairless, with raised yellow dots and raised large rounded leaf-scars.

Leaves alternate, 5--8 in (13--20 cm) long, hairless, compound, with long slender leafstalk of 2--4 in (5--10 cm) and 3 leaflets with long slender stalks of 1 1/4--3 in (3--7.5 cm), jointed above middle. Leaflets broadly ovate or rounded, 1 1/2--3 in (4--7.5 cm) long and 1 1/4--2 1/2 in (3--6 cm) wide, abruptly long-pointed at apex, rounded or blunt at base, the 2 lateral also unequal or 1-sided, not toothed on edges, slightly thickened, the upper surface shiny dark green and curved up slightly, with side veins inconspicuous, the lower surface dull green, brownish when dry, with gland dots visible under a lens.

Flower clusters (panicles) at leaf bases, 2 1/2--4 1/2 in (6--11 cm) long, branched, with several flowers on spreading stalks, mostly male and female on different plants (dioecious), 3/16 in (5 mm) long, with calyx of 4 minute finely hairy lobes and corolla of 4

Wood dull whitish yellow and soft, subject to blue-stain fungi and without growth rings.

Uncommon in wet forests at 2,000--5,000 ft (610--1,524 m) altitude.

Special area--Kokee.
Range--Kauai, Oahu, Maui, and Hawaii only.

A few additional varieties and forms of this variable species have been named. This genus is confined to Hawaii and has 3 other species of shrubs. Stone wrote a monograph about the genus (1962).

Zanthoxylum oahuense Hillebr.

greenish yellow petals overlapping in bud; male flowers with 4 stamens nearly 1/8 in (3 mm) long; female flowers have greenish pistil with ovary and stigma.

Fruits (follicles) 1/2 in (13 mm) long, elliptical, rough, dark brown, splitting open on 1 side. Seed 1, elliptical, 3/8 in (10 mm) long, slightly rough, shiny black.

Wood is light yellow, fine-textured, rather soft with a prominent figure on tangential faces resulting from parenchyma tissue.

This species occurs in wet forests at middle altitudes on Oahu.

Range--Oahu only, recorded from Konahuanui, Niu Valley, and Koolau Range, to about 2,500 ft (762 m) altitude.

Other common name--hea'e.
Botanical synonym--*Fagara oahuensis* (Hillebr.) Engler.

The generic name from Greek, yellow and wood, has been spelled also *Xanthoxylum*. Nine species have been described from the Hawaiian Islands, all endemic. They vary from shrubs to large trees and occur in both wet and dry forests. Leaves have a resinous citruslike odor.

MAHOGANY FAMILY (MELIACEAE*)

70. Chinaberry, pride-of-India

Melia azedarach L.

Chinaberry, or pride-of-India, is a popular ornamental tree planted for its showy cluster of pale purplish 5-parted spreading flowers and for the shade of its dense dark green foliage. It is further characterized by the bipinnate leaves with long-pointed saw-toothed leaflets and pungent odor when crushed, and by the clusters of nearly round golden yellow poisonous berries conspicuous when leafless.

Small to medium-sized deciduous tree often becoming 20--50 ft (6--15 m) tall and 1--2 ft (0.3--0.6 m) in trunk diameter, with crowded, abruptly spreading branches forming hemispherical or flattened crown. Bark dark or reddish brown, smoothish, becoming fur-

rowed. Inner bark whitish, slightly bitter and astringent. Twigs green, hairless or nearly so.

Leaves alternate, 8--16 in (20--40 cm) or more in length, bipinnate or partly tripinnate. Leaflets numerous short-stalked, paired along slender green branches of leaf axis but single at ends, lance-shaped to ovate, 1--2 in (2.5--5 cm) long and 3/8--3/4 in (1--2 cm) wide, short-pointed and mostly 1-sided at base, thin, hairless or nearly so, dark green on upper surface, and paler beneath.

Flower clusters (panicles) 4--10 in (10--25 cm) long at leaf bases, long-stalked and branched. Flowers showy fragrant, numerous on slender stalks, about

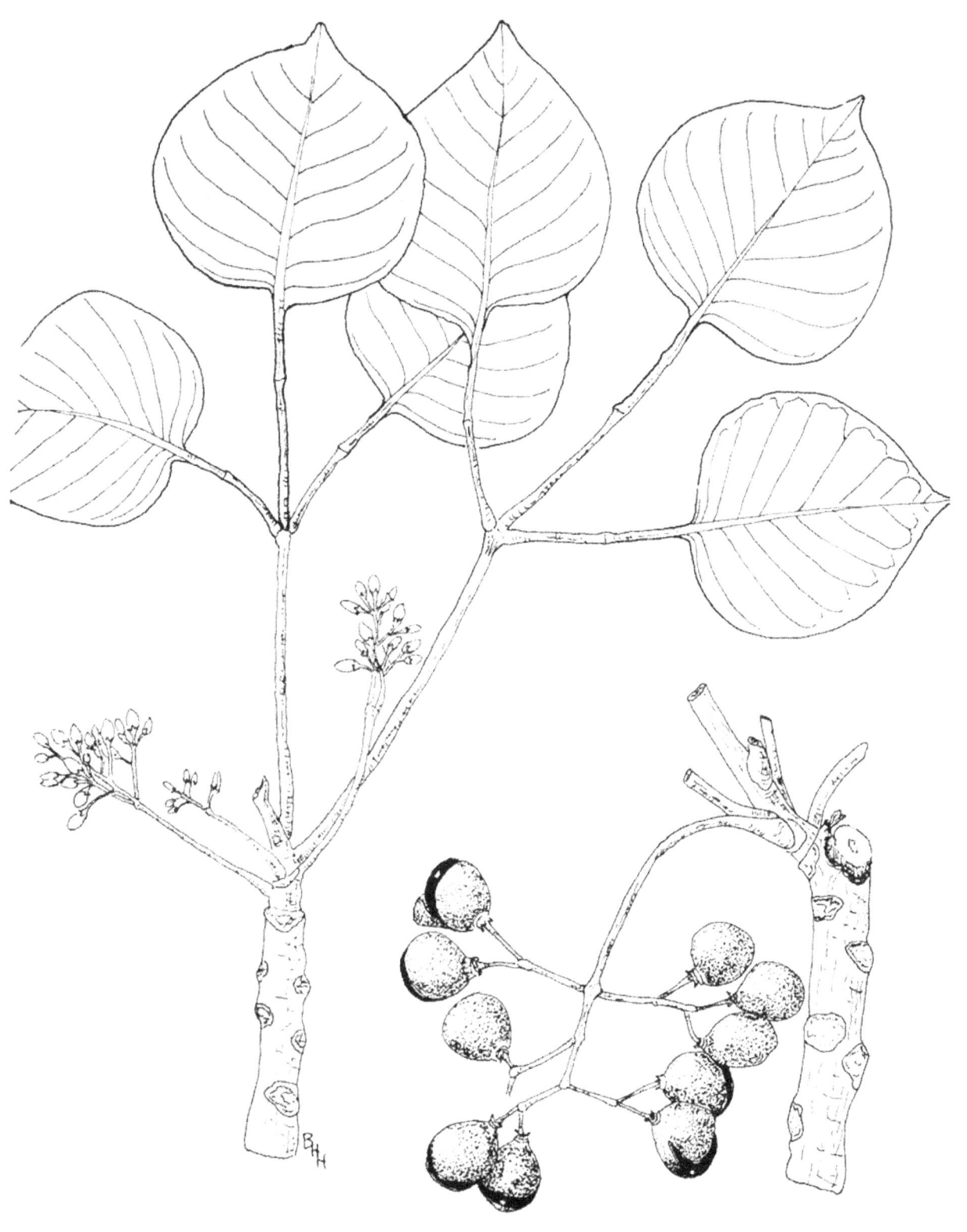

69. A'e *Zanthoxylum oahuense* Hillebr.
Twig with leaves and female flowers, and fruits (lower right), 1 X.

3/8 in (10 mm) long and 5/8--3/4 in (15--19 mm) wide. Calyx of 5 greenish sepals 1/16 in (1.5 mm) long; 5 pale purplish or lilac-colored petals 3/8 in (10 mm) long, narrow, spreading and slightly turned back; usually 10 stamens on narrow violet tube of 5/16 in (8 mm); and pale green pistil 5/16 in (8 mm) long with disk at base, 3--6 celled ovary, and long style.

Fruits or berries (drupes) about 5/8 in (15 mm) in diameter, smooth, but becoming a little shriveled, slightly fleshy. Stone hard, containing 5 or fewer narrow dark brown seeds 5/16 in (8 mm) long. These fruits are bitter, with poisonous or narcotic properties. Flowering from March to June in Hawaii, and the old slightly wrinkled yellow fruits generally present.

Sapwood yellowish white, heartwood light brown to reddish brown and attractively marked. Wood moderately soft, weak and brittle, and very susceptible to attack by dry-wood termites. Uses of the wood elsewhere include tool handles, cabinets, furniture, and cigar boxes. It has not been used in Hawaii.

Extensively planted around the world for ornament and shade. This attractive tree is easily propagated from seeds, cuttings, and sprouts from stumps. It grows rapidly but is short-lived, and the brittle limbs are easily broken by the wind.

This species is poisonous, at least in some parts, and has insecticidal properties. Leaves and dried fruits have been used to protect stored clothing and other articles against insects. Various parts of the tree, including fruits, flowers, leaves, bark, and roots, have been employed medicinally in different countries. The berries are toxic to animals and have killed pigs, though cattle and birds reportedly eat the fruits. An oil suitable for illumination was extracted experimentally from the berries. The hard, angular, bony centers of the fruits, when removed by boiling are dyed and strung as beads. In parts of Asia this is a sacred tree.

Commonly cultivated and naturalized through the Hawaiian Islands in lowlands. It is reported to be hardy up to 9,000 ft (2,743 m) altitude, especially in uplands of Kauai, Maui, and Hawaii. Birds apparently have spread the fruits. Introduced into Hawaii about 1839, according to Degener. It may be seen commonly along the roads of the Kona District of Hawaii and near Ulupalakua Ranch on Maui.

Special areas--Waimea Arboretum, Tantalus.

Champion--Height 75 ft (22.9 m), c.b.h. 18.5 ft (5.6 m), spread 96 ft (29.3 m). Koahe, South Kona, Hawaii (1967).

Range--Native of southern Asia, probably from Iran and Himalaya to China, but cultivated and naturalized in tropical and warm temperate regions of the world. Widely planted and escaped and naturalized locally in southeastern continental United States, California, Puerto Rico, and Virgin Islands.

Other common names--'inia, 'ilinia (Hawaii); chinatree, umbrella-tree, umbrella chinaberry, Indian-lilac, Persian lilac, beadtree; alelaila, lilaila, pasilla (Puerto Rico); lilac (Virgin Islands); paraíso (Spanish); lelah (Pohnpei).

Umbrella chinaberry, or Texas umbrella-tree, is a horticultural variety with compact crown of erect radiating branches and drooping foliage. The Hawaiian name 'inia is a corruption of the word India, according to Degener.

71. Australian toon

Toona ciliata M. Roem.*

Large tree introduced in 1918 from Australia for forest plantations, characterized by large compound leaves with 10--20 paired narrowly ovate unequal-sided leaflets, apparently deciduous, and oblong seed capsules splitting into 5 parts. About 50 ft (15 m) tall and 1 ft (0.3 m) in trunk diameter, possibly becoming larger. Bark light gray, smoothish, finely fissured. Inner bark with brownish outer layer, inner layer pink and white streaked, bitter. Twigs stout, brown, hairless, with large half-round leaf-scars.

Leaves alternate, even pinnate, 12--16 in (30--41 cm) long, hairless, with slender round light green axis. Leaflets paired, mostly 10--20, on slender stalks of 3/8 in (1 cm). Leaflet blades narrowly ovate, 2 1/4--7 in (6--18 cm) long and 1--2 1/2 in (2.5--6 cm) wide, long-pointed at apex, blunt and unequal at base, slightly curved and unequal-sided, not toothed on edges, thin, hairless, above slightly shiny green, beneath paler.

Flower clusters (panicles) 12--16 in (30--41 cm) long, slender, branched, spreading from near ends of twigs. Flowers many, very short-stalked, white, slightly fragrant, cylindrical, 1/4 in (6 mm) long, composed of 5 tiny brownish green sepals, corolla of 5 white oblong petals, 5 alternate stamens within corolla, and pistil with conical 5-celled ovary and short style.

Fruits (seed capsules) oblong, 3/4 in (2 cm) long, 3/8 in (1 cm) in diameter, with tiny style at apex, brown to blackish, splitting open from apex into 5 parts and exposing 5-angled light brown axis. Seeds several, 5/8 in (15 mm) long, narrowly oblong, whitish, much flattened, 2-winged.

Sapwood is pale brown and the heartwood reddish brown, strongly scented, lightweight (sp. gr. 0.35) soft, resistant to dry-wood termites, but not resistant to decay in ground. The wood has been tested at the Forest Products Laboratory and found to be similar to red alder (*Alnus rubra*) in most properties. Its shrinkage in drying is, however, quite high for its density. The wood is similar also to that of the related Spanish-cedar (*Cedrela* spp.). Elsewhere used for furniture, cabinetwork, and construction. So far very little wood has been used in

70. Chinaberry, pride-of-India   *Melia azedarach* L.*
Flowers, leaf, fruits, 2/3 X (P. R. v. 1).

Hawaii. It is rather soft for furniture and paneling but is so utilized in Australia.

The brown 5-parted seed capsules and their 5-angled axes have been used in Hawaii in dry fruit and flower arrangements.

Australian toon is found in forest plantations in moist areas throughout the Hawaiian Islands. It is suitable also as a shade tree. This was the most commonly planted tree in the Waiakea Forest Reserve near Hilo during the 1960's. Now, there are over 2,300 acres (931 ha). The best stand, on Round Top Drive in Honolulu was planted in 1924.

Special areas--Wahiawa, Tantalus.
Champion--Height 117 ft (35.7 m), c.b.h. 11.3 ft (3.4 m), spread 72 ft (21.9 m). Kohala Forest Reserve, Muliwai, Hawaii (1968).
Range--Native of Himalaya region of tropical Asia from India to China and through Indonesia to Australia. Introduced elsewhere for forest plantations.
Other common names--Australian redcedar, Burma toon; tun (Puerto Rico).
Botanical synonym--*Cedela toona* Roxb.

The Hawaiian tree is the southern variety native in eastern Australia, *Toona ciliata* var. *australis* (F. Muell.) C. DC. (*Cedrela australis* F. Muell.).

SPURGE FAMILY (EUPHORBIACEAE)

72. Kukui, candlenut-tree

Aleurites moluccana (L.) Willd.**

Kukui, the State tree of Hawaii is well known and is recognized from a distance by its silvery green or grayish foliage. This large spreading tree is common in moist lowland mountain forests through the Hawaiian Islands. The large long-stalked leaves are mostly 3- or 5-lobed and have 5 or 7 main veins from base. This species was apparently introduced by the early Hawaiians for the large nutlike elliptical, hard, oily seeds, for which many uses have been found.

Large evergreen forest tree to 80 ft (24 m) tall and 3 ft (0.9 m) in trunk diameter, sometimes larger, with broad spreading or irregular crown. Of largest size in narrow valleys, where trunk is tall and straight. Smaller, with twisted trunks and long branches, in exposed sites. Bark gray brown, smoothish with many thin fissures. Inner bark with dark red outer layer and brown within, tasteless, with thin whitish slightly bitter sap or latex. Twigs stout, greenish when young, becoming brown. Young leaves, young twigs, and flowering branches are densely covered with tiny whitish or rusty brown star-shaped and scaly hairs, which produce the distinctive gray color of foliage.

Leaves alternate, with stout leafstalks of 3--6 in (7.5--15 cm) or more, often longer than blades, yellow green hairy, with 2 dot glands at top above. Blades mostly 4--8 in (10--20 cm) long and wide, broadly ovate, with 3 or 5 (sometimes 7) long-pointed lobes or none, base nearly straight, thin, upper surface green and becoming hairless, lower surface light green with star-shaped hairs along veins.

Flower clusters (panicled cymes) terminal, much forked, hairy, 3 1/2--6 in (9--15 cm) long and broad, bearing many white flowers about 3/8 in (10 mm) long, mostly male and few female toward base (monoecious). Male flowers many, consisting of rounded hairy calyx 1/8 in (3 mm) long, splitting into 2--3 lobes, corolla of 5 white petals 5/16 in (8 mm) long, and 15--20 stamens. Female flowers few, composed of calyx, corolla, and pistil with hairy round 2-celled 2-ovuled ovary and 2 styles each 2-forked.

Stone fruits rounded, greenish to brown, 1 1/2--2 in (4--5 cm) in diameter, borne singly on stout stalks, leathery and slightly fleshy, not splitting open. Seeds 1--2, elliptical, about 1 in (2.5 cm) long, with hard, rough black shell.

The wood is white, lightweight (sp. gr. 0.35), soft, and of fine to coarse texture, the fine-textured type having the appearance of holly (*Ilex*) wood. It is usually colored by blue stain fungi before conversion to lumber and is not resistant to decay or insects. The wood is not currently utilized but was used by the Hawaiians for lightweight canoes and fishnet floats.

Because of its low durability, kukui is an excellent host for the edible bracket fungus pepeiao akua (*Auricularia polytricha*). During the 1800's, Chinese immigrants developed an industry of growing pepeiao on felled kukui logs for local consumption and shipment to China. A large amount of kukui was destroyed to support this industry.

Kukui was made the official tree of the State of Hawaii because of "the multiplicity of its uses to the ancient Hawaiians for light, fuel, medicine, dye, and ornament, as well as the distinctive beauty of its light-green foliage which embellishes many of the slopes of our beloved mountains." The State Legislature took this action in 1959.

Hawaiians had many uses for the big seeds, which are borne in large quantities--as many as 75--100 pounds (34--45 kg) annually--by a large tree. The seed shells, black when mature and white earlier, were made into leis and now into costume jewelry and curios. After roasting and shelling, the oil seeds were strung on a piece of coconut midvein for torches or candles, as the English name indicates. Oil pressed from the seeds was burned in stone lamps and, mixed with soot, used as paint. It has been extracted commer-

71. Australian toon   *Toona ciliata* M. Roem.*
Flowers, leaf, and fruits (lower left), 2/3 X.

cially for use as a drying oil in paints and varnishes, and for medicines. Long ago, as many as 10,000 gallons (37,879 liters) of oil were exported annually, but the high cost of labor, even in early times, made the industry unprofitable. The oil cake served as fertilizer and as cattle food. The raw seeds are reported to be toxic or purgative and should be eaten only in moderation. However, roasting or cooking apparently reduces the danger. The Hawaiians ate the roasted kernels with seaweed (limu) and salt as a condiment called 'inamona, which tastes somewhat like peanuts.

The whitish sap or latex, like a gum or resin, served as a folk remedy and was painted on tapa or bark cloth to make it more durable and waterproof. The Hawaiians obtained from the green fruit covering a black dye for tattooing and from root bark another for painting canoes. An infusion of bark and water was a fish net preservative. Additional uses were recorded by Degener (1930, p. 193--199).

Kukui is common as a wild tree in moist lowland forests from sea level to 2,200 ft (671 m) altitude through the Hawaiian Islands. Also planted as a shade and ornamental tree. Kukui, along with koa, was one of the first trees planted widely by the Division of Forestry as watershed cover. The extensive stands in the gullies of the Honolulu Watershed Forest Reserve resulted from those plantings begun in 1904. The Division of Forestry records the planting of 16,000 kukui through the islands.

Special areas--Keahua, Waimea Arboretum, Foster, Tantalus, Haleakala, City, Volcanoes.

Champion--Height 67 ft (20.4 m), c.b.h. 10.6 ft (3.2 m), spread 59 ft (18 m). Kapapala, Pahala, Hawaii (1968).

Range--Native probably of Malaysian region and named for the Moluccan Islands, the exact home uncertain. Widely spread by the early inhabitants through the Pacific Islands to Hawaii. Introduced elsewhere through the tropics and becoming naturalized. Uncommon in Puerto Rico and Virgin Islands.

Other common names--tutui, candlenut; nuez, nuez de India (Puerto Rico); lumbang (Guam); Sakan (Palau); lama (Am. Samoa).

73. Hame

Antidesma platyphyllum Mann

This handsome small evergreen native tree with shiny leaves and abundant dark purple, slightly flattened, pea-sized fruits, is widely distributed through the islands. A large shrub or small tree 20--30 ft (6--9 m) high, with trunk to 1 ft (0.3 m) and open crown of few nearly erect branches. Bark whitish gray, smoothish to deeply furrowed, the inner bark fibrous, pink, and bitter. Twigs slightly zigzag, with minute pressed hairs when young, raised rounded leaf scars, and very small rounded hairy buds.

Leaves alternate in 2 rows, becoming hairless, with short reddish leafstalks 1/8--3/8 in (3--10 mm) long. Blades elliptical, 3--5 in (7.5--13 cm) long and 1 1/2--2 1/2 in (4--6 cm) broad, short- to long-pointed at apex and rounded, blunt, or slightly notched at base, not toothed on edges, slightly thick and succulent, curved up on both sides of midvein, light green, shiny above, beneath slightly shiny with reddish midvein.

Flower clusters (panicles) on twigs back of leaves, 2--4 in (5--10 cm) long, with few slender finely hairy branches. Flowers male and female on different plants, small. Male flowers nearly stalkless, consisting of calyx about 1/16 in (1.5 mm) long with 4--5 lobes, 4--5 long stamens, and minute nonfunctional pistil. Female flowers short-stalked, composed of finely hairy calyx with 5--8 lobes and hairless pistil with 1-celled ovary and 3 very short styles.

Fruits (drupes) many, elliptical, nearly 3/8 in (1 cm) long, slightly flattened, with calyx and styles, shiny, turning from green to dark red, juicy. Seed 1, slightly flattened.

Wood is reddish brown, fine-textured, and hard. It takes a fine polish and is suitable for cabinetwork but is not found in commercial quantities. It is reported that the wood is resistant to marine borers or shipworms. The Hawaiians used the logs as anvils for beating the fibrous bark of the native shrub olonā (*Touchardia latifolia* Gaud.).

The reddish fruits are edible, sweet, and have a juice that stains hands and clothes. Mixed with kamani oil (*Calophyllum inophyllum*), the fruit made a bright red dye for tapa, particularly the tapa used for malos (loincloths).

Common in wet and dry forests, especially at 1,500--3,000 ft (457--914 m) altitude, widespread through the islands.

Special area--Kokee.

Champion--Height 52 ft (15.8 m), c.b.h. 6.7 ft (2 m), spread 23 ft (7 m). Kaupulehu, Kailua-Kona, Hawaii (1968).

Range--Hawaii only.

Other common names--ha'a, mehame, hamehame, mehamehame, ha'āmaile.

A second species is ha'a or mehame, *Antidesma pulvinatum* Hillebr. It is distinguished by the dull green leaves notched or heart-shaped at base and with tufts of hairs in vein angles and small black fruits less than 1/4 in (6 mm) long. The crown is rounded and symmetrical. This small tree is found on Oahu, Molokai, Maui, and Hawaii, in dry areas, especially the aa (rough) lava fields at low altitudes. It is reported as common on the lava fields of South Kona, Hawaii, especially at Kapua at 2,000 ft (610 m) altitude. Known only from Hawaii. Intermediate plants found together with both species on the Island of Hawaii apparently are hybrids (*Antidesma kapuae* Rock).

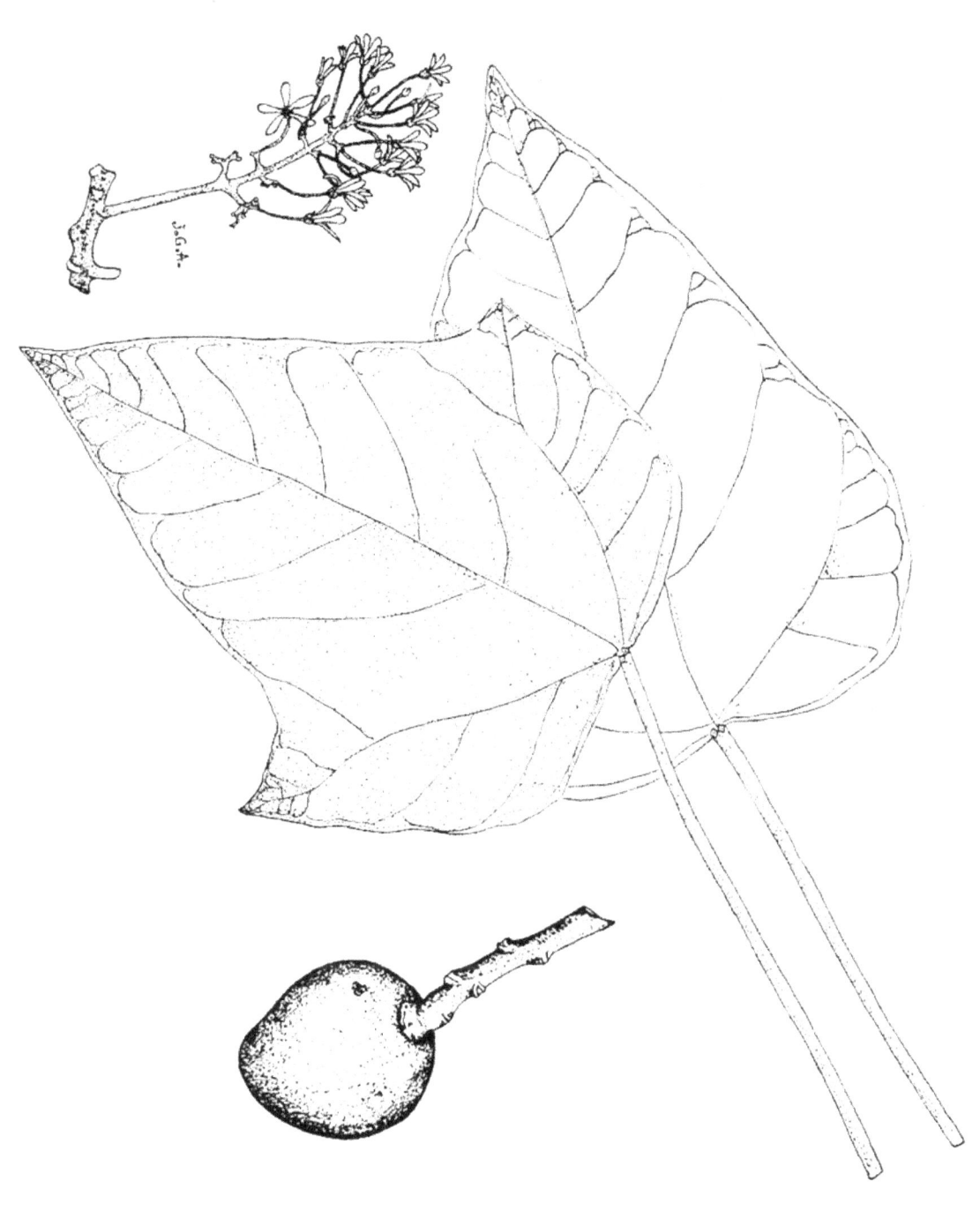

72. Kukui, candlenut-tree *Aleurites moluccana* (L.) Willd.**
Flowers (upper left), leaves, fruits (below), 2/3 X (P. R. v. 2).

Antidesma, which has flattened seeds and smooth-edged leaves close together, is sometimes confused with *Xylosma*, which has prominent lenticels in the bark, round seeds, and toothed, well-spaced leaves.

CASHEW FAMILY (ANACARDIACEAE)

74. Mango, manako *Mangifera indica* L.*

Mango is well known by its large elliptical or egg-shaped yellow or pinkish fruits with edible flesh and large seed in a mass of fibers. The tree is also a handsome ornamental and shade tree with very dense rounded crown of large narrow dark green leaves, drooping in showy red-brown clusters when first produced, and with large clusters of small yellow green to pink flowers. Also, the wood has many uses.

Medium-sized to large evergreen tree frequently attaining 20--65 ft (6--20 m) in height with stout trunk 2--3 ft (0.6--0.9 m) in diameter. Bark brown, smoothish, with many thin fissures, thick, becoming darker, rough, and scaly or furrowed. Inner bark light brown and bitter. Whitish sap exudes from cut twigs, and resin from cuts in the trunk. Twigs stout, pale green, and hairless.

Leaves alternate, hairless, with leafstalks 1/2--1 1/2 in (1.3--4 cm) long, swollen at base. Blades lance-shaped or narrowly oblong, 6--12 in (15--30 cm) long and 1 1/2--3 in (4--7.5 cm) wide, long-pointed at both ends or short-pointed at base, curved upward from midvein with many straight side veins and sometimes with edges a little wavy, leathery, shiny dark green above, paler beneath.

Flower clusters (panicles) large showy terminal, 6--8 in (15--20 cm) or more in length, with reddish hairy branches. Flowers numerous, 5-parted, about 1/4 in (6 mm) wide, short-stalked finely hairy fragrant, partly male and partly bisexual (polygamous). Calyx yellow green, 1/16 in (1.5 mm) long, deeply 5-lobed; corolla of 5 spreading petals more than 1/8 in (3 mm) long, pink, turning reddish; 5 stamens, 1 fertile and 4 shorter and sterile, borne on a disk; and in some flowers a pistil with 1-celled ovary and slender lateral style.

Fruits (drupes) hanging on long stalks, large, aromatic, mostly 3--4 1/2 in (7.5--11 cm) long, larger in improved varieties, slightly narrowed toward blunt apex and a little flattened, with smooth thin skin, soft at maturity, with thick juicy yellow or orange flesh. Seed case 2 1/2--3 1/2 in (6--9 cm), flattened, with long ridges and grooves, containing 1 seed. Flowering mainly in winter and spring and maturing fruits about 6 months later from spring to fall. If rains are prolonged during flowering period, a fungus (*Colletotrichum gloeosporioides*) destroys the flowers, and a poor fruit crop results.

The wood is lustrous blond without distinct sapwood. It is however, frequently stained during drying and may be mottled with darker spots. The heartwood when it forms in very old trees is dark brown. Wood is hard, moderately heavy (sp. gr. 0.57), tough, strong, and medium-textured and has straight to wavy grain, often with a pronounced curly or fiddleback figure. It shrinks very little in drying, so seasons well and stays in place despite humidity changes. Machining characteristics of Puerto Rican wood are as follows: planing, shaping, and turning are fair; boring, mortising, and resistance to screw splitting are good; and sanding is poor. The wood works easily, but grain irregularities cause tearouts. It is susceptible to attack by dry-wood termites and is not resistant to decay.

In Hawaii the wood has been employed for furniture, paneling, carved and turned bowls and trays, and gunstocks. Elsewhere, it has been used for flooring, construction, boxes and crates, carts, plywood, dry cooperage, and meat chopping-blocks. In French Oceania and the Cook Islands, most canoes are made from mango. Beautiful furniture has been made from a variety grown near Hilo that invariably has pronounced curly grain. The First Methodist Church in Hilo has a spectacular display of mango paneling on the wall behind the altar.

Mango is one of the most popular fruits through the tropics. Though usually eaten raw, mangos are also cooked or made into preserves, jelly, juice, or chutney. Numerous improved varieties with larger and less fibrous fruits have been developed. Many of these have several embryos (polyembryonic) and breed true from seed. Others with 1 embryo (monoembryonic) must be propagated vegetatively by budding or grafting. The most popular fruit varieties in Hawaii are the Hayden, Pirie, and Shibata. The Pirie was introduced in 1899 and the Hayden in 1930. Green mango fruit is quite popular and is often pickled.

Mango is an excellent hardy shade tree. It is also among the important honey plants, secreting quantities of nectar, and the flowers reportedly are edible. Livestock eat the fruits. The seeds, flowers, bark, leaves, and resin have been employed medicinally, and the bark and leaves yield a yellow dye.

Some people have skin sensitive to the resinous sap of the fruit peel and foliage, developing a rash similar to the rash from poison-ivy, which is in the same plant family. Climbing trees to gather fruits is hazardous because of the brittle branches.

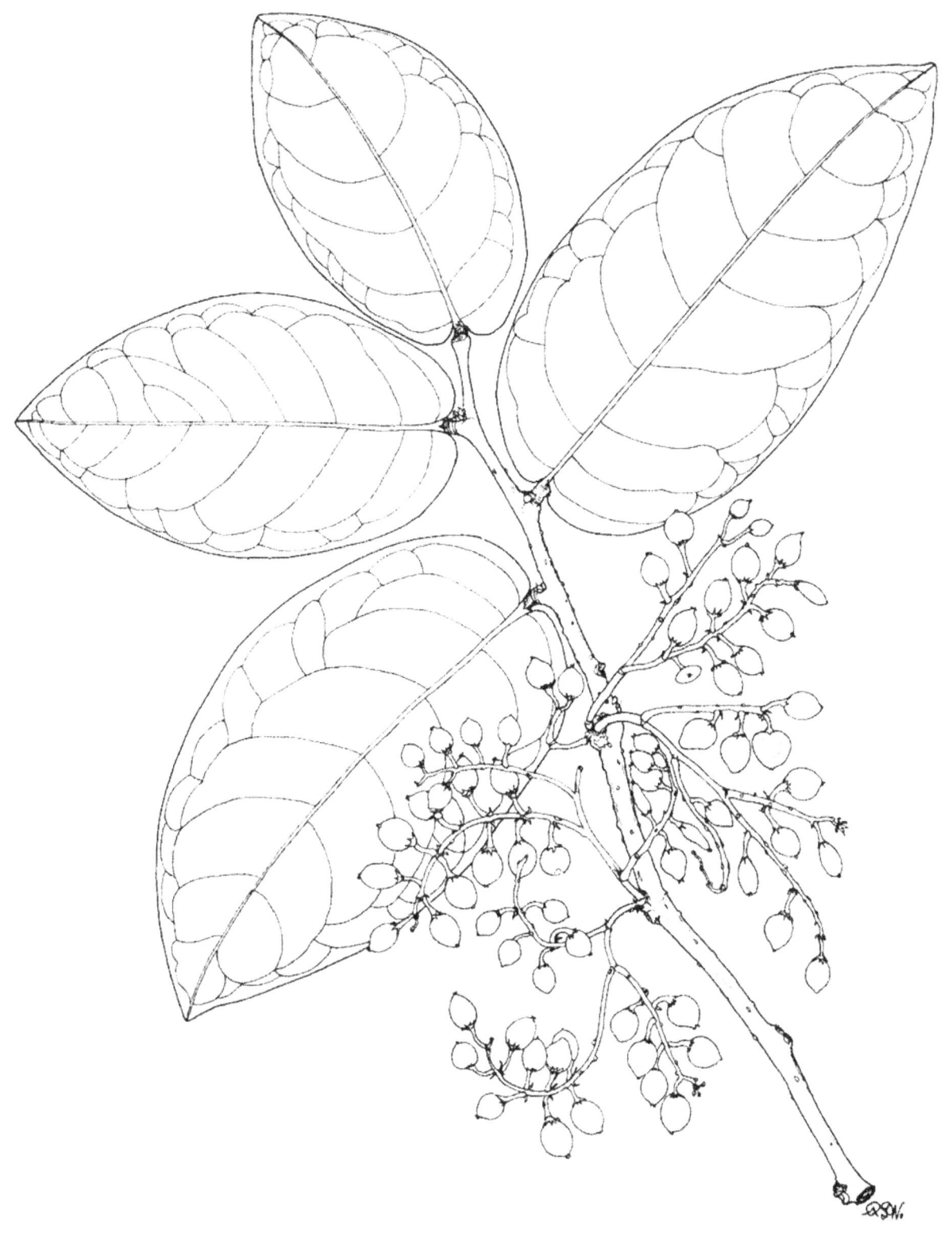

73. Hame   *Antidesma platyphyllum* Mann
Fruiting twig, 1 X.

Mango has been cultivated for more than 4,000 years in India, where more than 500 varieties are known. It was introduced into Hawaii probably prior to 1825 by Don Francisco Paula y Marín (1774-1837). Don Marín, Spanish-born friend and advisor of the Hawaiian king, imported fruits such as the pineapple and other plants from many parts of the world. The first introductions were from three different countries--Chile, the Philippines, and China--and more than 40 improved varieties followed later. The Mediterranean fruit fly and mango flies, which damage the developing fruits, came too!

In Hawaii, mango is planted and naturalized mainly in the lowlands through the islands. Huge old common mango trees are found occupying overgrown home sites in all the wetter valleys. There are estimated to be about 4 million board feet of mango sawtimber in Hawaii.

Special areas--Keahua, Waimea Arboretum, Foster, Tantalus.

Champion--Height 71 ft (21.6 m), c.b.h. 24.7 ft (7.5 m), spread 70 ft (21.3 m). Rainbow Falls State Park, Hilo, Hawaii (1968).

Range--Native to tropical India, probably from India east to Vietnam. Widely planted as a fruit tree and naturalized in tropical regions. Cultivated in southern Florida, where it is naturalized locally, and in southern California. Common in Puerto Rico and Virgin Islands.

Other common names--iedel (Palau); manga (Yap); kangit (Truk and Pohnpei).

75. Neneleau, Hawaiian sumac

Rhus sandwicensis Gray

Small deciduous native tree forming thickets in lowlands, with large compound leaves and wavy toothed leaflets paired except at end, and with whitish almost tasteless sap. To 15--25 ft (4.5--7.6 m) high and 4--12 in (0.1--0.3 m) in trunk diameter, spreading by stems from creeping roots. Bark brown gray, smooth. Inner bark whitish within green outer layer, bitter. Twigs stout, light green, with rusty brown pressed hairs and with hairy rounded buds above U-shaped leaf-scars.

Leaves alternate, large, pinnate, 12--18 in (30--46 cm) long, with yellow green hairy axis round and enlarged at base, not winged. Leaflets mostly 11--15, paired and almost stalkless except 1 at end, lance-shaped or oblong, 2--4 in (5--10 cm) long and 1--2 in (2.5--5 cm) wide, long-pointed at apex, rounded and unequal-sided at base, wavy toothed, thin, with many nearly straight side veins raised beneath, above dull green and almost hairless, beneath paler and finely hairy, red when young and again turning red before falling.

Flower clusters (panicles) terminal, erect, very large, 6--12 in (15--30 cm) long, much branched, finely hairy. Flowers very numerous, crowded, small, about 1/8 in (3 mm) long and broad, short-stalked, pale yellow, composed of calyx of 5 hairy green sepals united at base, corolla of 5 petals spreading and turned back, 5 stamens, and pistil with ovary and 2--3 short styles.

Fruit (drupe) egg-shaped, more than 1/8 in (3 mm) long, flattened, reddish, hairy, 1-seeded.

The wood is described as yellowish gray with dark resinous streaks, lightweight, coarse-textured, and tough. It has been used for saddle trees on Hawaiian ranches. Formerly, it served for ox yokes and plows.

The bark has been used locally for tanning goat skins. According to Degener, a keg of bark was shipped to Boston in 1868. Again in 1918, commercial use was considered but was abandoned because of a fungal disease that killed some plants. The shrubs are showy and ornamental.

Neneleau is found in the lowland forest zone at 600--2,000 ft (183--6,190 m) altitude or above. Common on the island of Hawaii and uncommon and in scattered or isolated thickets in Kauai, Oahu, Molokai, and Maui. It is common along highways near Hilo, Hawaii, and along the Hamakua coast.

Special area--Waiakea.

Range--Hawaii only.

Other common name--neleau.

Botanical synonyms--*Rhus chinensis* var. *sandwicensis* (Gray) Deg. & Greenwell, *R. semialata* Murr. var. *sandwicensis* (Gray) Engler.

This is the only native Hawaiian representative of its family. Several introduced species are better known. The Hawaiian plants have been treated also as a variety of the Asiatic species Chinese sumac, *Rhus chinensis* Mill., which ranges from Japan through southern China to India. That species differs in that the leaf axes are winged and it has larger red fruits 5/16 in (8 mm) in diameter, reported to be edible.

76. Christmas-berry

Schinus terebinthifolia Raddi*

This attractive introduced ornamental shrub or small tree has become widely naturalized and is a weed. It has pinnate leaves with narrowly winged axis and 5, 7, or 9 shiny leaflets, clusters of small white flowers, and many bright red poisonous fruits 3/16 in (5 mm) in diameter. The aromatic resinous sap produces a rash or dermatitis in some persons.

Evergreen shrub or small tree to 25 ft (7.6 m) high, with trunk 8 in (0.2 m) in diameter, often gnarled. Bark gray, smooth or becoming furrowed into long narrow flat ridges. Twigs light brown, finely hairy when young, with

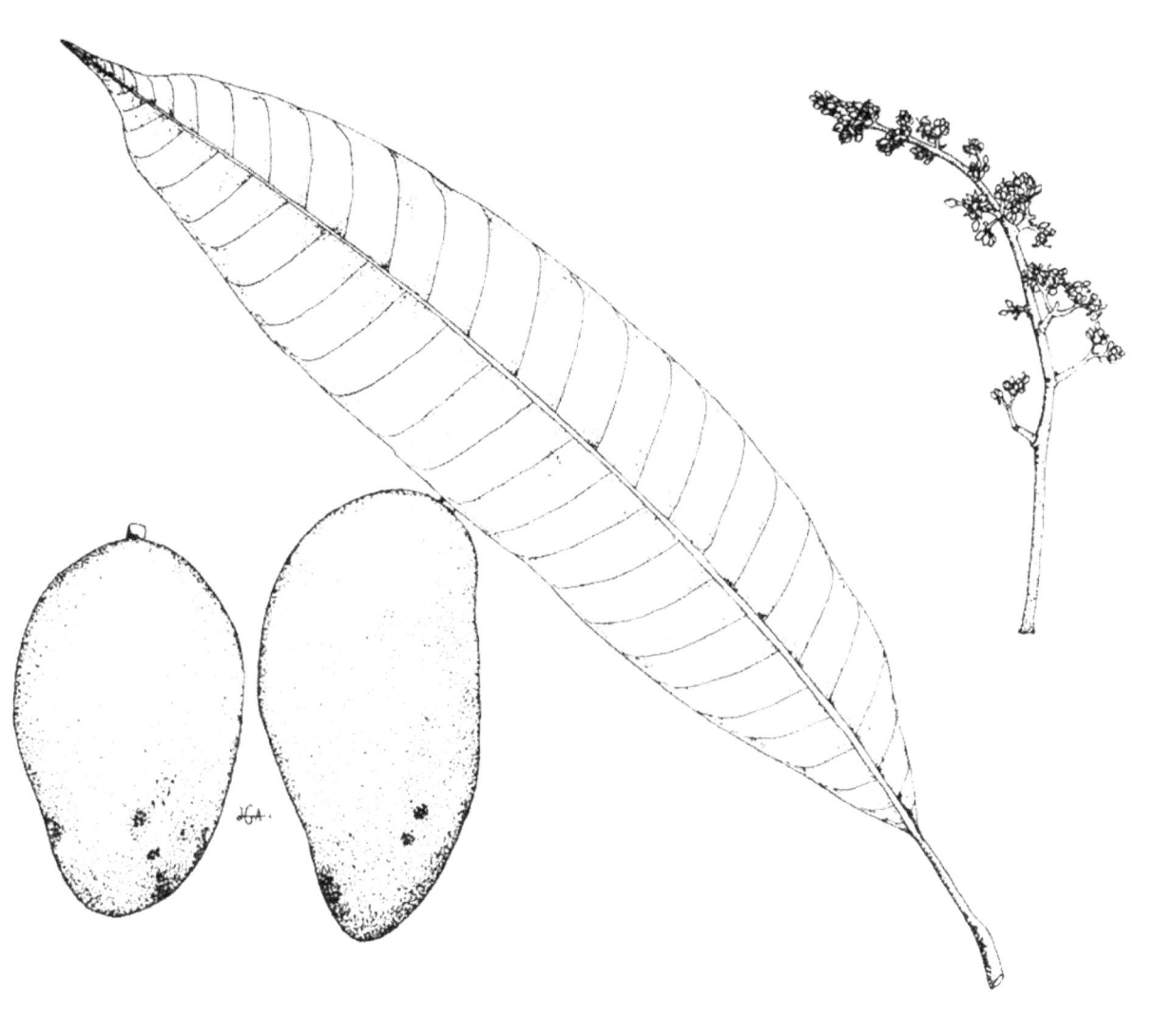

74. Mango, manako *Mangifera indica* L.*
Leaf, fruits, flowers, 2/3 X (P. R. v. 1)

75. Neneleau, Hawaiian sumac *Rhus sandwicensis* Gray
Flowering twig, 1/2 X; flowers (lower right), 2 X and 4 X (Degener).

76. Christmas-berry Schinus terebinthifolia Raddi*
Flowering twig (above), fruiting twig (below), 1 X (P. R. v. 2).

many raised dots (lenticels). Sap aromatic, resinous, suggesting turpentine, turning blackish upon exposure.

Leaves alternate pinnate, 3--6 in (7.5--15 cm) long, with narrowly winged green finely hairy axis of 1--3 in (2.5--7.5 cm) and mostly 5, 7, or 9 (3--13 or more in varieties) stalkless leaflets paired except at end. Leaflets elliptical or oblong, 1--2 in (2.5--5 cm) long and 1/2--3/4 in (1.3--2 cm) wide, the largest at end to 3 in (7.5 cm) by 1 in (2.5 cm), short-pointed at both ends, often with inconspicuous small blunt teeth toward apex, slightly thickened, hairless or nearly so, upper surface shiny green with several straight side veins, and lower surface dull light green.

Flower clusters (panicles) mostly at base of upper leaves, 1--4 in (2.5--10 cm) long, much branched, composed of many short-stalked flowers, partly male and female on different plants (dioecious). Flower about 1/8 in (3 mm) long and broad consists of calyx of 5 tiny pointed green sepals; corolla of 5 spreading white petals less than 1/8 in (3 mm) long; 10 stamens attached at base of large ring-shaped disk; and pistil with rounded ovary, short style, and dot stigma.

Fruits (drupes) many in dense clusters, bright red, with calyx at base and dot stigma at apex, with aromatic resinous brown pulp, slightly bitter. Seed 1, elliptical, light brown, less than 1/8 in (3 mm) long. Female plants produce abundant fruits which mature mostly in autumn and remain attached until December.

The hard, heavy, reddish brown wood is utilized in Hawaii only as firewood and chipped mulch from trees cut during land clearing.

Fruiting branches are picked for making Christmas wreaths and decorations. Classed as a honey plant.

Christmas-berry was introduced into Hawaii before 1911 as an ornamental. Later, it was planted extensively through the islands for the bright red berries and shiny evergreen foliage used in Christmas decorations. It has escaped widely in dry lowlands and now is classed as an undesirable weed. It is very common in Ka'u and North Kona on Hawaii, southeastern Maui, and on Oahu near Mokuleia.

Special areas--Waimea Arboretum, Foster, Tantalus, Haleakala, City, Volcanoes.

Champion--Height 39 ft (11.9 m), c.b.h. 12.5 ft (3.8 m), spread 46 ft (14.0 m). Waimea Village, S. Kohala, Hawaii (1968).

Spreading rapidly as a weed in southern Florida and known there as "Florida-holly." Planted also as an ornamental in southern California, southern Arizona, Puerto Rico, and Virgin Islands.

Range--Native of southern Brazil, Paraguay, and Argentina. Introduced northward to southern border of continental United States and in Old World tropics.

Other common names--Christmas-berry tree, wilelaiki, nani-o-Hilo (Hawaii); Brazil peppertree (continental United States); pimienta de Brazil (Puerto Rico).

The Hawaiian name wilelaiki, according to Neal, refers to Willie Rice, a politician, who often wore a hat lei of the berries.

Some people are sensitive to the resinous sap, which produces a rash or dermatitis. Pollen may cause sneezing and headaches, and the fruits are poisonous.

KARAKA FAMILY (CORYNOCARPACEAE*)

77. Karaka

Corynocarpus laevigatus J. R. & G. Forst.*

Medium-sized evergreen ornamental introduced tree with large elliptical leaves, small whitish flowers, and orange plumlike edible fruit with very poisonous seed. Tree 20--50 ft (6--15 m) tall, with trunk to 2 ft (0.6 m) in diameter, with rounded crown often flowering as a shrub. Bark on small trunks gray, smoothish. Inner bark whitish, slightly bitter. Twigs stout, hairless, with raised half-round leaf scars.

Leaves alternate, hairless, with short stout leafstalks 3/8--3/4 in (1--2 cm) long. Blades elliptical, 3 1/2--6 in (9--15 cm) long and 1 1/4--2 1/2 in (3--6 cm), wide, rounded or blunt at apex, short- to long-pointed at base, not toothed, thick and leathery, above shiny dark green with inconspicuous veins, beneath dull and paler.

Flower clusters (panicles) terminal, branched, 3--4 in (7.5--10 cm) long. Flowers many in groups of 3, short-stalked, greenish, small, about 1/4 in (6 mm) wide, composed of 5 rounded sepals, corolla of 5 elliptical finely-toothed petals united at base, 5 short stamens attached at base of corolla and opposite petals, 5 lobes (staminodia) between stamens, and pistil with 1-celled ovary, narrow style, and dot stigma.

Fruit (drupe), elliptical, often unequal sided, 1--1 1/2 in (2.5--4 cm) long, rarely to 2 1/2 in (6 cm), hairless, shiny, turning from dark green to yellow to orange, fleshy. Seed 1, large, very bitter and very poisonous.

The sweetish fruit is reported to be edible, but the seed is very poisonous, containing a cyanogenic glucoside, corynocarpin. It is recorded that the Maoris of New Zealand ate the starchy seeds after long cooking and soaking in salt water.

A handsome ornamental or park tree. Grown also in southern California.

77. Karaka *Corynocarpus laevigatus* J. R. & G. Forst.*
Flowering twig, 1/2 X; corolla with stamens (lower right), 3 X (Degener).

Scattered and naturalized in moist soils in the islands. Introduced before 1891 in Kokee region of Kauai and established there. Brought to Molokai before 1912. Afterwards seeds were distributed to other islands. Wild hogs spread the seeds also.

Karaka is common only on Kauai, where the Division of Forestry planted almost 5,000 trees between 1925 and 1937. It is spreading rapidly and is expected to become a noxious weed in the future. On Molokai, it is presently confined to one small patch and will likely be eradicated before it spreads.

Special area--Kokee.
Range--Native of New Zealand.
Other common names--karakanut, New Zealand karakanut, karaka-tree.

The genus has about 5 species in the southwestern Pacific region from New Guinea to Australia (Queensland) and New Zealand, New Hebrides, and New Caledonia. It is placed alone in its family, which is probably related to the cashew family (Anacardiaceae). The descriptive generic name is Greek meaning "club and fruit."

HOLLY FAMILY (AQUIFOLIACEAE)

78. Kāwa'u, Hawaiian holly

Ilex anomala Hook. & Arn.

Hawaii has 1 native tree species related to American holly of eastern continental United States, which is in the same genus. This handsome evergreen tree has elliptical leathery leaves, many small white flowers, and clusters of small rounded black fruits.

A medium-sized tree to 30 ft (9 m) high and 1 ft (0.3 m) in trunk diameter, often a small shrub, with irregular crown, hairless throughout. Bark light to dark gray, smooth. Inner bark with outer green layer, light yellow with brown streaks, bitter. Twigs stout, slightly angled, light green, with raised half-round leaf-scars. End buds more than 1/8 in (3 mm) long, composed of pointed scales which form a ring around twig to mark end of season's growth.

Leaves alternate but partly crowded, with light green leafstalks 1/4--3/4 in (6--20 mm) long, flattened above. Blades elliptical, mostly 1 1/2--3 1/2 in (4--9 cm) long and 1--2 1/2 in (2.5--6 cm) wide, thick, stiff, brittle, and leathery, rounded or blunt at both ends, turned under at edges (rarely with teeth, except on seedlings), above shiny dark green with network of fine veins slightly sunken, beneath dull light green with inconspicuous veins.

Flower clusters (cymose panicles) 1--3 in (2.5--7.5 cm) long at leaf bases, the long stalk and branches flattened. Flowers many, short-stalked, from rounded greenish buds, male and female on different plants (dioecious), about 1/2 in (13 mm) broad. Calyx less than 1/8 in (3 mm) long, greenish, with 4--5 rounded lobes; corolla white, sometimes pink-tinged, composed of short tube and 6--12 widely spreading rounded lobes 1/8 in (3 mm) long; stamens as many as corolla lobes, attached in notches, white, in female flowers short and not functioning; and pistil with rounded yellow green 12--20-celled ovary, no styles, and as many crowded short stigmas as cells, in male flowers small and not functioning.

Fruits (drupes) common on twigs back of leaves, rounded but slightly flattened, about 5/16--3/8 in (8--10 mm) in diameter, shiny black, smooth, with calyx and stigmas remaining, bitter. Nutlets 10--20, 1/8 in (3 mm) long.

Wood whitish, or grayish yellow with lighter colored sapwood, hard. It was prized by the Hawaiians for canoe trimmings and to make the anvil on which bark was beaten into bark cloth or tapa. It has also been used for saddle trees.

Common and widespread in open wet forests through the 6 larger islands, to 6,500 ft (1,981 m) altitude.

Special areas--Kokee, Haleakala, Volcanoes.
Champion--Height 45 ft (13.7 m), c.b.h. 3.9 ft (1.2 m), spread 32 ft (9.8 m). Honaunau Forest Reserve, Kailua-Kona, Hawaii (1968).
Range--Hawaiian Islands only.
Other common names--kā'awa'u, 'aiea.

The name kāwa'u has been applied occasionally to *Styphelia*, *Mezoneuron*, and *Zanthoxylum* also.

Botanical synonyms--*Ilex sandwicensis* (Endl.) Loess., *I. hawaiensis* S. Y. Hu.

This very variable species and a few close relatives in Polynesia are classed as the most primitive for the genus of more than 350 species of wide, mostly tropical distribution. The Marquesan and Tahitian trees are scarcely distinguishable from those of Hawaii.

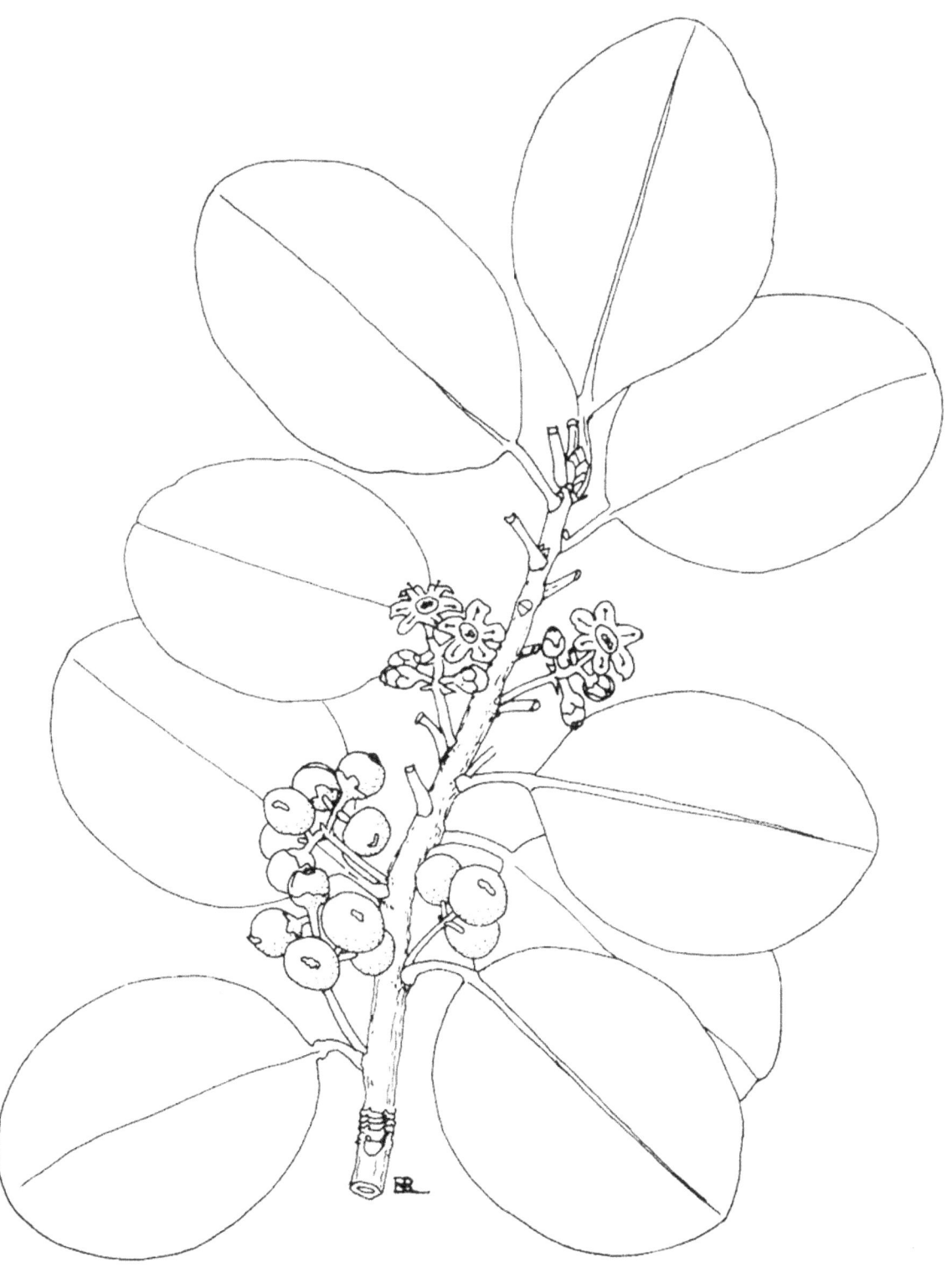

78. Kāwa'u, Hawaiian holly *Ilex anomala* Hook. & Arn.
Twig with flowers and fruits, 1 X.

BITTERSWEET FAMILY (CELASTRACEAE)

79. **Olomea**   *Perrottetia sandwicensis* Gray

This species, widespread in wet forests through the Hawaiian Islands, is the only native species of its family. It is an evergreen shrub or small tree to 23 ft (7 m) high, with finely toothed pinkish tinged elliptical leaves. The bark is gray, smoothish to finely fissured; inner bark gray, slightly bitter. Branches sometimes long and drooping. Twigs green or reddish, hairless or nearly so, ending in small hairy buds of 1/8 in (3 mm), composed of tiny pointed scales or stipules that fall early.

Leaves alternate, with slender pinkish leafstalks of 3/8--1 1/4 in (1--4 cm). Blades elliptical or ovate, 2--5 in (5--13 cm) long and 1--3 in (2.5--7.5 cm) wide, short- to long-pointed at apex, blunt or rounded at base, finely toothed on edges, thin, becoming hairless, upper surface slightly shiny green with pinkish midrib and curved side veins, light green beneath.

Flower clusters (panicles) at leaf bases, 1 1/2--4 in (4--10 cm) long, much branched. Flowers many, very small, of 1 or both sexes (polygamo-dioecious), short-stalked, 1/16 in (1.5 mm) long and broad, greenish red, composed of 5 pointed reddish tinged sepals; 5 orange green pointed petals; 5 stamens from a disk, alternate with petals, much longer and spreading widely (short in female flowers); and pistil with 2-celled ovary and short 2-forked style.

Fruit (berry) round and slightly flattened, 3/16 in (5 mm) in diameter, bright red, juicy, with sepals at base and black style at apex. Seeds 2--4, 1/16 in (1.5 mm) long, rounded, shiny green, smooth, sticky.

The wood is described as golden brown with reddish tint, moderately hard, and straight grained.

Hawaiians made fire by friction by rotating a piece of this hard wood on a piece of the soft wood of No. 87, hau (*Hibiscus tiliaceus*).

The trees are attractive in October and November, bearing numerous clusters of drooping red berries.

Fairly common in moist wet forests at 1,000--6,000 ft (305--1,829 m) altitude, throughout the islands.

Special areas--Kokee, Haleakala, Volcanoes.
Range--Hawaii only.
Other common names--waimea, pua'a olomea.

This genus honors George Samuel Perrottet (1793-1870), Swiss-born French botanical explorer, and contains about 15 species scattered from tropical America to Asia.

SOAPBERRY FAMILY (SAPINDACEAE)

80. **'A'ali'i** *(color plate, p. 35)*   *Dodonaea viscosa* Jacq.

Native evergreen much-branched shrub or small tree with sticky yellow green leaves and rounded dry fruits with 3--4 papery wings, in the broad sense, here regarded as one variable species through the Hawaiian Islands and beyond. Sometimes a small tree to 30 ft (9 m) and 3 in (7.5 cm) in trunk diameter, with rounded crown. Bark dark brown, finely fissured. Twigs slender, slightly angled, light brown, hairless.

Leaves alternate, 1 1/2--3 in (4--7.5 cm) long and 3/8--5/8 in (1--1.5 cm) wide, narrowly lance-shaped or reverse lance-shaped (oblanceolate) and broader toward short-pointed apex, tapering to long-pointed base and very short stalk, not toothed on edges, slightly thickened, finely hairy when young, yellow green, sticky and resinous especially when young, slightly shiny, paler beneath.

Flower clusters (panicles) at ends of twigs, about 1 in (2.5 cm) long, finely hairy, sticky. Flowers partly male, female, and bisexual on the same plant or different plants (polygamous), about 1/4 in (6 mm) long, with 5 hairy sepals and no petals. Male flowers have 10 stamens. Female flowers have pistil with ovary, short style, and 4 dot stigmas.

Fruit (capsule) swollen, dry, rounded, 3/8--5/8 in (1--1.5 cm) long, notched at both ends, with 3--4, sometimes 2, papery round wings, of colors ranging from yellow green to pink to brown to dark maroon, spreading up to 3/4 in (2 cm) across, finely hairy on edges, sometimes viscous or sticky, 3--4-celled. Seeds 4--1, elliptical, blackish, 1/8 in (3 mm) long.

The wood is yellow brown and the small heartwood is black when present. It is very hard and heavy and said to be durable. It was sometimes used for house posts and spears.

The leaves were used by the Hawaiians for medical purposes. Like hops, the flowers were served to impart a bitter flavor, and also were used as a tonic. The attractively colored fruit is also used in leis for the hair. A red dye was made from the capsules.

Throughout the Hawaiian Islands, including Niihau, especially in dry regions at about 10--7,700 ft (3--2,347 m) altitude. Most easily seen in the national parks on Hawaii and Maui and the Waimea Canyon area of Kauai. The tallest specimens, about 30 ft (9 m) high, are along the Mauna Loa Strip Road in Hawaii Volcanoes National Park. The tree-sized plants are mostly in

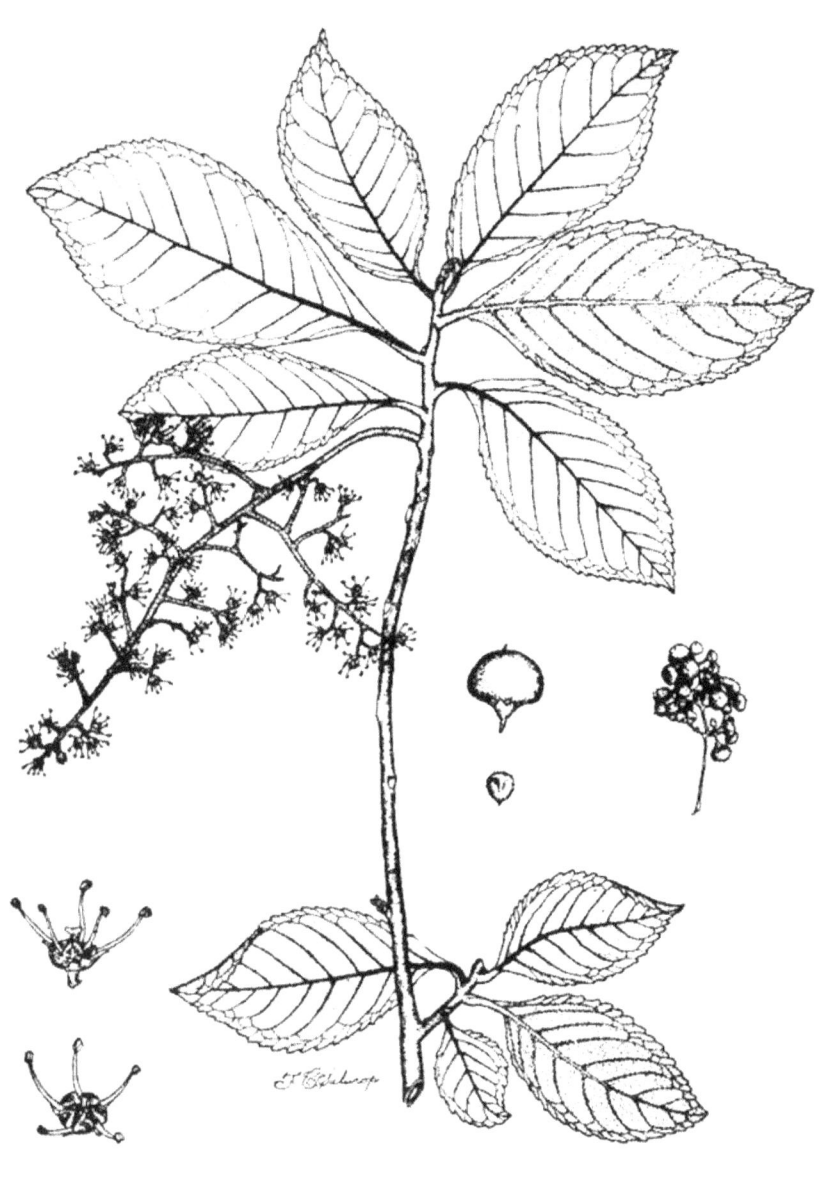

79. Olomea   *Perottetia sandwicensis* Gray
Flowering twig, 1/2 X; flowers (lower left), 2 X; fruit, 2 X, seed, 4 X, and cluster of fruits (right) 1/2 X (Degener).

the upper elevation forests of Hawaii and Maui, but are also observed occasionally in Waianae and Koolau Ranges on Oahu.

Special areas--Haleakala, Volcanoes.

Range--Widespread through tropics of both hemispheres, including Puerto Rico and Virgin Islands, Florida, and Arizona.

Other common names--'a'ali'i-ku ma kua, 'a'ali'i ku makani; lampuaye (Guam); mesechelangel (Palau).

81. Āulu

This medium-sized evergreen tree native to Oahu and Kauai is a relative of the widespread species No. 82, wingleaf soapberry, Sapindus saponaria L., but differs in that the leaves are simple rather than compound. A tree to 60 ft (18 m) high and 1 1/2 ft (0.5 m) in trunk diameter. Bark gray, smoothish, slightly fissured. Inner bark light orange streaked. Twigs light brown, finely hairy when young, becoming hairless, with tiny cracks, with small dark brown pointed hairy side buds.

Leaves alternate, simple, hairless, with long slender leafstalks 3/4--2 in (2--5 cm) long, flattened above and enlarged at brown base. Blades narrowly ovate or oblong, 3--6 in (7.5--15 cm) long and 1 1/4--2 3/4 in (4--7 cm) wide, blunt at base and extending slightly down leafstalks, gradually narrowed to bristle point at apex, not toothed on edges, slightly thickened and leathery. Upper surface slightly shiny dark green with fine slightly curved side veins and sides curved up from light yellow midvein; lower surface dull light green with raised light yellow midvein.

Flower clusters (panicles) terminal and at base of uppermost leaves, 2--6 in (5--15 cm) long, much branched, the long wide-spreading branches dark brown, finely hairy. Flowers very numerous, male and female, short-stalked, slightly bell-shaped, about 1/4 in

82. Wingleaf soapberry, mānele

This handsome tree is of special interest as 1 of 2 tree species native to both Hawaii and the continental United States (southern Florida). (The other is No. 80, 'a'ali'i, Dodonaea viscosa Jacq.). It is characterized by pinnate leaves with usually 6--12 paired elliptical to lance-shaped dull green leaflets and axis slightly winged when young and by the shiny brown ball-like 1-seeded berries 5/8--3/4 in (15--19 cm) in diameter.

A small to large deciduous tree becoming 80 ft (24.4 m) tall in Hawaii and as much as 6 ft (1.8 m) in trunk diameter, larger than elsewhere, with enlargements or buttresses at base, and with compact crown. Bark light brown or gray, smoothish and warty, becoming finely fissured, shedding in large scales and exposing smooth dark layer. Inner bark light orange brown, slightly bitter and astringent. Twigs stout, light gray with raised reddish brown dots (lenticels), finely hairy when young.

Botanical synonyms--Dodonaea eriocarpa Sm., D. sandwicensis Sherff, D. stenocarpa Hillebr.

Three other species with many named varieties and forms recorded from Hawaii are united here under a single very variable species.

This species and No. 82, wingleaf soapberry or mānele, Sapindus saponaria, are the only tree species native in both Hawaii and continental United States.

Sapindus oahuensis Hillebr.

(6 mm) wide, light greenish yellow. Male flowers with calyx of 5 overlapping rounded light green hairy sepals 1/8 in (3 mm) long, corolla of 5 spreading greenish white hairy petals 3/16 in (5 mm) long, 8 stamens on rounded disk, and tiny pistil. Female flowers have similar calyx and corolla, 8 minute sterile stamens on a disk, and pistil with elliptical ovary slightly 2--3-lobed and 2--3-celled and dot stigma.

Fruit an elliptical berry (coccus), sometimes 2, 3/4--1 in (2--2.5 cm) long, shiny and leathery. Seed 1, elliptical black, 1/4--3/4 in (12--20 mm) long, rough.

Wood light brown, hard, presumably similar to that of the species described next.

The seeds were used in home remedies as a cathartic and were strung in leis. Like those of the common soapberry, they may be poisonous.

Fairly common, scattered in dry forests at 200--2,000 ft (61--610 m) altitude.

Special areas--Wahiawa, Bishop Museum.

Range--Oahu and Kauai only.

Other common names--kaulu, Oahu soapberry. The name lonomea is used on Kauai.

The trees of Kauai, known by the Hawaiian name lonomea, have been treated also as a separate species, Sapindus lonomea St. John (1977b).

Sapindus saponaria L.

Leaves alternate pinnate, 8--16 in (20--40 cm) long. Leaflets stalkless or nearly so, 2 1/2--5 in (6--13 cm) long and 3/4--1 1/2 in (2--4 cm) wide, long or short-pointed at apex, base short-pointed or blunt and often oblique and unequal with side toward leaf apex broader, not toothed on edges, thin, upper surface dull green and hairless, lower surface slightly paler and sometimes soft hairy.

Flower clusters (panicles) terminal and lateral, to 4--8 in (10--20 cm) long, larger elsewhere, very numerous small 5-parted whitish flowers 3/16 in (5 mm) across, mostly male but some female or bisexual (polygamous). Male flowers have 5 spreading unequal sepals about 1/16 in (1.5 mm) long, outer 2 smaller, whitish and tinged with green; 5 white hairy rounded petals smaller than sepals; 8 light yellow stamens more than 1/16 in (1.5 mm) long on a light green disk; and minute brown nonfunctional pistil. Female flowers have sepals, petals,

80. 'A'ali'i *Dodonaea viscosa* Jacq.
Twig with fruits, 1 X.

81. Āulu *Sapindus oahuensis* Hillebr.
Twig with female flowers, male flowers (upper right), fruit (lower left), 1 X.

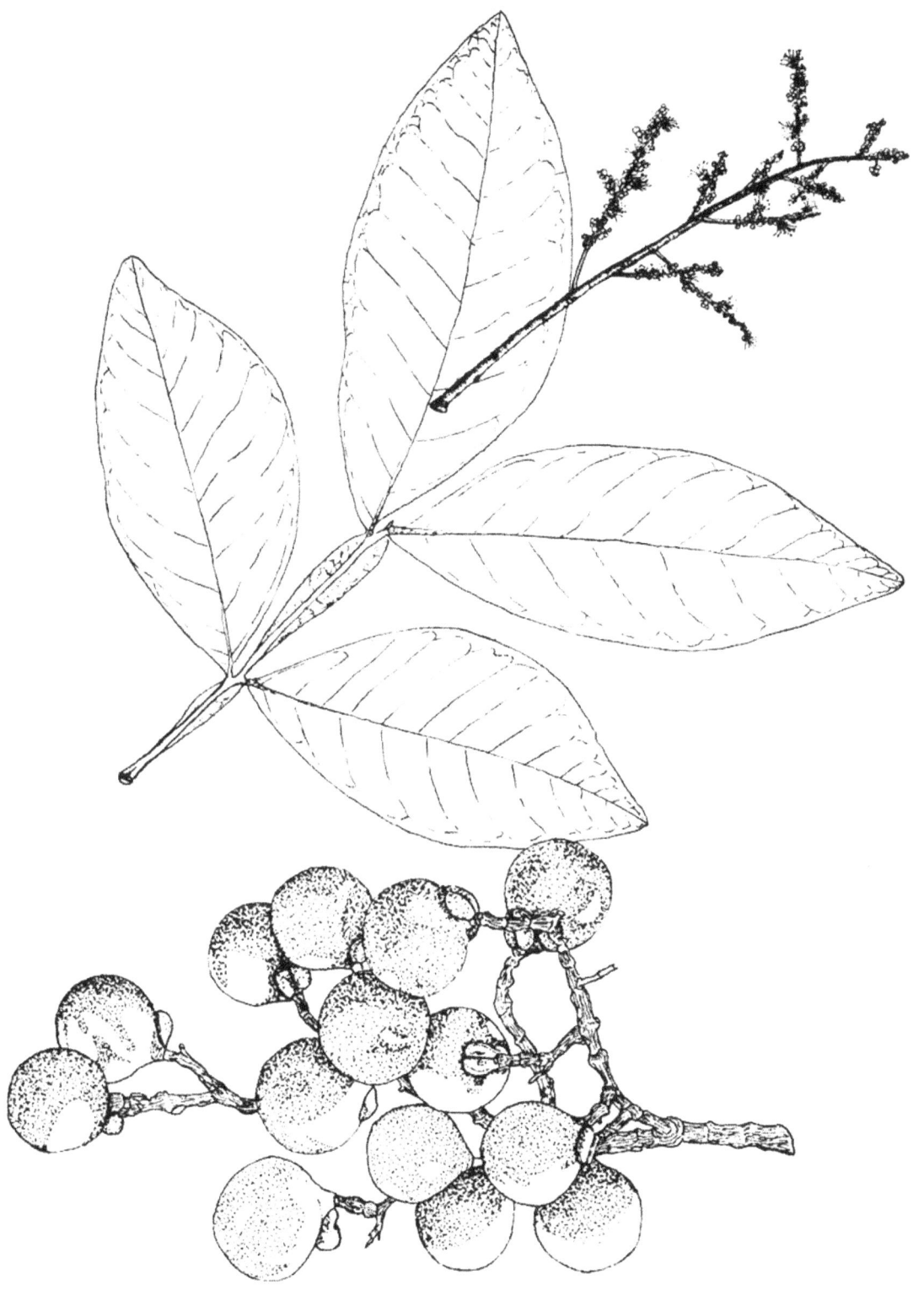

82. Wingleaf soapberry, mānele   *Sapindus saponaria* L.
Flowers, leaf, and fruits (below), 1 X (P. R. v. 1).

shorter stamens, and greenish pistil more than 1/16 in (1.5 mm) long with 3-celled ovary and slender style.

Fruits (berries or cocci) in clusters on hard woody stalks, 1 (sometimes 2 or 3) developing from a pistil and others disklike at base, with leathery shiny brown skin and yellow sticky bitter poisonous flesh, clear or translucent. Seed 1, round black, 3/8--1/2 in (10--13 mm) in diameter, poisonous.

Sapwood whitish and heartwood yellow or light brown. Wood hard and heavy (sp. gr. 0.8), coarse-textured, and not durable when exposed. Elsewhere employed for posts and in carpentry.

The scientific and common names refer to the use elsewhere of the fleshy fruit as a substitute for soap. When cut up, the fleshy part, containing about 30 percent saponin, produces abundant suds in water.

The seeds are used in leis in Hawaii. In tropical America, crushed seeds serve as a fish poison when thrown into a stream. An insecticide has been made from ground seeds, and medicinal oil extracted. Also elsewhere, the hard round seeds have been used as beads in necklaces and rosaries as well as marbles and formerly, as buttons.

A common shade tree in tropical America and classed as a honey plant. Infusions of the roots and leaves have been prepared for home remedies.

Mānele is native in the middle forest zone at 3,000--4,500 ft (914--1,372 m) altitude on the island of Hawaii, for example, Mauna Loa and Puu Waawaa. The trees of largest size are accessible and easily seen in Kipuka Puaulu near Kilauea Volcano within Hawaii Volcanoes National Park. There are also some at Ulupalakua on Maui. Another form is planted in Hawaii as a shade tree.

Special areas--Waimea Arboretum, Wahiawa, Volcanoes, Kipuka Puaulu, Ala Moana Park.

Champion--Height 106 ft (32.3 m), c.b.h. 10.1 ft (3.1 m), spread 84 ft (25.6 m). Hawaii Volcanoes National Park, Hawaii (1968).

Range--Widespread in tropical America from northern Mexico to Brazil and Argentina and through West Indies including Puerto Rico and Virgin Islands. Also in Florida and at 2 coastal localities in Georgia, the range extended northward partly by prehistoric Indians and partly by cultivation. Native to Hawaii and other Pacific Islands including the Marquesas and Society Islands to New Caledonia. Introduced into Old World tropics.

The native Hawaiian trees found in 1909 by Joseph F. Rock seemed different from the introduced trees of another form in Honolulu and were named *Sapindus thurstonii* Rock. Soon afterwards he concluded that the native trees, which have deciduous foliage, were the same as the evergreen species widespread on the American continent. The segregate was revived by St. John (1977b).

The separate or disjunct distribution of this tree species in continental America and also Hawaii and other Pacific Islands is unexplained. However, Degener (1930, p. 202) observed that the dried berries have an air space between the outer wall and seed formed by the shrinking flesh and that they will float in water. Also when removed from the fruit, at least half of the seeds will float. Thus, long distance transportation by ocean currents may occur. Seeds are often found in beach drift on various islands. Rock found that the Hawaiian trees attain a larger size, both in height and particularly in trunk diameter, than those anywhere on the mainland.

Other common names--a'e, soapberry; jaboncillo (Puerto Rico, Spanish).

Botanical synonym--*Sapindus thurstonii* Rock.

BUCKTHORN FAMILY (RHAMNACEAE)

83. **Kauila** *(color plate, p. 36)* *Alphitonia ponderosa* Hillebr.

Medium-sized to large handsome evergreen tree of dry forests characterized by rusty hairs on twigs, leafstalks, flower stalks, and under surface of the young ovate leaves, by lateral clusters of small greenish flowers spreading like a 5-pointed star, and by rounded fruits ringed near middle. To 50--80 ft (15--24 m) high with straight trunk 8--24 in (0.2--0.6 m) in diameter, or only a shrub on exposed ridges. Bark whitish gray, rough and furrowed. Twigs with raised half-round leaf scars and rusty hairy buds composed of minute leaves.

Leaves alternate, with leafstalks 1/2--1 in (13--25 mm) long. Blades ovate, 2--6 in (5--15 cm) long, long-pointed at apex, rounded at base, not toothed on edges, slightly thickened and leathery, with curved parallel side veins, shiny dark green and hairless above, and beneath dull light green with rusty hairy raised veins.

Flower clusters (cymes) at leaf bases, shorter than leaves, with widely forking hairy branches. Flowers several, of one or both sexes (polygamous), short-stalked, about 1/4 in (6 mm) across. The short cuplike base (hypanthium) bears 5 spreading pointed hairy sepals more than 1/16 in (1.5 mm) long; 5 narrow spoon-shaped petals half as long and partly enclosing the 5 opposite stamens; and pistil with 2--3-celled ovary covered by broad rounded disk and with short 2--3-forked style.

83. Kauila *Alphitonia ponderosa* Hillebr.
Twig with flowers and fruits, 2/3 X; flower and fruit (lower left), 3 X (Degener).

Fruit (drupe) about 5/8 in (15 mm) in diameter, with ring formed by cuplike base, containing 2--3 stones. Seeds oblong, shiny, with red covering.

The wood has a beautiful cherry red or dark red color with wide light yellowish brown sapwood. It has distinct growth rings and is diffuse porous, fine-textured, very hard, strong, and durable. One of the heaviest native woods, it sinks in water. It was highly valued by the Hawaiians and served for tools in the absence of metals. The many uses included hut beams, mallets for beating tapa cloth, spears 13--20 ft (4--6 m) long, javelins, and the o'o or digging stick for cultivating fields. The lintels above the windows of the Hawaiian Mission Printing House, built in 1821, were of this wood and were quite sound when removed for restoration in 1972.

Rounded polished rods of the wood became hairpins for women.

Widespread in the lower dry forest on the leeward side of the 6 large Hawaiian Islands, sometimes on exposed ridges and on aa lava fields at 800--4,100 ft (244--1,250 m) elevation in koa forest. Rare except on Kauai.

Special areas--Kokee, Volcanoes.
Champion--Height 62 ft (18.9 m), c.b.h. 7 ft (2.1 m), spread 54 ft (16.5 m). Kokee State Park, Kauai (1968).
Range--Known only from Hawaiian Islands.
Other common name--o'a.
Botanical synonym--*Alphitonia excelsa* auth., not (Fenzl) Reiss. ex Endl.

St. John (1977a) has distinguished 6 varieties, each restricted to a separate island.

ELAEOCARPUS FAMILY (ELAEOCARPACEAE)

84. Kalia

Elaeocarpus bifidus Hook. & Arn.

Medium-sized evergreen tree of wet forests of Kauai and Oahu, with ovate hairless leaves, small greenish flowers, and shiny dark brown stone fruit. To about 30 ft (9 m) tall and 1 ft (0.3 m) in trunk diameter. Bark dark gray, rough, thin, fibrous. Branches and twigs slender, drooping hairless, ending in gummy or varnished bud.

Leaves alternate, hairless, with slender leafstalks 1 1/4--4 in (4--10 cm) long. Blades ovate, 3--6 in (7.5--15 cm) long and 1 1/4--3 1/4 in (4--8 cm) wide, long-pointed at apex, short-pointed at base, wavy toothed on edges, thin, shiny green above.

Flower clusters (racemes) at leaf bases, unbranched, 1 1/4--3 1/4 in (3--8 cm) long. Flowers 5--8 on stalks of 3/8 in (1 cm), almost 3/8 in (1 cm) long, pale greenish yellow, narrowly bell-shaped, composed of 4--5 thick pointed sepals 5/16 in (8 mm) long, 4--5 thick, narrowed and notched petals as long as sepals, 14--21 stamens on a disk, much shorter than petals, and pistil with egg-shaped 2--3-celled ovary, short style, and 2--3-forked stigma.

Stone fruit (drupe) elliptical and olive-shaped or rounded, about 3/4--1 in (2--2.5 cm) long, shiny dark brown, with thick hard stone, 1-seeded.

Wood whitish, soft, fine-textured, and straight-grained. Wood of a different species, *Elaeocarpus joga* Merrill from Guam, in tests was found suitable for interior parts in furniture.

Hawaiians formerly made rope from the fibrous bark. The larger branches served for house rafters, and the slender ones for thatching rods.

Some flower clusters are deformed by mites and become bright red as if showy flowers. The trees are attacked by mistletoes also.

Common in wet forests at 300--4,000 ft (91--1,219 m) on Kauai and on both ranges of Oahu. Reported from Niihau in 1832.

Special areas--Kokee, Waimea Arboretum.
Range--Known only from Kauai and Oahu.

This genus and family are also united within the linden (basswood) family (Tiliaceae).

MALLOW FAMILY (MALVACEAE)

85. Koki'o ke'oke'o, native white hibiscus

Hibiscus arnottianus Gray

Small tree or tall shrub native of Oahu and sometimes cultivated, with showy large fragrant funnel-shaped flowers about 5--6 in (13--15 cm) across the 5 spreading elliptical white petals. Evergreen, 10--30 ft (3--9 m) high, with several trunks 3 in (7.5 cm) or more in diameter, with dense crown, hairless throughout (a variety hairy). Bark gray, smooth; inner bark light green,

almost tasteless. Twigs green when young, becoming gray with raised half round leaf scars.

Leaves ovate, 2--4 in (5--10 cm) long and 1 1/2--3 in (4--7.5 cm) broad, thin, green, and slightly shiny on both surfaces, lower surface often finely hairy, with 5 main veins, from rounded base, apex blunt, edges finely wavy

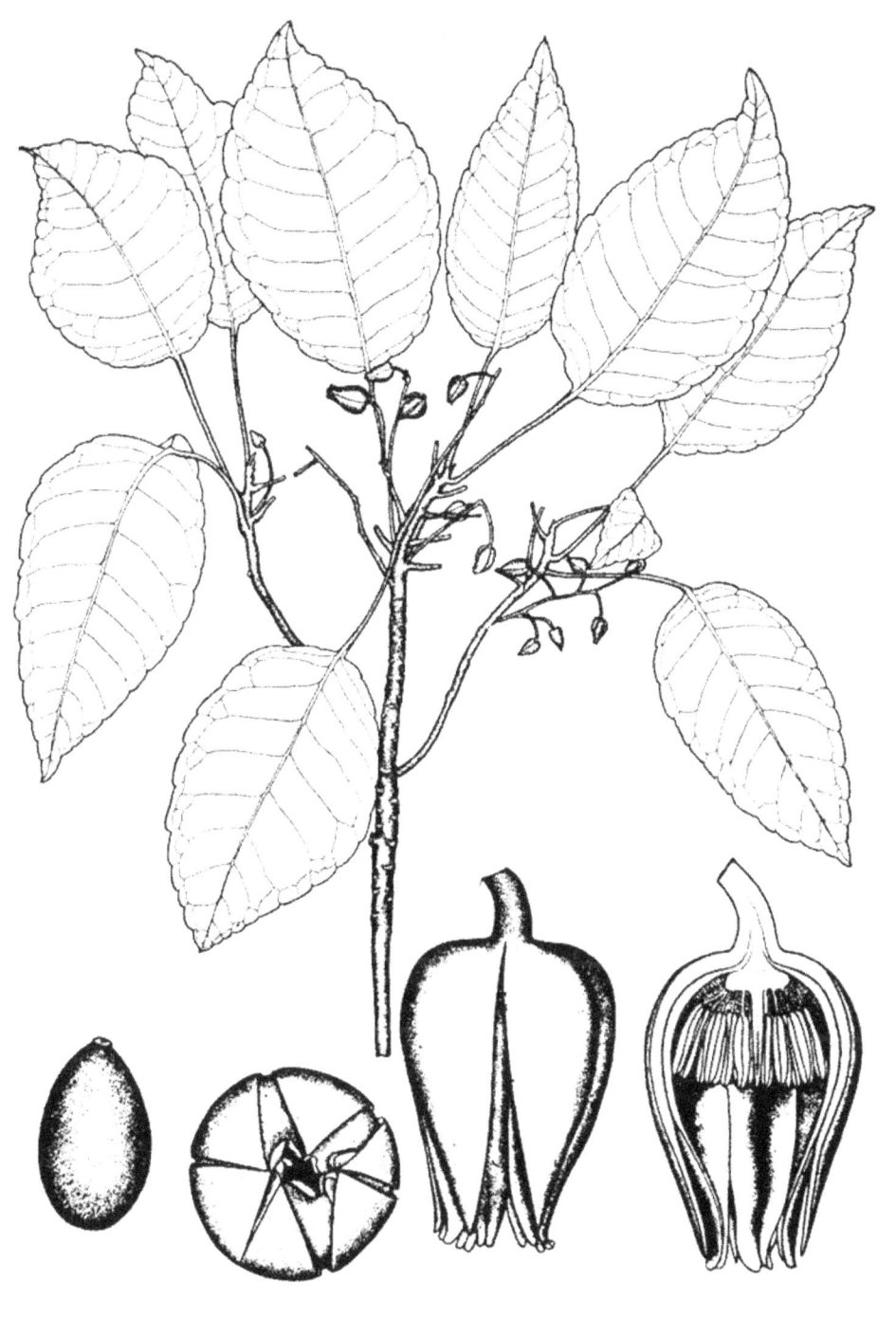

84. Kalia *Elaeocarpus bifidus* Hook. & Arn.
Flowering twig (above), 1/2 X; seed (lower left), 1 X; flowers (below), 5 X (Degener).

or straight; leafstalks 3/4--1 1/4 in (2--4 cm); stipules long-pointed, shedding early.

Flowers single on jointed stalks at 1 or 2 uppermost leaves. Calyx tubular, cylindrical, light green, 2 cm long, 5-toothed and split on a side, above 5--7 narrow, curved green scales; petals 5, white, sometimes pinkish, 3--4 1/2 in (7.5--11 cm) long, united at base into narrow tube 1 in (2.5 cm) long; many threadlike stamens along the upper part of a long dark red or white column extending from tube 3 1/4--4 in (8--10 cm), enclosing pistil; ovary cylindrical, 5-celled with several ovules in each cell, the slender style with 5 exposed erect dark red branches ending in brown dot stigmas.

Seed capsules oblong, 1 in (2.5 cm) long, thin-walled, 5-celled, enclosed by calyx. Seeds 3/16 in (5 mm) long, brown hairy.

Native only on Oahu and Wailau Valley, Molokai, in wet forests at 1,000--3,000 ft (305--914 m) altitude. Originally common in mountains near Honolulu.

Special areas--Lyon Arboretum, Waimea Arboretum, Maui Zoological and Botanical Gardens, Wahiawa Botanic Garden, Kapiolani Rose Garden near Honolulu Zoo, and Manuka State Park on Hawaii.

Other common names--pā-makini, Punaluu hibiscus (variety with large heart-shaped hairy leaves). A common name for *Hibiscus* is pua aloalo, which refers to the flower. Aloalo refers to the whole plant.

This species named in 1854 honors its discoverer, George Arnold Walker Arnott, (1799-1868), Scottish botanist. It is mentioned in old Hawaiian songs and legends.

Several other mostly shrubby species of *Hibiscus*, with flowers of assorted colors, are native in Hawaii. *H. arnottianus* is a source of numerous horticultural varieties, which are popular ornamentals. Its flowers last longer, 2 days instead of 1.

White Kauai hibiscus, *Hibiscus waimeae* Heller, also with showy white flowers, is native to Kauai and common in Waimea Canyon at 2,000--3,000 ft (610--914 m). This small tree reaches 30 ft (9 m) and 1 ft (0.3 m) in trunk diameter. It is easily cultivated and can be seen at most special areas listed above.

86. Blue mahoe, Cuban-bast

Hibiscus elatus Sw.*

Mahoe is an introduced forest and shade tree related to No. 87, hau, *Hibiscus tiliaceus*, (and sometimes considered a variety) but is a taller tree with larger flowers 3--5 in (7.5--13 cm) long and broad, also funnel-shaped and yellow, and with blunt pointed egg-shaped seed capsules, densely hairy with hairy seeds.

Large evergreen introduced tree to 80 ft (24 m) high with tall straight trunk 16 in (0.4 m) in diameter or larger. Bark gray, smooth to finely fissured. Inner bark fibrous, whitish to whitish green, and slightly bitter. Twigs stout, green when young, with rings at nodes. Young twigs, leafstalks, lower leaf surfaces, calyx, and seed capsules densely covered with minute gray star-shaped hairs.

Leaves alternate with slender round leafstalks 2 1/2--4 in (6--10 cm) long and with 2 large oblong short-pointed hairy light green basal scales (stipules) 1 1/4 in (3 cm) long, shedding early and leaving ring scar. Blades heart-shaped and nearly round, about 5--7 in (13--18 cm) long and broad, abruptly short- or long-pointed at apex and heart-shaped or notched at base, with straight or finely wavy edges, with mostly 9 main veins from base, slightly thickened, upper surface green and hairless, lower surface gray hairy with 1--3 narrow glands near base of main veins.

Flowers 1--3 borne at leaf bases at ends of twigs on stout green stalks of 1/2 in (13 mm), with light green hairy basal cup (involucre) 3/4 in (2 cm) long with 9 long-pointed lobes. Calyx 1 1/2--2 in (4--5 cm), light green hairy, tubular with 5 narrow long-pointed spreading lobes. Petals 5, yellow with large dark red spot at base inside, 3 1/2--5 in (9--13 cm) long, narrow elliptical spreading, united at base. Stamens numerous on whitish column united with corolla at base. Pistil has densely hairy 5-celled ovary, long style, and 5 rounded stigmas. Flowers opening and closing same day, petals withering and turning to orange and red.

Seed capsules egg-shaped, 1--1 1/2 in (2.5--4 cm) long, blunt-pointed, densely yellow brown hairy, splitting into 5 parts, calyx and involucre shedding. Seeds many, hairy.

Sapwood light brown, heartwood bluish, greenish, or streaked when freshly cut. Wood varies from soft, fine-textured, with silky luster to hard, of medium to coarse texture, and rather dull. Where this species is native, the attractive colored wood has been prized for cabinetwork, furniture, interior trim, and gunstocks. Other uses include construction, railroad crossties, and shingles.

The fibrous bark of young trees makes good ropes. Formerly, the lacelike inner bark was used for tying bundles of Havana cigars and was called Cuba bark. An infusion of the mucilaginous leaves and young twigs has served in home remedies.

This species is being tested in Puerto Rico in experimental forest plantations. It grows rapidly, reaching a height of 60 ft (18 m) in 10 years. A red-flowered form is known there.

Planted as a fast growing handsome shade tree in southern Florida, because of its resistance to salt spray.

Introduced in Hawaii about the middle of the present century as a forest and shade tree in moist lowland zones.

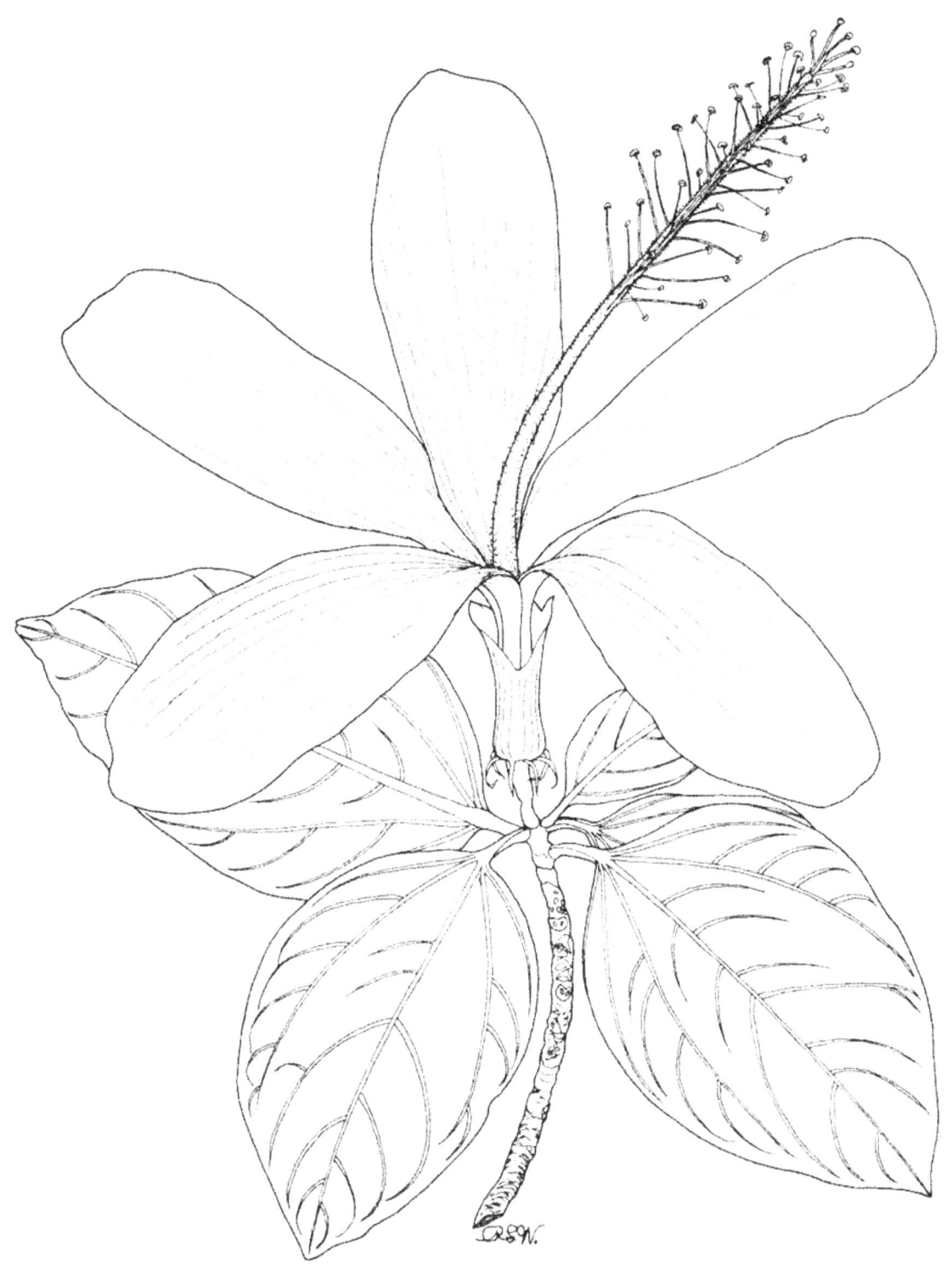

85. Koki'o ke'oke'o, native white hibiscus *Hibiscus arnottianus* Gray
Flowering twig, 1 X.

Special areas--Waimea Arboretum, Tantalus, Waiakea.

Champion--Height 82 ft (25.0 m), c.b.h. 8.4 ft (2.6 m), spread 61 ft 918.6 m). Lihue Sugar Co., Lihue, Kauai (1968).

Range--Native only in upland parts of Cuba and Jamaica.

87. Hau, sea hibiscus *(color plate, p. 37)*

Hibiscus tiliaceus L.

This common small tree of lowlands, especially shores, through the islands, is characterized by large funnel-shaped bright yellow flowers 3--3 1/2 in (7.5--8 cm) long and broad, usually with dark red "eye spot" inside, and by long-stalked heart-shaped and nearly round leaves with mostly 7 or 9 main veins from base, whitish gray hairy beneath.

Small evergreen native tree 13--33 ft (4--10 m) high, with short crooked trunk to 6 in (15 cm) in diameter and with broad crown of widely spreading or crooked branches, or a shrub with many prostrate branches forming dense thickets. Bark gray or light brown, smooth; inner bark fibrous. Twigs stout, with rings at nodes, becoming brown and hairless. Young twigs, leafstalks, lower leaf surfaces, calyx, and seed capsules densely covered with minute whitish gray star-shaped hairs.

Leaves alternate with leafstalks of 2--5 in (5--13 cm) and with 2 large short-pointed whitish hairy basal scales (stipules) 1--1 1/2 in (2.5--4 cm) long, shedding early and leaving a ring scar. Blades 4--7 in (10--18 cm) long and broad, sometimes larger, abruptly short- or long-pointed at apex and heart-shaped at base, rarely wavy toothed on edges, slightly thickened and leathery, shiny yellow green and hairless on upper surface, lower surface with 3 narrow glands near base of main veins.

Flower clusters (panicles) at or near ends of twigs, branching. Flowers many, few in each cluster, each with whitish hairy stalk of 3/4--2 in (2--5 cm) and gray green hairy basal cup (involucre) 3/4 in (2 cm) long usually with 9--10 narrow pointed lobes. Calyx 1--1 1/4 in (2.5--3 cm) long, gray green hairy, tubular with 5 narrow long-pointed lobes. Petals 5, yellow, usually with dark red spot at base inside, 2 1/2--3 1/2 in (6--9 cm) long, rounded but broader on one side, with tiny star-shaped hairs on outer surface, united at base. Stamens numerous on column about 2 in (5 cm) long united with corolla at base. Pistil has densely hairy conical 5-celled ovary, long slender style, and 5 broad stigmas. Flowers opening and closing same day, the petals withering and turning to orange and later to red.

Seed capsules elliptical, 1--1 1/4 in (2.5--3 cm) long, long-pointed, gray green hairy, splitting into 5 parts and breaking open the calyx and involucre which remains attached. Seeds, 3 from each cell, brownish black, 1/8--3/16 in (3--5 mm) long, hairless. Flowering and fruiting probably through the year.

Sapwood whitish and heartwood dark greenish brown. Wood moderately soft and porous, and moderately heavy (sp. gr. 0.6). It has been used sparingly by Hawaii's craftwood industry for carved and turned bowls and bracelets. The wood so used is mostly sapwood, so that the mottled dark heartwood inclusions give a marble-like appearance. Freshly cut wood has an odor similar to coconut.

Hawaiians used the wood for outriggers of canoes, floats for fish nets, long spears for games, and for cross sticks of kites. Fires were started by friction by rubbing a pointed stick of a hardwood such as No. 79, olomea (*Perrottetia sandwicensis*), against a grooved piece of the much softer hau wood. Hau is preferred by local Boy Scout troops for earning fire starting merit badges.

An important use of the tough fibrous inner bark, here and wherever this species grows, is for ropes and cords. Several long strips are braided together depending upon the strength needed. It was beaten into tapa or bark cloth and used for mats. The "grass" skirts exported for hula dancers from Samoa and elsewhere in the Pacific are actually made of hau fiber. The same material is used to strain the beverage kava in Samoa. Elsewhere, in times of famine the bark, roots, and young leaves were eaten. Flowers, roots, and bark served in folk remedies.

Hau is planted through the tropics as an ornamental for the showy flowers and as a shade tree. Branches can be trained over trellises to form arbors. Easily propagated by cuttings and started in fence rows as living fenceposts. The long spreading branches form roots upon contact with the ground, making dense thickets and in coastal swamps aiding in building the land. Classed as a honey plant. A weed in pastures, rangelands, and waste places (Haselwood and Motter 1966).

Common in lowlands and especially on beaches through the Hawaiian Islands to about 1,500 ft (457 m) altitude.

Special areas--Keahua, Waimea Arboretum, Tantalus, Haleakala, Volcanoes.

Range--Seashores through tropics, native probably in Old World. Common and widespread through Pacific Islands and regarded as native in Hawaii. Naturalized in

The common name mahoe is from the Spanish, majagua. Blue refers to the staining common to the wood of this tree.

Other common names--majó, emajagua excelsa (Puerto Rico); majagua, majagua azul (Cuba); blue mahoe (Jamaica).

Botanical synonyms--*Paritium elatum* (Sw.) G. Don, *Pariti tiliaceum* var. *elatum* (Sw.) Deg. & Greenwell.

86. Mahoe, Cuban-bast *Hibiscus elatus* Sw.*
Flowering twig, fruits (right), 2/3 X (P. R. v. 2).

New World, including Florida, Puerto Rico, and Virgin Islands.

Other common names--linden hibiscus, mahoe; emajagua (Puerto Rico, Spanish); pago (Guam, N. Marianas); ermall (Palau); gaal (Yap); kilife (Truk); kalau (Pohnpei); lo (Kosrae, Marshalls); fau (Am. Samoa).

Botanical synonyms--*Paritium tiliaceum* (L.) St.-Hil., Juss., & Camb., *Pariti tiliaceum* (L.) Britton.

Minor variations have been observed. Examples are forms with double flowers, white petals with maroon dot at base, and pure yellow petals. One variation introduced into Hawaii from Guadalcanal is an erect tree to 66 ft (20 m) tall.

The English name mahoe is a corruption of the Spanish common name majagua or emajagua. That American Indian word is applied in tropical America to several unrelated trees with useful fibrous bark.

Additional uses by the early Hawaiians as well as legends have been recorded by Degener (1930, p. 213--219).

88. Milo, portiatree *(color plate, p. 38)*

Thespesia populnea (L.) Soland. ex Correa**

Milo is common along shorelines through the islands and is also planted for ornament and shade, but it is said to have been introduced by the early Hawaiians. It is recognized by the large bell-shaped flowers, like hibiscus, with 5 overlapping pale yellow petals, single at leaf bases, by the dark gray, rounded but flattened, hard dry fruits that usually do not split open, and by the long-pointed shiny heart-shaped leaves usually with 7 main veins from base.

An evergreen medium-sized tree 20--30 ft (6--9 m) in height, with straight trunk 8--24 in (0.2--0.6 m) in diameter, and dense crown. Long spreading or nearly horizontal lower branches of crowded plants form dense thickets. Bark gray or light brown, smoothish or slightly fissured, becoming thick and rough. Inner bark yellowish, tough and fibrous. Twigs stout, green and covered with very small brown scales when young, becoming gray. Leafstalks, blades, flowerstalks, calyx, and fruits have scattered tiny brown scales also.

Leaves alternate, with long leafstalks of 2--4 in (5--10 cm). Blades heart-shaped, 4--8 in (10--20 cm) long and 2 1/2--5 in (6--13 cm) broad, long-pointed, not toothed on edges, slightly thickened and leathery, usually with 7 main veins from base, shiny dark green on upper surface, paler beneath, becoming nearly hairless.

Flowers single at leaf bases, opening 1 at a time, on stout stalks of 1/2--2 in (1.3--5 cm). Calyx cup-shaped, green, about 3/8 in (10 mm) high and 1/2 in (13 mm) across, remaining at base of fruit, with 3--5 narrow green scales (bracts) 1/2 in (13 mm) or more in length on outside, falling from bud. Petals 5, broad rounded oblique, 2 in (5 cm) or more in length, pale yellow, usually with maroon spot at base, with tiny star-shaped hairs on outer surface. Stamens many on column 1 in (2.5 cm) long joined at petals at base. Pistil has 5-celled ovary with slender style and 5 broader stigmas. Flowers opening and closing the same day, petals withering and turning to purple or pink.

Fruits (seed capsules) rounded but flattened, about 1 1/4 in (3 cm) in diameter and 3/4 in (2 cm) high, slightly 5-ridged, dark gray, hard, woody and dry, with calyx at base, usually remaining attached and not splitting open. Seeds several, elliptical, 3/8 in (1 cm) long, brown hairy. Flowering from early spring to late summer.

The beautiful wood of milo, with light brown sapwood and reddish brown to chocolate brown heartwood, takes a fine polish. It is moderately heavy (sp. gr. 0.6), easy to work, has a low shrinkage in drying and is durable. It is classed as resistant to attack from drywood termites and is used elsewhere in boatbuilding and cabinetwork. The Hawaiians carved it into beautiful bowls, such as calabashes for poi. It is presently an important craftwood, used for turned bowls and carved figures (tikis). The wood contains an oil that retards drying of oil-base varnishes but does not affect lacquers.

Elsewhere, rope has been made from the tough fibrous bark. It is reported that flowers and young leaves are mildly poisonous, though they also have been eaten. The seeds have been employed medicinally.

Several milo trees were planted around the house of King Kamehameha at Waikiki. In some Pacific Islands this species was regarded as sacred and was cultivated around temples.

Occasionally planted in the tropics as a street tree and ornamental, it produces dense shade and much leaf litter; it is also used as a living fence. In West Indian islands, where cotton is an important crop, this species is eradicated because it is a host of the cotton stainer, a red insect that stains fibers of growing cotton.

In Hawaii, common along sandy shores and borders of brackish marshes through the islands. Also planted around houses, formerly more than now. A particularly large tree is on the Ward Avenue side of Thomas Square in Honolulu. The trees are hardy in dry coastal areas but more common on windward shores where protected by reefs. The rounded but flattened dark brown seed capsules usually present in the upper crown distinguish milo from No. 87, hau, *Hibiscus tiliaceus*, which has elliptical long-pointed gray green seed capsules.

Special areas--Keahua, Waimea Arboretum, Haleakala, City, Volcanoes.

Champion--Height 42 ft (12.8 m), c.b.h. 9 ft (2.7 m), spread 69 ft (21 m). Kekaha, Kauai (1968).

Range--Widely distributed on tropical shores through the world, native in Old World. Transported by floating fruits and seeds. Naturalized elsewhere along

87. Hau, sea hibiscus *Hibiscus tiliaceus* L.
Flower, leaf, and seed capsule, 2/3 X (P. R. v. 1).

shores of southern Florida, through West Indies including Puerto Rico and Virgin Islands, and continental tropical America from Mexico to Brazil and Chile.

Other common name--seaside mahoe; emajagüilla, otaheita (Puerto Rico); haiti-haiti (Virgin Islands); kilulo (Guam); banalo (N. Marianas); badrirt (Palau); bangbeng (Yap); polo (Truk); pone (Pohnpei); panu (Kosrae); milo (Marshalls, Am. Samoa).

Common names in Puerto Rico and Virgin Islands are derived from Tahiti, a Pacific island where this species is native.

Some dissemination of this species from island to island through the tropics by ocean currents apparently is natural. Seeds in the lightweight fruits can germinate after floating a year in seawater.

CHOCOLATE FAMILY (STERCULIACEAE)

89. Melochia

Melochia umbellata (Houtt.) Stapf*

Introduced small weedlike tree of roadsides and waste places, to 50 ft (15 m) high and 6 in (15 cm) in trunk diameter, with large ovate, long-pointed, finely sawtoothed, soft hairy leaves. Bark smooth, light gray. Twigs stout, light green, turning brown. Twigs, stipules, leaves, flowers, and fruits with soft gray hairs, partly star-shaped. Buds large, of stipules folded together, the 2 large half round stipules less than 3/8 in (1 cm) long.

Leaves alternate, with long slender stalk 2 3/4--4 in (7--10 cm) long. Blades broadly ovate, large, 5--6 1/2 in (13--16.5 cm) long and 5 1/2 in (14 cm) wide, abruptly long-pointed, heart-shaped at base, finely sawtoothed, thin, soft hairy, dull green above and gray green beneath.

Flower clusters (panicles) erect from base of upper leaves, large, 4--6 in (10--15 cm) long, branched. Flowers many, few grouped at ends of short stalks, small, about 1/4 in (6 mm) long, 5-parted, composed of bell-shaped 5-toothed greenish to brownish calyx, 5 elliptical usually pink petals, 5 stamens on a tube, and pistil with deeply 5-angled ovary and 5 threadlike styles.

Fruit a small oblong capsule 5/16 in (8 mm) long, gray or brown, deeply 5-angled, 5-celled, splitting open, with calyx and petals at base. Seeds 2 in each cell, small, brown, about 1/4 in (6 mm) long, including long wing.

The sapwood is white and heartwood pale brown. A low-density hardwood (sp. gr. 0.5) with moderately large shrinkage in drying, it appears to have potential for use as a utility furniture species such as for drawer sides and backs. It has an attractive figure imparted by zones of darker color in the growth rings.

A weedy tree, reported as planted elsewhere for shade, because of its rapid growth. It is a honey plant and forms thickets in clearings and forest borders.

Introduced about 1925, it has been planted extensively on Hawaii and sparingly on Kauai and Oahu. It was one of the species included in aerial seeding of a burn on the Panaewa and lower Waiakea Forest Reserves near Hilo in 1928, which resulted in a large population there. Trees may be seen commonly along the volcano road near the turnoff to Kulani, along the Stainback Highway, and near Hilo Airport.

Special area--Waiakea.

Range--Native from India to New Guinea and Philippines. Introduced beyond in southern Asia and Pacific Islands.

Botanical synonym--*Melochia indica* (Gmel.) Kurz.

MANGOSTEEN FAMILY (GUTTIFERAE*)

90. Kamani

Calophyllum inophyllum L.**

This handsome tree found along or near seashores is identified by cream-colored resinous sap or latex in bark, paired thick and stiff, elliptical shiny dark green leaves with closely spaced parallel veins, white flowers clustered among leaves, and round ball fruits 1--1 1/2 in (2.5--4 cm) in diameter. Apparently introduced by the early Hawaiians, rather than native.

Medium-sized to large evergreen tree 40--60 ft (12--18 m) high and 1--1 1/2 ft (0.3--0.5 m) in trunk diameter or larger, with broad spreading crown of irregular branches. Bark light gray or brown, smoothish, becoming slightly cracked into shallow broad furrows and long flat ridges. Inner bark pink, fibrous, and bitter. Twigs stout, green, turning to brown, hairless, with cream-colored or light yellow resinous sap or latex. Bud about 1/4 in (6 mm) long, narrow and pointed, dark brown, composed of minute leaves without scales.

Leaves opposite in 4 vertical rows on twig, hairless, with light green leafstalks 5/8--3/4 in (1.5--2 cm) long, stout and flattened above. Blades elliptical, 4--8 in (10--20 cm) long and 2 1/2--3 1/2 in (6--9 cm) wide, slightly notched at rounded apex, rounded at base, slightly turned up from yellow green midvein and concave, with narrow whitish line along straight border;

88. Milo, portiatree   *Thespesia populnea* (L.) Soland. ex Correa**
Twig with flowers and fruits, 1 X (P. R. v. 1).

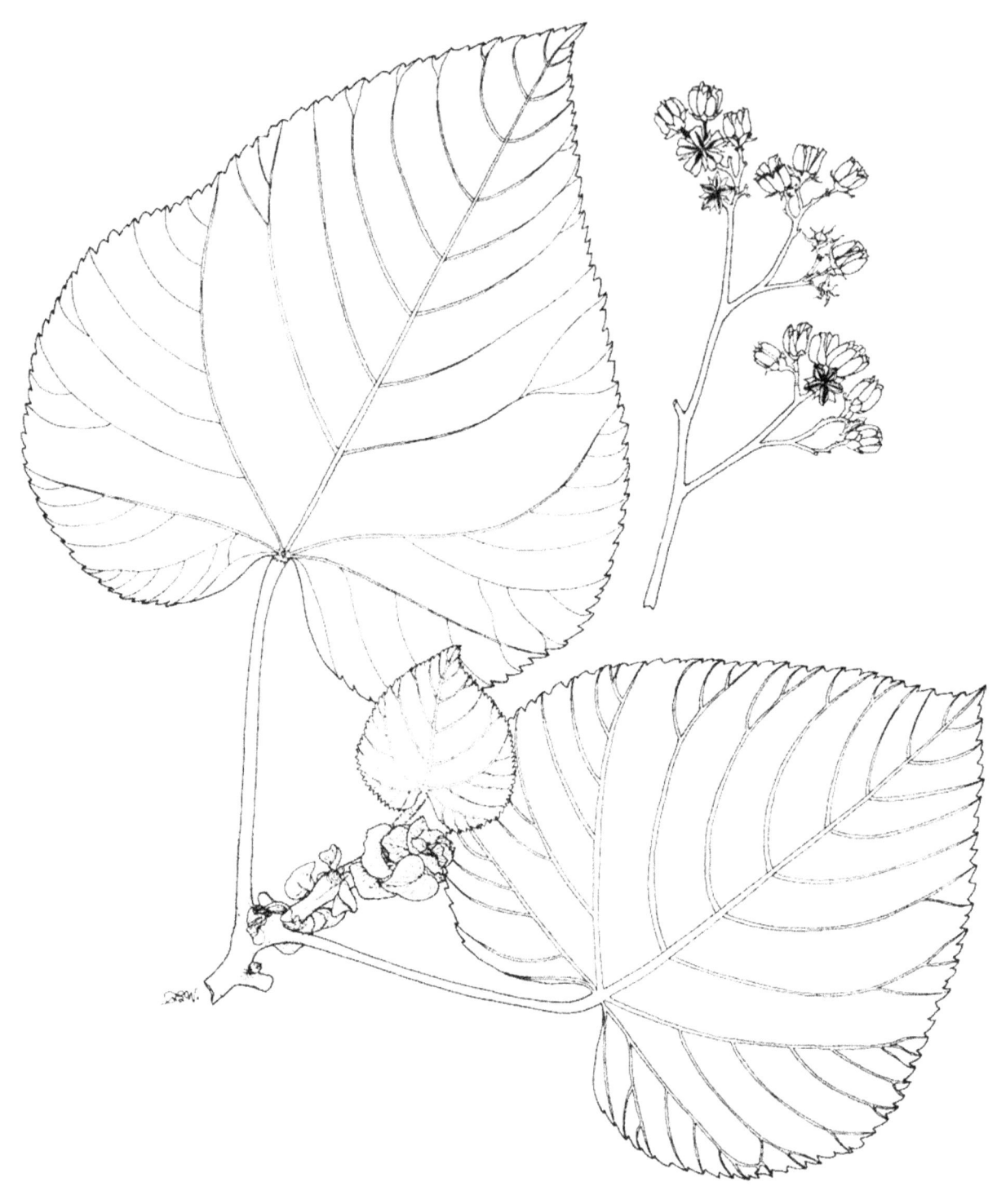

89. Melochia *Melochia umbellata* (Houtt.) Stapf*
Leafy twig and fruits (upper right), 2/3 X.

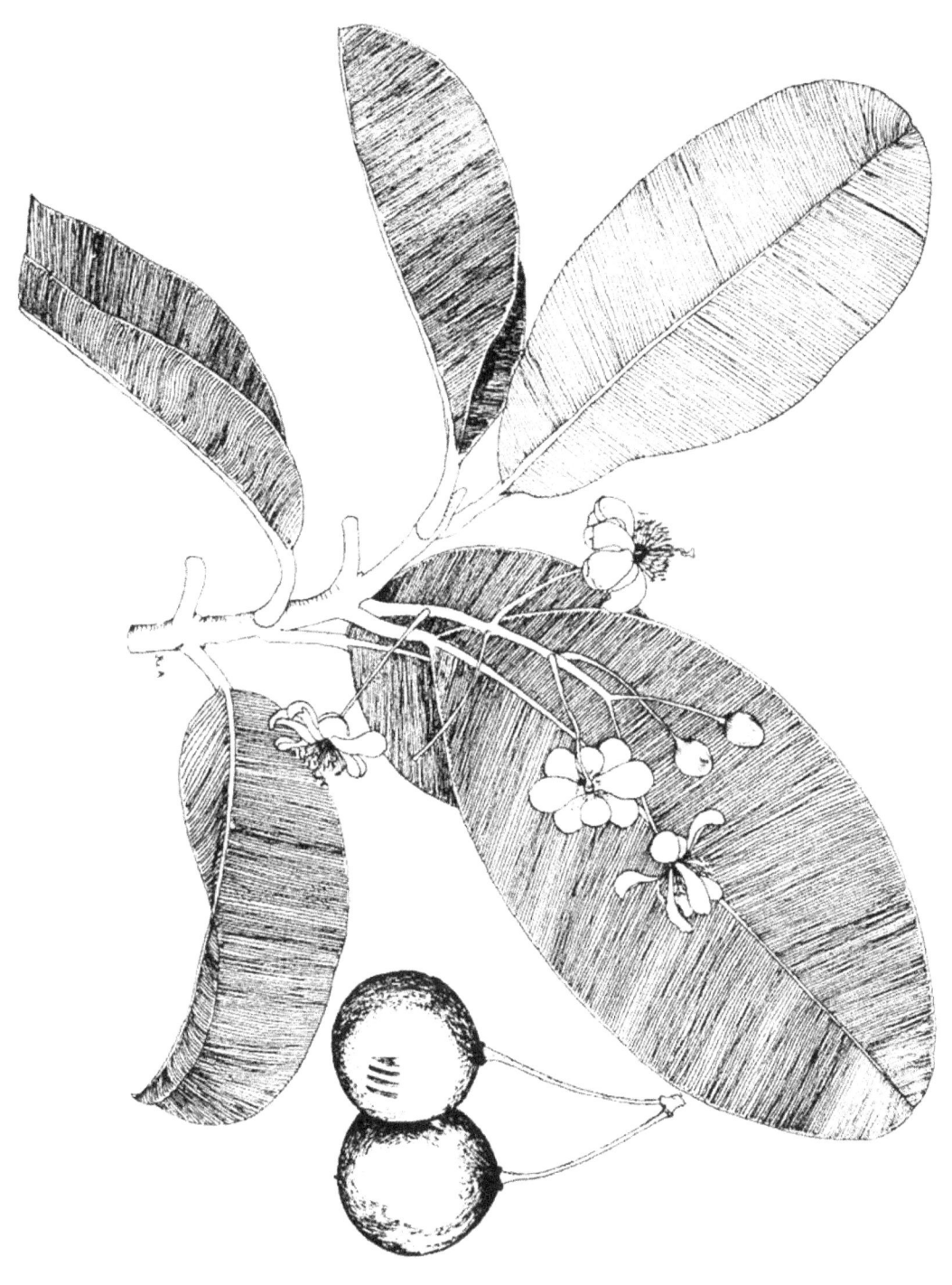

90. Kamani *Calophyllum inophyllum* L.**
Flowering twig, fruits (below), 2/3 X (P. R. v. 2).

thick and stiff, with numerous very fine straight parallel side veins scarcely visible until dried, upper surface shiny dark green, lower surface yellow green and slightly shiny.

Flower clusters (racemes) 2--6 in (5--15 cm) long at leaf base, with 4--15 fragrant white flowers about 1 in (25 mm) across on long stout stalks to 1 1/2 in (4 cm). Calyx of 4 rounded concave white sepals 1/4--3/8 in (6--10 mm) long, in 2 pairs; corolla of 4--8 elliptical to oblong concave spreading petals 1/2--5/8 in (13--15 mm) long; many stamens about 5/16 in (8 mm) long, with orange anthers and white filaments slightly united at base; and pistil with round red 1-celled ovary containing 1 ovule, slightly curved style, and disk stigma.

Fruits (drupes) few in cluster, round balls, light green, becoming yellow or brown, with thin pulp. Seed 1 large brown, round, 3/4--1 1/4 in (2--3 cm) in diameter, including shell and kernel.

The sapwood is white and heartwood reddish brown. When fresh cut, the heartwood is distinctly white and red. The wood has a pronounced figure on tangential faces imparted by parenchyma tissue and interlocked grain. It is of moderate density (sp. gr. 0.6), with a relatively large shrinkage in drying which can result in considerable warping of lumber. It is subject to tearouts in machining due to its irregular grain, but turns well except for a peculiar tendency to snag chisels. It is not resistant to decay or termites and is currently used in Hawaii only occasionally for carved and turned craftwood products. Elsewhere, it is used for general construction, cabinetmaking, boat-building, railroad crossties, and similar purposes. Hawaiians made bowls from the wood. The doors of the main floor of Iolani Palace in Honolulu have panels of kamani veneer.

The aromatic latex or resin has served in folk remedies and as tacamahaca resin of commerce.

The thick dark green oil extracted from the seeds has been employed medicinally. Formerly, it was used as ointment for skin and hair and for burning in lamps. Reportedly, the seeds are poisonous.

Regarded as sacred, this tree was grown around temples in the South Sea Islands. Planted in the tropics as an ornamental or shade tree for the dense shiny dark green foliage and fragrant flowers. The species is slow growing. It is a common street tree in Puerto Rico and has been introduced into southern Florida.

Large groves occur near the coasts through the Hawaiian Islands. One in Halawa Valley, Molokai, was noted by early explorers and apparently has been replaced by younger trees. This species is utilized as an ornamental along coasts because of its resistance to salt spray. It is planted along Dillingham Boulevard, near the Dole Pineapple Cannery, along Kailua Road, along Lunalilo Home Road at Koko Marina, and many other places on Oahu as a street tree. Particularly common between Isaac Hale Beach Park and Opihikao in the Puna District of Hawaii.

Special areas--Waimea Arboretum, Iolani.

Champion--Height 59 ft (18.0 m), c.b.h. 18.5 in (5.6 m), spread 81 ft (24.7 m). Malama-ki Forest Reserve, Puna, Hawaii (1968).

Range--Native of East Indian region of southeast Asia and Pacific islands but widely planted through the tropics.

Other common names--kamanu, Alexandrian-laurel, beautyleaf; daog (Guam, N. Marianas); btaches (Palau); biyuch (Yap); rakich (Truk); isou (Pohnpei); eet (Kosrae); lueg (Marshalls); fetau (Am. Samoa).

True kamani sometimes has been confused with an introduced tree of shores, No. 94, false kamani or tropical-almond, *Terminalia catappa*, with which it is usually planted. The two species are rather distinctive in leaf size, leaf color, and general appearance. A large kamani tree with a name tag grows in Iolani Palace grounds near the gazebo (bandstand). Next to it is a false kamani, also labeled.

FLACOURTIA FAMILY (FLACOURTIACEAE)

91. Maua, xylosma

Xylosma hawaiiense Seem.

Handsome small deciduous native tree mostly of dry forests, with shiny broadly elliptical leaves, edges straight or finely wavy toothed, and small greenish or reddish flowers, male and female on different trees. To 60 ft (18 m) tall and 1 1/2 ft (0.5 m) in trunk diameter, usually smaller and often only 15 ft (4.6 m) high, with spreading rounded crown of slightly drooping branches. Bark gray, smoothish, sometimes warty, becoming thick, rough, and furrowed into small scaly plates. Inner bark light yellow or orange within green outermost layer, bitter. Twigs hairless, dark red and slightly angled when young, becoming gray brown with raised dots and raised half-round leaf scars. End bud 1/8 in (3 mm) long, rounded, brown, scaly.

Leaves alternate in 2 rows, hairless, the young leaves reddish, bronze green, or copper-colored with red veins. Leafstalks slender, 3/8--3/4 in (1--2 cm) long, dark red. Blades broadly elliptical, 2--4 in (5--10 cm) long and 1 1/4--3 in (3--7.5 cm) wide, slightly thickened, apex short-pointed, blunt, rounded, or slightly notched, base

91. Maua, xylosma *Xylosma hawaiiense* Seem.
Fruiting twig (left), twig with male flowers (upper right), twig with female flowers (lower right), 1 X.

rounded to short-pointed, edges straight or finely wavy toothed, above shiny dark green, beneath slightly shiny green.

Flower clusters (racemes) at bases of new leaves or back of leaves, 1/2--1 in (13--25 mm) long, unbranched. Flowers mostly male and female on different trees (dioecious), several, greenish or reddish, about 1/4 in (6 mm) long and broad, without petals, on slender pinkish stalks. Male flowers with cuplike calyx of 4--5 rounded finely hairy or hairless sepals less than 1/8 in (3 mm) long and above a disk many threadlike spreading stamens 1/4 in (6 mm) long with dotlike anthers, sometimes with pistil. Female flowers with 4--5 sepals and pistil composed of elliptical 1-celled ovary, sometimes a short style, and 2--4 flattened stigmas.

Fruit (berry) rounded or elliptical, about 1/2 in (13 mm) long, bluish, blackish, or reddish, slightly shiny, with calyx at base and stigmas at apex, slightly fleshy or nearly dry, bitter, astringent, and not edible. Seeds 1--2, elliptical, about 1/4 in (6 mm) long, brown.

Wood is reddish brown with light and dark banding resembling growth rings, heavy, hard, brash, but easily worked.

Borer insects attack and kill the branches.

Widely distributed and common locally in dry forests through the islands, especially on leeward sides, at 800--4,500 ft (244--1,372 m) altitude. Windswept and stunted where exposed. Found on aa (rough) lava fields.

Special areas--Kokee, Volcanoes, Kipuka Puaulu.
Champion--Height 58 ft (17.7 m), c.b.h. 5.1 ft (1.6 m), spread 43 ft (13.1 m). Hoomau Ranch, Honomalino, Hawaii (1968).
Range--Hawaiian Islands only.
Botanical synonym--*Drypetes forbesii* Sherff.

MEZEREUM FAMILY (THYMELAEACEAE)

92. 'Ākia
Wikstroemia oahuensis (Gray) Rock

Plants of 'ākia, genus *Wikstroemia*, are easily recognized, though further identification of the species, about 12, found through the Hawaiian Islands is not so easy. Most species are evergreen shrubs, and only a few reach tree size. They have blackish or gray, often reddish brown, very tough bark with strong fibers, which served early Hawaiians as rope. These reputedly poisonous plants formerly were pounded into pulp and thrown into water to stupefy fish to aid in their capture. Plants of 'ākia are known by the mostly small narrow paired leaves, slender very tough twigs with strong fibers difficult to break, with raised triangular leaf-scars, small narrowly tubular 4-lobed greenish to yellow fragrant flowers, and small orange stone fruits.

This species varies from a low shrub of 2--4 ft (0.6--1.2 m) to a small tree to 25 ft (7.6 m) high and 6 in (15 cm) in trunk diameter. Bark dark gray or blackish, smoothish, very tough, fibrous, and bitter. Twigs paired and partly forking by 2, light green when young, later dark brown, almost hairless, with raised triangular leaf-scars and enlarged nodes.

Leaves opposite, hairless, with short light green leafstalks of less than 3/8 in (1 cm) long. Blades narrowly elliptical, 1 1/4--2 3/4 in (3--7 cm) long and 1/2--1 in (1.3--2.5 cm) wide, short-pointed at both ends, not toothed, slightly thick and leathery, upper surface dull green with side veins fine and inconspicuous, and lower surface dull light green.

Flower clusters (like umbels) terminal, with short stalk less than 1/4 in (6 mm) long, which is curved down and persistent. Flowers several, short-stalked, very narrow, 3/8 in (10 mm) long and less than 1/8 in (3 mm) wide, light greenish yellow, without corolla, composed of tube (hypanthium), calyx with 4 spreading lobes turned back, 8 tiny stamens, 4 within tube and 4 in the throat opposite lobes, and pistil with elliptical 1-celled ovary, very short style, and larger round stigma.

Fruits (drupes) 1 or 2, oblong or egg-shaped, 1/2 in (13 mm) long and 5/16 in (8 mm) in diameter, enclosed by tube, with stigma at apex, from light yellow to orange, fleshy, bitter. Seed 1, egg-shaped, 5/16 in (8 mm) long, pointed, blackish.

The wood of another species (*W. sandwicensis* Meisn.) is whitish and very soft, subject to blue stain.

Scattered in understory of wet forests of Oahu, for example, Niu Valley, Mt. Konahuanui, and Poamoho Ridge, to 4,600 ft (1,402 m) altitude).

Range--Kauai, Oahu, Molokai, Lanai, Maui.
Other common names--asasa, false 'ohelo.
Botanical synonyms--*Wikstroemia basicordata* Skottsb., *W. degeneri* Skottsb., *W. elongata* Gray, *W. eugenioides* Skottsb., *W. haleakalensis* Skottsb., *W. isae* Skottsb., *W. lanaiensis* Skottsb., *W. leptantha* Skottsb., *W. macrosiphon* Skottsb., *W. palustris* Hochr.; *W. recurva* (Hillebr.) Skottsb., *W. sellingii* Skottsb., *W. vaccinifolia* Skottsb.

Plants of 'ākia are extremely poisonous if eaten, according to Degener (1930), though harmless to the touch. Root and bark were ingredients of a deadly drink for suicide or for execution of criminals by order of a chief. However, Arnold (1944) concluded that plants of this genus were not as poisonous as commonly reputed and possibly might not be toxic to humans. Baldwin

92. ʻĀkia   *Wikstroemia oahuensis* (Gray) Rock
Flowering twig (above), fruiting twig (below), 1 X.

(1979) also reported questionable toxicity, but recommended that the fruits not be eaten.

The narcotic substance of this genus and a few unrelated plants served also as a fish poison. The bark and leaves were pounded into powder with stones on rocks near a tidal pool or stream. This powder was placed in double handfuls in the fibrous sheath of a coconut leaf or in a twisted bunch of grass. Then the material was quickly inserted under a rock or in crevices where fish were expected. Within 10 minutes, the fish would swim about aimlessly or float on their sides and could be caught easily. Fortunately, fish so caught were edible.

MANGROVE FAMILY (RHIZOPHORACEAE*)

93. Mangrove

Rhizophora mangle L.*

This species naturalized on protected muddy seashores is easily recognized by the mass of peculiar branching curved and arching stilt roots that enable the trees to spread in shallow salt and brackish water and form dense, impenetrable thickets at tide level. Each fruit supports an attached odd cigarlike seedling that elongates hanging down.

Small evergreen tree to 33 ft (10 m) high and 8 in (0.2 m) in trunk diameter, reported to reach a height of 75 ft (23 m) elsewhere. Bark gray or gray brown, smooth and thin on small trunks, becoming furrowed and thick on larger ones. Inner bark reddish or pinkish, with slightly bitter and salty taste. Twigs stout, gray or brown, hairless, ending in conspicuous narrow pointed green bud 1--2 in (2.5--5 cm) long, covered with 2 green scales (stipules) around pair of developing leaves and making ring scar around twig when shedding.

Leaves opposite, crowded at end of twig, hairless, with slightly flattened leafstalks of 1/2--7/8 in (13--22 mm). Blades elliptical, 2 1/2--4 in (6--10 cm) long, blunt-pointed at apex and short-pointed at base, slightly rolled under at edges, slightly leathery and fleshy with side veins not visible, shiny green above, yellow green beneath.

Flowers usually 2--4 together at leaf base on forked green stalk altogether 1 1/2--3 in (4--7.5 cm) long, slightly fragrant, pale yellow, about 3/4 in (2 cm) across. The bell-shaped pale yellow base (hypanthium) less than 1/4 in (6 mm) long bears 4 widely spreading narrow pale yellow sepals almost 1/2 in (13 mm) long, leathery and persistent; 4 narrow petals 3/8 in (10 mm) long, curved downward, whitish but turning brown, white woolly or cottony on inner side; 8 stamens; and pistil of 2-celled ovary mostly inferior but conical at apex, with 2 ovules in each cell, slender style, and 2-lobed stigma.

Fruits dark brown, conical, about 1 1/4 in (3 cm) long and 1/2 in (13 mm) in diameter, with enlarged curved sepals, remaining attached. The single seed germinates inside fruit, forming long narrow first root (radicle) green except for brown enlarged and pointed end, to 1/2 in (13 mm) in diameter. When about 8--12 in (20--30 cm) long, the heavy seedling falls into the mud or water. It may be carried by water and ocean currents before becoming firmly rooted. Flowering and fruiting continue through the year.

The sapwood is light brown, heartwood reddish brown or dark brown. The wood is hard, very heavy (sp. gr. 0.9--1.2), durable in the soil but susceptible to attack by dry-wood termites.

Elsewhere, it is used as roundwood for posts and poles and is excellent for fuel and charcoal. Wood in larger sizes has been employed also for marine piling and wharves, shipbuilding, and in cabinetwork. The bark is important commercially in tanning leather, and the leaves are rich in tannin also. A dye and medicines have been obtained from the bark. Fishermen in Puerto Rico preserve their lines with an extract from the roots.

Mangrove forests on depositing shores aid in extending the shoreline, holding the black mud in place and gradually advancing on the side toward the ocean. Where native, this species with its stilt roots growing in shallow water extends farther seaward than mangroves of a few other plant families.

Planted and naturalized in salt marshes of Kauai, Oahu, Molokai, Lanai, and Hawaii. Trees can be seen from the bridge at Heeia, as well as the shoreline reefs of Kaneohe Bay and Leeward Molokai.

Champion--Height 61 ft (18.6 m), c.b.h. 3.7 ft (1.1 m), spread 23 ft (7.0 m). Kohana-iki, Kailua-Kona, Hawaii (1968).

Range--Widely distributed on silt shores of Atlantic Coast of tropical America from Florida, Bermuda, and Bahamas through West Indies and from northeastern Mexico to Brazil. The same or closely related species also on Pacific Coast from northwestern Mexico to Peru, on coasts of western Africa, and in Melanesia and Polynesia.

Other common names--red mangrove, common mangrove, American mangrove; mangle, mangle colorado (Puerto Rico, Spanish).

Botanical synonyms--*Rhizophora mangle* var. *samoensis* Hochr., *R. samoensis* (Hochr.) Salvoza.

Where necessary to distinguish other unrelated mangroves of seashores, this species is known as red mangrove.

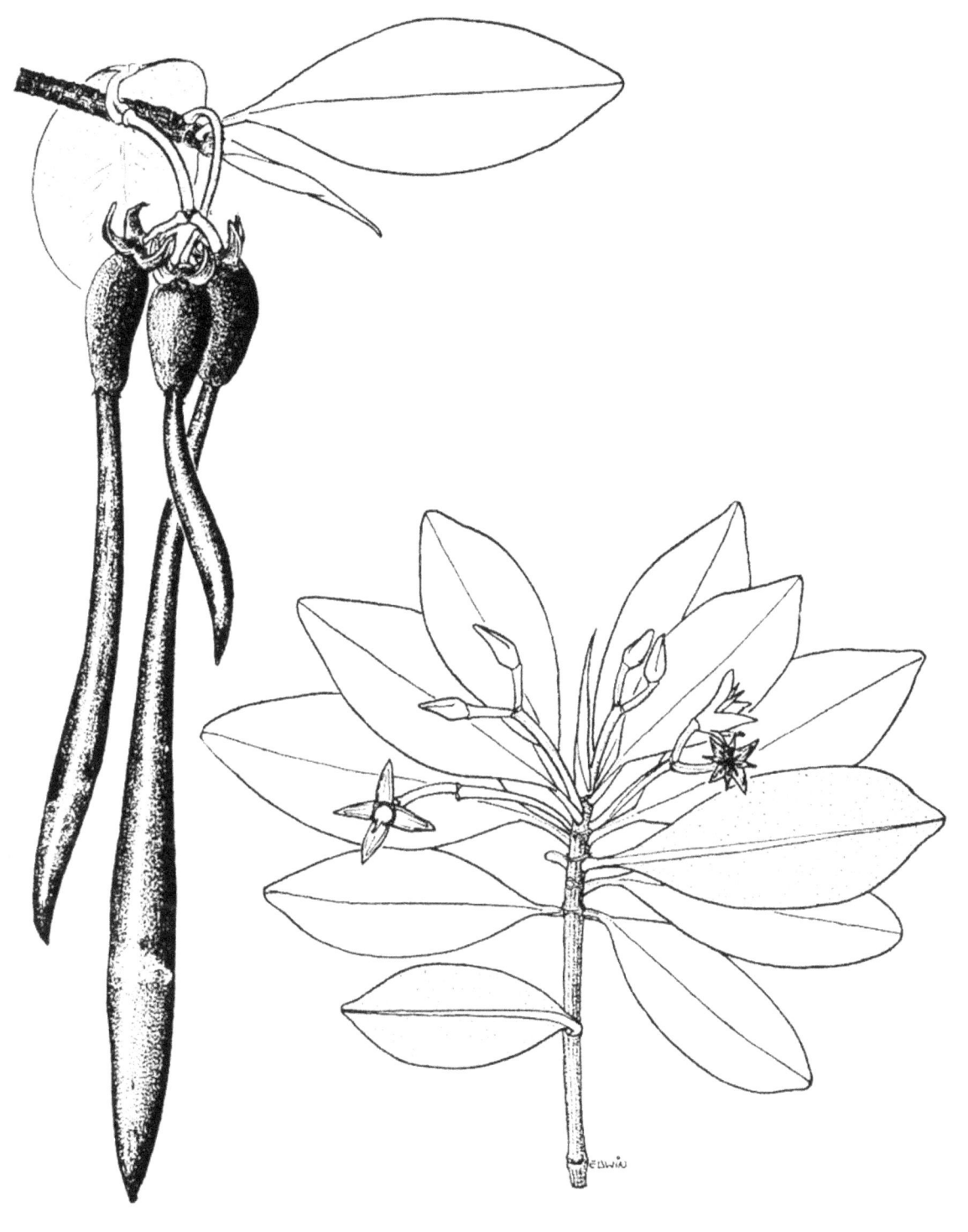

93. Mangrove *Rhizophora mangle* L.*
Fruiting twig (left), flowering twig (right), 2/3 X (P. R. v. 1).

Red mangrove was first introduced to Hawaii in 1902, according to Degener (1933-1986), to hold the soil in the mudflats of southwest Molokai. It has become thoroughly naturalized there. Twenty years later, this and other mangroves were planted in salt marshes of Oahu. Classed also as a weed, often spreading into fish ponds, where eradication is difficult (Hasselwood and Motter 1966).

COMBRETUM FAMILY (COMBRETACEAE*)

94. Tropical-almond, false kamani*Terminalia catappa* L.*

This tree from the East Indies is widely planted for shade, ornament, and nuts around the world, especially along tropical sandy seashores, and has become naturalized. It is characterized by horizontal branches in circles or tiers at different levels on the trunk, large leathery leaves broadest toward apex (obovate), turning reddish before falling, and the elliptic slightly flattened greenish or reddish fruits about 2 in (5 cm) long containing a large edible seed or nut inside the hard husk.

Usually a small to medium-sized tree 30--50 ft (9--15 m) high and 1 ft (0.3 m) in trunk diameter, but sometimes much larger in diameter and with slight buttresses, evergreen except in areas with a marked dry season. Bark gray, smoothish, thin, becoming slightly fissured. Inner bark pinkish brown, slightly bitter and astringent. Twigs brown, finely hairy when young, slender but swollen at leaf scars and nodes.

Leaves alternate but crowded near ends of twigs, with stout finely brown hairy leafstalks of 3/8--3/4 in (1--2 cm). Blades 6--11 in (15--28 cm) long and 3 1/2--6 in (9--15 cm) broad, abruptly short-pointed or rounded at apex and gradually narrowed toward rounded base, not toothed on edges, slightly thickened, upper surface shiny green or dark green and hairless, and lower surface paler and often finely brown hairy.

Flower clusters (narrow racemes) at leaf bases are 2--6 in (5--15 cm) long. Flowers numerous small greenish white, 3/16--1/4 in (5--6 mm) across, mostly short-stalked, with slightly unpleasant odor, mostly male and a few bisexual flowers near base (polygamous). Both kinds have a greenish white or light brown hairy calyx with cup-shaped tube and 5 or 6 pointed spreading lobes 1/16 in (1.5 mm) long and bearing twice as many small stamens near base. Bisexual and female flowers, which are stalkless, have a slender style and a narrow basal tube (hypanthium) 3/16 in (5 mm) long, brownish green and finely hairy, resembling a stalk but containing the inferior 1-celled ovary.

Fruits (drupes) about 1 in (2.5 cm) broad, pointed, slightly flattened and with 1 or 2 narrowly winged edges, light brown at maturity. The thin outer layer is slightly sour and can be eaten. Inside the hard fibrous husk there is a light brown, thick hard stone containing an oily seed or nut about 1 1/4 in (3 cm) long and 3/8 in (1 cm) broad, somewhat like the true almond. Flowering and fruiting nearly through the year.

The heartwood is reddish brown when first cut, becoming pale brown with age and has a subdued figure imparted by dark banding at the terminus of each growth ring. The sapwood is lighter in color. The wood is hard, moderately heavy (sp. gr. 0.59), moderately strong, tough, medium to coarse-textured, and with irregular and often interlocked grain. It is very susceptible to attack by dry-wood termites. Rate of air-seasoning is rapid, and amount of degrade is moderate. Machining characteristics are as follows: planing is very poor; shaping, boring, and mortising are fair; turning is poor; and sanding and resistance to screw splitting are good.

The wood has been used occasionally as a substitute of No. 90, kamani (*Calophyllum inophyllum*), in Hawaii's craftwood trade. However, it is considered inferior in appearance and is very susceptible to attack by lyctus and ambrosia beetles when stored in large pieces awaiting carving. Common uses elsewhere are for posts and fuel. This attractive wood if carefully handled in machining would be suitable for millwork, furniture, veneer, and cabinetwork. It has been recommended for boatbuilding, general construction, bridge timbers, crossties, flooring, and boxes and crates.

The bark, roots, astringent green fruits, and leaves contain tannin and have been used in tanning. A black dye serving for ink has been obtained from bark, fruits, and foliage also. An oil has been extracted from the seeds.

This species is extensively planted and is naturalized along seashores, being hardy and salt tolerant, though reportedly not resistant to hurricanes or strong winds. It is an attractive roadside tree for its peculiar branching and reddish tinged old leaves and grows rapidly.

Introduced very early in Hawaii, probably before 1800, and now naturalized at low altitudes, mainly near shores.

Special areas--Waimea Arboretum, Foster, Iolani.

Champion--Height 45 ft (13.7 m), c.b.h. 21.3 ft (6.5 m), spread 76 ft (23.2 m). Haili Church, Hilo, Hawaii (1968).

Range--Native of East Indies and Oceania but planted and naturalized through the tropics, including southern Florida and Puerto Rico and Virgin Islands.

Other common names--kamani haole, kamani'ula, umbrella-tree (Hawaii); India-almond (continental Unit-

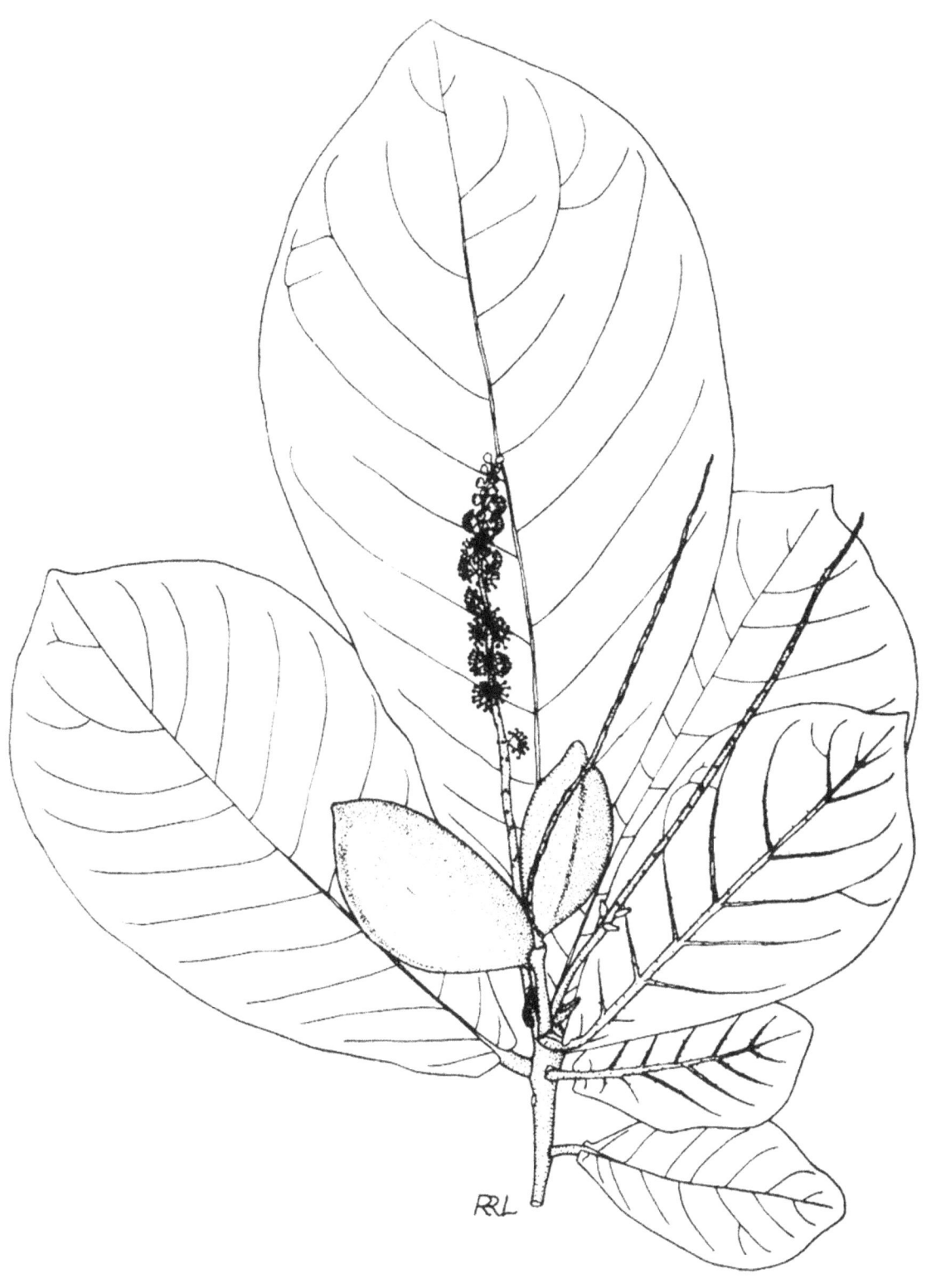

94. Tropical-almond, false kamani *Terminalia catappa* L.*
Twig with flowers and fruits, 2/3 X (P. R. v. 1).

ed States); almendro (Puerto Rico, Spanish); almond, West-Indian-almond (Virgin Islands); talisai (Guam, N. Marianas); miiche (Palau); kel (Yap); as (Truk); thipwopu (Pohnpei); srofaf (Kosrae); kotal (Marshalls); talie (Am. Samoa).

95. Jhalna

Terminalia myriocarpa Heurck & Muell.-Arg.*

Large evergreen tree introduced for forestry tests, characterized by paired large narrowly elliptical toothed leaves with 1 or 2 glands on leafstalk, many small light yellow flowers in large slender drooping clusters, and 2-winged pinkish tinged yellow fruits. To 80 ft (24 m) high, the trunk 2 ft (0.6 m) in diameter, enlarged at base with slight buttresses and surface roots. Bark gray, smoothish. Twigs light green, turning brown, finely hairy when young, becoming hairless.

Leaves paired (opposite), leafstalks 3/8 in (1 cm) long, light green, with 1 or 2 rounded greenish glands near apex. Blades narrowly elliptical, 4--7 1/2 in (10--18 cm) long and 1 3/4--3 in (4.5--7.5 cm) wide, long-pointed at apex, rounded or short-pointed at base, finely wavy toothed, thin, becoming hairless, above slightly shiny green with many slightly curved parallel side veins, beneath dull light green.

Flower clusters (panicles) very large, 12 in (30 cm) or more in length, terminal, with many long spreading drooping slender, slightly flattened branches. Flowers many, male and bisexual in same cluster (polygamous), small, 1/8 in (3 mm) long, with cup-shaped light yellow 5-toothed calyx. Male flowers short-stalked, with 10 stamens. Bisexual flowers with short stamens, stalklike base (hypanthium) finely hairy and containing inferior ovary, style, and dot stigma.

Fruits (drupes) many, crowded and stalkless on long slender branches, 1/8 in (3 mm) long, with narrow finely hairy body and 2 broad pinkish tinged yellow wings more than 1/2 in (13 mm) across.

The heartwood is pale brown, with darker brown zones at the end of each growth ring giving the wood an attractive figure. This heavy hard wood takes an excellent finish and is widely used in its native Assam for furniture and general construction. Wood is not durable but very permeable to preservatives and so easily treated.

The first recorded plantings of this tree were in 1928 by the Division of Forestry on Oahu and Kauai. Between then and 1960, more than 26,000 trees were planted in the forest reserves on all islands. The trees have shown strong response to site and are quite variable in size. The most impressive planting is at Kalopa State Park on Hawaii. In other locations, the trees are often much smaller. Best suited to sites with rainfall of 100--150 in (2,540--3,810 mm) annually at 1,000--3,000 ft (305--914 m) elevation.

Special areas--Pepeekeo, Kalopa.
Champion--Height 125 ft (38.1 m), c.b.h. 19.9 ft (6.1 m), spread 110 ft (33.5 m). Kohala Forest Reserve, Muliwai, Hawaii (1968).
Range--Native of India and from Nepal eastward through Assam to Burma.
Other common name--hollock (Assam, India).

MYRTLE FAMILY (MYRTACEAE)

96. Lanceleaf gum-myrtle

Angophora costata Domin*

Large evergreen tree introduced in forest plantations, related to *Eucalyptus* but differing in the leaves paired (opposite) and in the 5 separate white petals. To 80 ft (24 m) high and 2 ft (0.6 m) in trunk diameter. Bark gray smooth, flaking off and becoming slightly mottled and gray brown, showy, becoming brown and scaly on very large trunks. Branches light gray, smooth to slightly scaly. Twigs long slender drooping, brown, hairless.

Leaves opposite or mostly so, hairless, drooping on slender leafstalks 1/2--3/4 in (13--20 mm) long, grooved above. Blades lance-shaped, 4--6 in (10--15 cm) long, 1--1 1/2 in (2.5--4 cm) wide, short-pointed at base and tapering to a very long point slightly curved to one side, slightly turned under at edges, slightly thick, dull light green with pale yellow midvein and many very fine straight parallel side veins.

Flower clusters (panicles) terminal, large, branched. Flowers many, crowded, with short hairy stalks less than 1/4 in (6 mm) long, composed of conical basal cup (hypanthium) 1/4 in (6 mm) long and broad, with 5 ridges and 5 teeth corresponding to calyx, with fine gray hairs and gland hairs; 5 rounded white petals 1/8 in (3 mm) long on rim of cup; many white threadlike stamens 3/8 in (10 mm) long spreading from rim about 3/4 in (2 cm) across; and pistil with inferior ovary, slender style, and dotlike stigma.

Fruits cuplike or like capsules, rounded, about 1/2 in (13 mm) in diameter, with 10 long ridges, blackish, hard and thick-walled, opening by a lid, 3-celled.

Wood light brown or gray, very hard, very heavy (sp. gr. 0.9), would be suitable for heavy timbers except that it contains many gum veins (gum-filled shakelike

95. Jhalna *Terminalia myriocarpa* Heurck & Muell.-Arg.*
Twig with flowers and fruits, 2/3 X.

separations in the wood), which make it essentially useless except as pulpwood or fuelwood.

The Division of Forestry has planted almost 20,000 trees in the forest reserves, primarily on Oahu and Maui, but also on Kauai and Hawaii. On Oahu there is a stand shortly beyond the start of Trail Number 2, Manoa Cliffs Trail on Tantalus Mountain. This species is also extensively planted on the land of Waiawa and Waiau above Pacific Palisades and Waimano Home. On Maui it is growing well on Upper Borge Ridge in the Makawao Forest Reserve. The tree has showy bark with a striking resemblance to Eucalyptus maculata and has potential as an ornamental.

Special areas--Aiea, Tantalus.

Champion--Height 132 ft (40.3 m), c.b.h. 9.6 ft (2.9 m), spread 54 ft (16.5 m). Honaunau Forest Reserve, Kailua-Kona, Hawaii (1968).

Range--Native of Australia.

Botanical synonym--Angophora lanceolata Cav.

97-109. Eucalyptus, eucalypt

Eucalyptus*

The genus Eucalyptus is probably the best known of all Hawaii's introduced forest trees. Recognition of the group is easy, but further identification is less simple, because of the very large number of species. This generic description is followed by a key, based mainly on bark and other vegetative characters. Thirteen species are described and illustrated as numbers 97--109.

Eucalyptus contains about 500 species of trees, nearly all confined to Australia (there are a few in Indonesia to the Philippines). Among them are very large trees and the world's tallest hardwoods, up to 318 ft (97 m) in height. Because of their large size and rapid growth, many species of eucalyptus have been propagated widely in forest plantations through tropical and subtropical regions. Wood quality varies greatly within the genus.

The generic name from two Greek words meaning "well" and "covered" refers to the distinctive character of the flower, the lid (operculum). That part, usually conical and often thick, covers the flower in bud and apparently represents the calyx. (The corolla is lacking or united with calyx.) The cup-shaped or bell-shaped base (hypanthium) of the open flower bears around the rim numerous spreading threadlike white, yellow, and sometimes red stamens and encloses the pistil with inferior ovary mostly 3--5-celled and slender style. Flowers are borne mostly in stalked clusters (umbels) commonly at base of leaves.

The fruit is a seed capsule slightly enlarged from flower base, hard and woody, opening at top by as many pores as cells and often with as many projecting pointed teeth (valves). There are many tiny seeds including a larger number very small and nonfunctional.

The leaves vary in shape with age of plant, from juvenile, commonly opposite, stalkless and broad, to adult, alternate, stalked, and often narrow. Foliage commonly is aromatic or resinous, and trunks of some species yield a gum or resin.

Differences in bark character provide groupings of species useful in identification. The gums, or most common group, have smooth thin bark that peels off in patches or layers. The stringy-barks have thick fibrous bark, while the iron-barks have thick hard furrowed bark that is often black. Bloodwoods have uniform scaly bark.

The 13 species of Eucalyptus in this handbook include the most widely tested and most successful in forest plantations in Hawaii. However, many other species have been introduced Bryan and Walker (1966) listed 77, but the total count is now more than 90.

Several technical publications on this genus in Australia may be consulted for reference and further identification. Titles by the following authors are listed under Selected References: Hall et al. (1975) (contains key to most species listed); Blakely (1965), Johnston and Marryatt (1965), Kelly (1959), Maiden (1902-24; 1903-33) (source of 3 drawings reproduced here); Mueller (1879-84) (the classic 10-volume monograph and source of 7 drawings reproduced here); Penfold and Willis (1961), Pryor (1976), Pryor and Johnson (1971).

96. Lanceleaf gum-myrtle   *Angophora costata* Domin*
Twig with flowers and fruits, 1 X (Maiden).

KEY TO SPECIES OF EUCALYPTUS

BARK ON TRUNK MOSTLY SMOOTH AND PEELING IN LARGE PIECES

    Trunk smooth or becoming smooth

        Bark bright green with pink or red brown blotches where recently peeled--100. Bagras eucalyptus, *Eucalyptus deglupta*

        Bark gray, dimpled; leaves with strong odor of lemon--99. lemongum eucalyptus, *E. citriodora*

    Trunk with some bark rough or peeling

        Lower trunk short to first branches, crooked; basal bark thin, strips or plates of thin old bark peeling above

            Old bark shed in long strips, partly persistent, smooth bark bluish gray, usually twisted grain in trunk; fruit top-shaped--101. bluegum eucalyptus, *E. globulus*

            Old bark shed in plates, rarely persistent, smooth bark white and brown in patches; fruit with prominent valves raised and curved inward (see figure)--98. river-redgum eucalyptus, *E. camaldulensis*

        Lower trunk long to first branches, straight; basal bark thick, rough, trunk smooth bluish or greenish gray above

            Peeling bark often present in patches on upper trunk, "stocking" of rough bark often more than 10 ft (3 m) high on trunk; fruit with valves slightly curved inward whether flush or protruding; seedlings usually without lignotubers; juvenile leaves lance-shaped, wide at base--102. rosegum eucalyptus, *E. grandis*

            Peeling bark rarely any patches on upper trunk, "stocking" of rough bark less than 8 ft (2.5 m) high; fruit with valves spreading or curved out whether flush or protruding; seedlings usually with lignotubers; juvenile leaves narrow, linear--108. saligna eucalyptus, *E. saligna*

BARK ON TRUNK MOSTLY ROUGH AND FISSURED

    Rough bark absent from branches and sometimes upper trunk

        Rough bark gray brown, finely fissured, stringy--105. blackbutt eucalyptus, *E. pilularis*

        Rough bark brown, deeply and coarsely fissured, soft, breaking away in chunks--97. bangalay eucalyptus, *E. botryoides*

    Rough bark persistent on both trunk and branches

        Bark hard, difficult to depress with hand, containing hard gum deposits

            Bark black or nearly so, trunk erect--109. red-ironbark eucalyptus, *E. sideroxylon*

            Bark brown or gray brown, trunk crooked--104. gray-ironbark eucalyptus, *E. paniculata*

        Bark soft, easily pressed in with hand

            Bark long-fibered, pulling away in strings, stringy

97. Bangalay eucalyptus   *Eucalyptus botryoides* Sm.*
Twig with flowers and fruits, 1 X (Mueller).

Bark red on inner surface when pulled away--106. kinogum eucalyptus, *E. resinifera*

Bark yellow on inner surface when pulled away--103. tallowwood eucalyptus, *E. microcorys*

Bark short-fibered, pulling away in short chunks

Bark extremely soft and deeply fissured, spongy, reddish brown; leaves broader than most other eucalypts; fruit vase-shaped with stalk; flower bud with beaked lid (operculum)--107. robusta eucalyptus, *E. robusta*

Bark soft and less deeply fissured, pulled loose with difficulty, brown or gray brown; fruit cylindrical and stalkless; flower bud with blunt or rounded lid (operculum)--97. bangalay eucalyptus, *E. botryoides*

97. Bangalay eucalyptus

Eucalyptus botryoides Sm.*

Characterized by very rough and thick deeply furrowed bark and by several bell-shaped seed capsules clustered at end of short stalk at leaf base. The bark is similar to that of No. 107, robusta eucalyptus, *E. robusta*, Hawaii's most common eucalypt, but more finely fissured and grayer in color.

A large tree 80--150 ft (24--45 m) high and 2--3 ft (0.6--0.9 m) in trunk diameter in closed forests. Crown open, spreading, irregular. Bark gray or brown, very rough and thick, deeply furrowed into long narrow ridges and becoming slightly shaggy; on smaller branches smooth, green to light brown or salmon pink. Twigs slender, angled, greenish yellow.

Leaves alternate, with flattened greenish yellow leafstalks of 1/2--3/4 in (13--15 mm). Blades lance-shaped, 4--6 in (10--15 cm) long and 1--2 1/2 in (2.5--6 cm) wide, curved, tapering to long point, short-pointed at base, moderately thick and leathery, with fine straight parallel side veins, slightly shiny green above, dull light green beneath, horizontal. Juvenile leaves opposite for 3--4 pairs, then alternate, stalked, lance-shaped, 2--3.5 in (5--9 cm) long and 1--1 1/4 in (2.5--3 cm) wide, thin and wavy, green, paler beneath.

Flower clusters (umbels) less than 1 in (2.5 cm) long on short flattened stalk of 3/8 in (1 cm) at leaf base. Flowers 6--10, stalkless, more than 1/2 in (13 mm) across the many spreading white stamens nearly 3/8 in (10 mm) long. Buds cylindrical, 3/8--1/2 in (10--13 mm) long and 1/4 in (6 mm) wide, lid half-round or blunt, often with small rounded point.

Seed capsules 5--7 clustered, stalkless, bell-shaped or cylindrical, about 3/8 in (10 mm) long and 1/4 in (6 mm) wide, light brown, with sunken disk and 3--5 short valves at rim level.

Sapwood is cream-colored and heartwood reddish brown. The wood is medium to coarse textured with interlocked grain and color variation in the growth rings providing a prominent figure on flat-sawn faces. It is very heavy (sp. gr. 0.9) with a large shrinkage in drying, very strong if from the outer part of logs, otherwise very brash. The wood has been used only for fenceposts in Hawaii. Its density, coloration, and probable poor seasoning characteristics suggest that it would be marginal for other uses than fuelwood and fenceposts.

The bulk of the bangalay eucalyptus in Hawaii is on Maui where, in several large stands belonging to Haleakala Ranch Co., there are more than 1.5 million board feet, mostly in trees more than 20 in (51 cm) in diameter that were planted before 1920. It has been planted sparingly on the other islands.

Special area--Waiakea.

Range--Southeastern Australia in narrow coastal strip near sea level. Usually scattered.

Other common names--southern-mahogany, bangalay (Australia).

In Australia, planted for shade and windbreaks, hardy near shores.

98. River-redgum eucalyptus

Eucalyptus camaldulensis Dehn.*

This tree is frequently planted in windbreaks in Hawaii, it is characterized by its whitish and brown, smoothish bark and short crooked trunk. The half-round seed capsules about 1/4 in (6 mm) long and wide, have 3--4 prominent triangular valves raised and curved inward.

A large tree 80--120 ft (24--36 m) high. Trunk usually short and crooked among trees in Hawaii, 2--3 ft (0.6--0.9 m) in diameter. Crown open, spreading when planted in windbreaks. Bark mostly smoothish, light gray or buff, peeling in long strips or irregular flakes and exposing whitish inner layers. Twigs very slender, angled.

Leaves alternate, with leafstalks of 1/2--1 in (13--25 mm). Blades narrowly lance-shaped to lance-shaped, 2 1/2--12 in (5--30 cm) long and 3/8--3/4 in (1--2 cm) wide, often curved, tapering to long point, short-pointed at base, thin or slightly thickened, with

98. River-redgum eucalyptus *Eucalyptus camaldulensis* Dehn.*
Twig with flowers and fruits, 1 X (Mueller).

many fine straight side veins and vein inside margin, dull pale green on both surfaces or occasionally grayish, drooping. Juvenile leaves opposite for 3--4 pairs, then alternate, stalked, ovate to broadly lance-shaped, gray green.

Flower clusters (umbels) about 1 in (2.5 cm) long, on slender stalk of 1/4--3/4 in (6--19 mm) at leaf base. Flowers 5--10, each on slender stalk of 1/4--1/2 in (6--12 mm), 1/2--5/8 in (13--15 mm) across the many spreading white stamens. Buds 1/4--1/2 in (6--13 mm) long and 3/16 in (5 mm) wide, with half-round base and longer conical lid with long narrow or blunt beak.

Seed capsules clustered on slender stalk, half-round or egg-shaped, about 1/4 in (6 mm) long and wide, light brown, with wide raised disk and 3-4 prominent triangular valves 1/16 in (1.5 mm) long, raised and curved inward.

Wood red, fine-textured and interlocked or wavy grained. Gum veins and pockets common in wood grown in Australia. Heavy (sp. gr. 0.65), hard and durable, and resistant to termites. Not difficult to saw, but tends to warp in drying. In Hawaii used only for fenceposts because of the generally poor form. In Australia, it is used extensively for structural timbers where strength and durability are required, also for railroad crossties. There it is the only timber tree present in the interior. Thus, the wood is used for construction despite its density and difficulties of working.

It is classed as a good honey plant in Australia. It flowers throughout the year.

Introduced to Hawaii in the 1880's and first planted at Ulupalakua on Maui and Eucalyptus Ridge on Tantalus Mountain, Oahu; one of the most commonly planted eucalypts in Hawaii. Primarily used as a windbreak tree. The Division of Forestry had planted 429,000 trees by 1960, and many more were planted by private landowners. Although it will grow well in wet forest conditions, this species has been used most extensively in semi-arid sites where it achieves relatively good growth. Recorded from Niihau.

Special area--Tantalus.

Range--Widespread in Australia, mainly along streams and in flood plains, it often forms pure forests. Of all eucalypts, the most widely distributed as a wild tree. Extensively planted in arid areas throughout the world. Very popular in Israel for forestation.

Other common names--river redgum, redgum, Murray redgum, river-gum (Australia).

Botanical synonym--*Eucalyptus rostrata* Schlecht.

99. Lemon-gum eucalyptus

Eucalyptus citriodora Hook.*

Easily recognized by the strong lemon odor of crushed foliage, this eucalypt grows better at lower altitude than most of the others introduced to Hawaii. A handsome large tree 80--160 ft (24--48 m) high, with straight trunk 2--4 ft (0.6--1.2 m) in diameter. Crown regular or irregular, thin. Bark smooth, gray, peeling in thin irregular scales or patches and becoming mottled, exposing whitish or faintly bluish inner layer with powdery surface, appearing dimpled; on large trunks dark gray. Twigs slender, slightly flattened, light green, brownish tinged.

Leaves alternate, with yellowish flattened leafstalks of 1/2--3/4 in (13--19 mm). Blades narrowly lance-shaped, 4--8 in (10--20 cm) long and 3/8--1 in (10--25 mm) wide, long-pointed at apex and short-pointed at base, thin, green on both surfaces, slightly shiny, with many fine parallel straight side veins scarcely visible. Juvenile leaves opposite for 3 or more pairs, then alternate, stalked, narrowly to broadly lance-shaped with wavy margins, 1 1/2--5 in (4--13 cm) long and 1/2--1 in (13--25 mm) wide, slightly bristly hairy, green above, purplish beneath.

Flower clusters (corymbs) lateral and terminal, to 2 1/2 in (6 cm) long, branched. Flowers many, 3--5 together on equal stalks (umbels), about 1/2 in (13 mm) across the many spreading white stamens. Buds short-stalked, egg-shaped, about 1/2 in (13 mm) long and 5/16 in (8 mm) wide, lid short, half-round with short point.

Seed capsules egg-shaped or urn-shaped, narrowed into short neck, 1/2 in (13 mm) long and 5/16--3/8 in (8--10 mm) wide, brown with scattered raised dots, with wide sunken disk and enclosed valves, opening by 3 narrow lines. Seeds few, irregularly elliptical, more than 1/8 in (3 mm) long, blackish, also many smaller ones nonfunctional.

Wood light brown to gray brown, very heavy (sp. gr. 0.85), very hard, strong and very tough; moderately durable to durable; straight or wavy grain. Works easily for a wood of its density. It has been used in Hawaii for boat framing, sugar mill conveyor belt slats, and general heavy construction. Also used on a small scale in Hawaii, and more extensively in South Africa, for handle stock. Relatively stable for a eucalypt, very slow drying, but maintains its shape well in drying without collapse or serious checking.

In Australia, considered a first-class saw timber with wide range of uses including general and heavy construction and tool handles.

In Hawaii, commonly seen planted in the lowlands. There is an avenue of these trees at the entrance to Ualakaa Park (Round Top) on Oahu. It is the principal tree bordering Wahiawa Reservoir. The Division of Forestry had planted 127,000 trees in the forest reserves by 1960. It was frequently used on the island of Hawaii to make boundaries of various land ownerships and fence lines within the reserves and long single rows of the tree can be seen running through the forest near

99. Lemon-gum eucalyptus   *Eucalyptus citriodora* Hook.*
Twig with flower buds (above) and fruits (lower right), 2/3 X.

Hilo. A large tree grows alongside H-1 Freeway in Moanalua Gardens.

Special areas--Wahiawa, Tantalus, Pepeekeo.

Champion--Height 98 ft (30.0 m), c.b.h. 16 ft (4.9 m), spread 90 ft (27.4 m). Kaumana Drive, Hilo, Hawaii (1968).

Range--Northeastern Australia (Queensland), from coast to more than 200 miles (322 km) inland. Usually mixed with other eucalypts.

Often planted as an attractive ornamental in Australia, but the crown is too sparse for shelterbelts. Adapted to summer rainfall climate and very extensively planted worldwide in warm monsoon tropics.

Strong lemon odor from the essential oil citronellal is produced by crushing the foliage. Leaves when distilled yield this oil, which is used as perfume for soap. On a warm still day in Hawaii, the smell near the trees can be almost overpowering.

Other common names--lemon-scented-gum, spotted-gum, lemon-scented iron-gum (Australia).

Botanical synonym--*Eucalyptus maculata* Hook. var. *citriodora* (Hook.) Bailey.

100. Bagras eucalyptus

Eucalyptus deglupta Blume*

The beautiful mottled bark of many colors distinguishes this tree and qualifies it as an ornamental for moist regions. It is, however, a rapidly growing tree that reaches enormous size and so must be used with caution as a garden or landscaping tree. The smoothish bark peels in long strips, exposing various shades of pink, purple, copper, brown, orange, and green. Further identified by the broad ovate mostly paired leaves and the numerous small flowers and small half-round fruits. Because of its very rapid growth and good wood, this tree is now used in low elevation forest plantations.

Where native, a large tree 50--200 ft (15--61 m) high, with straight clear trunk up to 8 ft (2.4 m) in diameter and with open crown. In Hawaii, to 150 ft (45 m) in moist wind-sheltered bottom land, its preferred site. One of the few eucalypts that occurs naturally in pure stands, this tree does well in plantations. Bark smoothish, thin, of many colors. Twigs slightly 4-angled.

Leaves mostly opposite, with leafstalks of 1/2 in (13 mm). Blades ovate, 2--6 in (5--15 cm) long and 1--2 3/4 in (2.5--7 cm) wide, long-pointed at apex, rounded or blunt at base, slightly thickened, shiny green, dull and paler beneath, side veins fine and curved. Juvenile leaves opposite, stalked, ovate to oblong lance-shaped, 2 in (5 cm) long and 1 1/2 in (4 cm) wide, long-pointed, thin.

Flower clusters (panicles) terminal and lateral, with flattened branches 2--4 in (5--10 cm) long. Flowers numerous, 3--7 in group (umbel) on slender stalks less than 1/4 in (6 mm) long, more than 1/2 in (13 mm) across the many spreading white stamens. Buds club-shaped, 3/16 in (5 mm) long and 1/8 in (3 mm) wide, with conical base and conical pointed lid of equal length.

Seed capsules short-stalked, half-round, 3/16 in (5 mm) long and broad, dark brown with thin disk and 3--4 pointed valves 1/16 in (1.5 mm) long, protruding and spreading.

Wood from trees up to 20 years old is pale reddish brown with density, appearance and working characteristics similar to red lauan (*Shorea negrosensis* Foxworthy) of the Philippines. Wood from older trees is said to be more dense, but the lightweight wood (sp. gr. 0.45) from young trees in Hawaii has a relatively low shrinkage and no serious growth stress problems in manufacturing. It is suited for a wide range of uses in furniture and construction.

Used elsewhere for construction, cabinetwork, and boat building.

This species was introduced to Hawaii at Wahiawa Botanic Garden in 1929 from the Philippines. Only 4,000 trees had been planted in the forest reserves before 1960, but since then *E. deglupta* has been used extensively, and several New Guinea and New Britain races have been introduced. Trees may be seen also at the Hilo State Office Building (where they had to be topped for safety 3 or 4 years after planting), next to Hamilton Library on the University of Hawaii campus, and at numerous other locations. The bark makes the tree easily recognizable.

Special areas--Keahua, Wahiawa, Foster, Waiakea.

Champion--Height 86 ft (26.2 m), c.b.h. 6.5 ft (2.0 m), spread 42 ft (12.8 m). Wahiawa Botanical Garden, Wahiawa, Oahu (1968).

Range--Philippine Islands (Mindanao), Moluccas, New Guinea, and New Britain. Wet lowland tropics or tropical rain forests.

Other common names--Mindanao-gum, New-Guinea-gum, amammanit eucalyptus; kamarere (Papua-New Guinea).

Botanical synonym--*Eucalyptus naudiniana* F. Muell.

This species ranges from New Britain northwestward through Indonesia to Mindanao in the Philippines. It achieves its best growth in pure stands that grow on flood plains along the rivers of New Britain. It is one of only two eucalypts (*E. urophylla* S. T. Blake is the other) not native to Australia and one of the few native to a tropical rain forest climate.

One of the world's fastest growing trees, reaching as much as 33 ft (10 m) in less than 18 months. In Hawaii, attaining a height of 100 ft (30 m) in 7 years. A planted stand near Port Moresby, New Guinea, had an average height of 180 ft (54 m) and diameter of 2 ft (0.6 m) at 14 years of age.

100. Bagras eucalyptus   *Eucalyptus deglupta* Blume*
Twig with flowers and fruits, fruits (lower right), 1 X.

101. Bluegum eucalyptus

Eucalyptus globulus Labill.*

One of the first eucalypts to be used extensively outside of Australia and perhaps the best known, particularly so in California, Spain, Portugal, and Argentina where it has been used both as a timber tree and an ornamental along city streets. Easily recognized by the very large white flowers single and almost stalkless at leaf bases, more than 2 in (5 cm) across the very numerous spreading white stamens, and by the large 4-angled warty seed capsules with whitish bloom. Crushed foliage has an odor like that of camphor.

A large tree in Hawaii 150--200 ft (46--61 m) high, with straight trunk 2--4 ft (0.6--1.2 m) in diameter and up to two-thirds of total height. Crown narrow and irregular, with drooping foliage. Bark smoothish, mottled gray, brown, and greenish, peeling in long strips, at base becoming gray, rough and shaggy, thick, and finely furrowed. Inner bark light yellow within thin green layer, fibrous, with slightly resinous or bitter taste. Twigs slender, angled, drooping, yellow green, turning dark red or brown.

Leaves alternate, with flattened yellowish leafstalks of 1/2--1 1/2 in (1.3--4 cm). Blades narrowly lance-shaped, 4--12 in (10--30 cm) long and 1--2 in (2.5--5 cm) wide, mostly curved, long-pointed at apex, short-pointed at base, dull green on both surfaces, thick and leathery, with fine straight side veins and vein inside margin, drooping. Juvenile leaves on 4-angled or winged twigs, opposite for many pairs, stalkless or clasping, ovate or broadly lance-shaped, 3--6 in (7.5--15 cm) long and 1 1/2--3 1/2 in (4--9 cm) wide, with bluish or whitish waxy bloom on lower surface.

Flowers 1, rarely 2--3, at leaf base on very short flattened stalk or none, more than 2 in (5 cm) across the very numerous spreading white stamens about 1/2 in (13 mm) long, with odor of camphor. Buds top-shaped, 1/2--5/8 in (13--15 mm) long and 1/2--1 in (13--25 mm) wide, 4-angled and very warty, bluish with whitish bloom, lid caplike and warty with central knob or point, and thin smooth pointed outer lid.

Seed capsules broadly top-shaped and 4-angled, 3/8--5/8 in (10--15 mm) long and 3/4--1 in (19--25 mm) wide, bluish with whitish bloom and with broad thick flat or convex disk extending over 3--5 blunt flat thick valves. Seeds many, irregularly elliptical, 1/8 in (3 mm) long, dull black, also many smaller nonfunctional.

The sapwood is white and the heartwood pale yellow brown. Wood heavy (sp. gr. 0.75) with a medium texture, and straight to interlocked grain. Occasional trees produce wood with a bird's-eye figure. Dense outer wood of logs is very strong and moderately durable. The inner wood is of lower density and apt to be brash. Growth stress is not as serious a problem in manufacture as with *E. saligna* and *E. robusta* but is nevertheless troublesome.

The wood is used in light and heavy construction. On Maui, the lumber has been used for home framing, flooring, and siding as well as for decorative interior panelling and furniture. Although it is a difficult wood to work and has a large movement in place with humidity changes, it is being used for normal hardwood purposes. A veneer plant in Hawaii successfully peeled bluegum and laid up plywood used for concrete forms. The most recent use of bluegum in Hawaii was for pulpwood chips, which were exported to Japan.

Plantations of this species are an important source of fuelwood in places such as the Andes where fuel is scarce. Trees grow very rapidly and after cutting form new trunks from stump sprouts. A large program of producing "biomass" by planting *E. globulus* and *E. saligna* for fuelwood as an alternate energy source is being investigated in Hawaii. The tree is a honey plant and is also used in windbreaks.

An essential oil named eucalyptol is distilled from the leaves and has medicinal use. This is the principal eucalyptus for oil. Leaf infusions have served elsewhere in home remedies.

Introduced to Hawaii in the 1870's or 1880's, this eucalypt was extensively planted between 1908 and 1920 on Haleakala Ranch on Maui, and on Kukaiau Ranch, Parker Ranch, and Kapapala Ranch on Hawaii. Until heavy logging began on Hawaii in 1973 to supply the chip market, there were 56 million board feet of timber on Hawaii. Maui has about 100 million feet. It was the fourth highest in timber volume of the tree species in Hawaii. Since 1973, about 40 million board feet have been cut on Hawaii and the stands cut are now growing as coppice. The coppice stands should be cut again when about 15 years old. This eucalypt has been planted on all the larger islands.

This is the common *Eucalyptus* of California. It is one of the most extensively planted eucalypts outside of Australia for forestry, shade, and ornament. It has been successful in different climates mainly subtropical or sometimes classed as warm temperature. Hybrids have originated in plantations elsewhere.

Special areas--Tantalus, Volcanoes.

Champion--Height 90 ft (27.4 m), c.b.h. 26.7 ft (8.1 m), spread 96 ft (29.3 m). Kukaiau Ranch, Honokaa, Hawaii (1968).

Range--Eastern Tasmania and very local in southern Victoria, and New South Wales, Australia. In pure stands on favorable sites or mixed with other eucalypts. Mediterranean or Atlantic climate.

Other common names--southern bluegum, bluegum, Tasmanian bluegum (Australia); Tasmanian blue eucalyptus.

The specific name meaning little ball refers to the appearance of the flower from a distance. This species

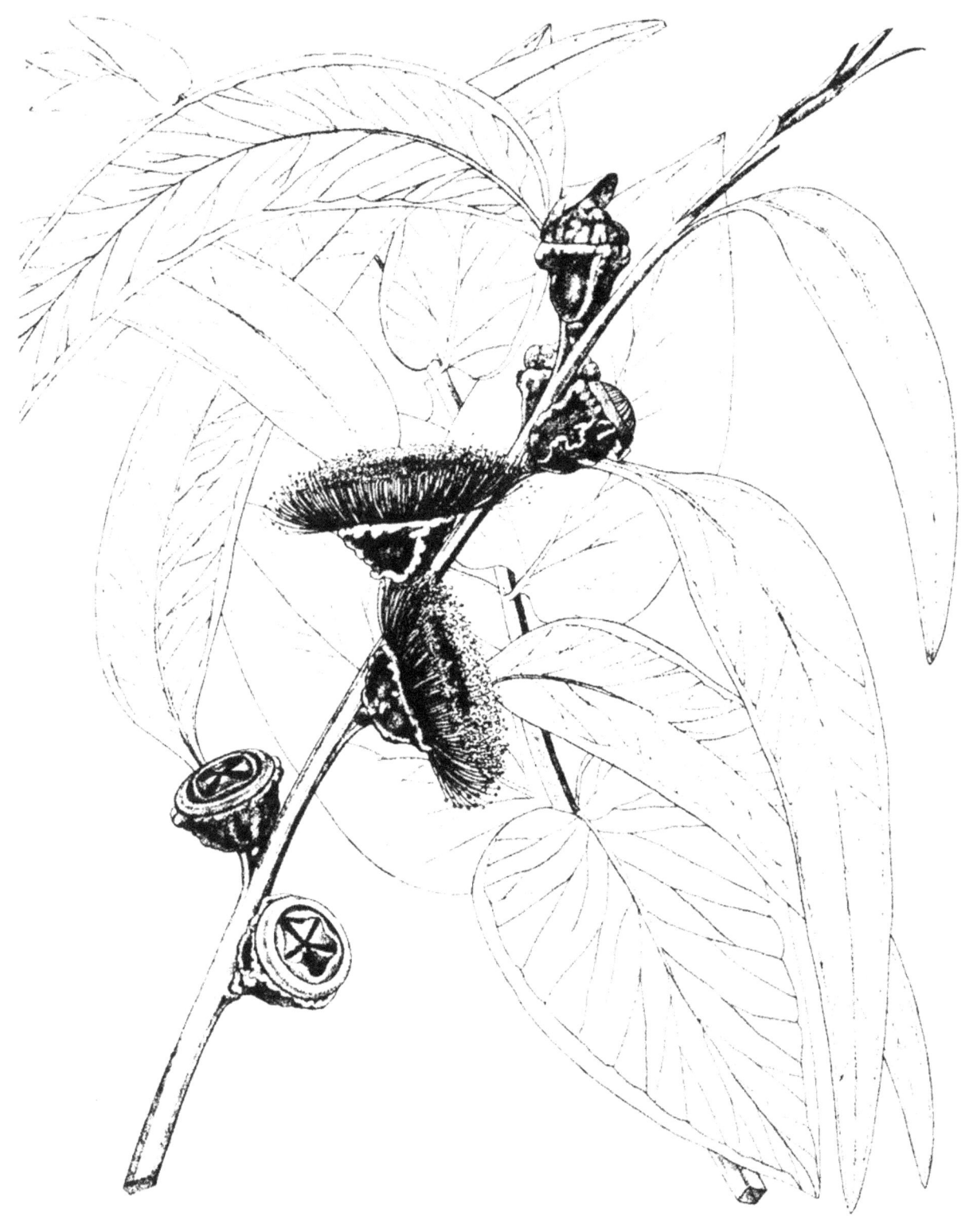

101. Bluegum eucalyptus *Eucalyptus globulus* Labill.*
Twig with flowers and fruits, twig with juvenile leaves (behind), 1 X (Mueller).

is the floral emblem of Tasmania and has nearly all its natural range within that island.

Eucalyptus globulus Labill. is now designated also as *Eucalyptus globulus* Labill. subsp. *globulus*. Two other subspecies *E. globulus* subsp. *bicostata* (formerly considered 2 species, *E. bicostata* and *E. stjohnii*) and *E. globulus* subsp. *maidenii* (F. Muell.) Kirkp. (formerly considered *E. maidenii*) are not described here (Food and Agriculture Organization of the United Nations 1979).

102. Rosegum eucalyptus

A slender tree 140--180 ft (42--54 m) high, similar to No. 108, saligna eucalyptus, *Eucalyptus saligna*, and hybridizing with that species. The two are compared in the key to species. Trunk 2--3 ft (0.6--0.9 m) in diameter, tall, of good form, unbranched for up to two-thirds of total height. Crown spreading and rather open when open grown. Crown small and compressed when in dense plantations.

Bark light gray, smooth, thick. Occasionally lowest 4--20 ft (1.2--6.0 m) of trunk has thicker light gray bark persisting as fairly regular small plates. Bark of branches and smaller trunks smooth, outer orange brown layers peeling in long narrow strips, exposing smooth greenish and greenish white inner layer. Twigs slender, angled, with whitish waxy coating.

Leaves alternate with slender leafstalks of 1/2--3/4 in (13--19 mm). Blades lance-shaped, 4--8 in (10--20 cm) long and 3/4--1 1/2 in (2--4 cm) wide, tapering to long narrow point at apex, blunt and slightly unequal at base, slightly wavy, shiny dark green above, paler beneath, thin, with fine regular veins, hanging obliquely or horizontal. Juvenile leaves opposite for 3--4 pairs, then alternate, short-stalked, oblong lance-shaped, 1--2 1/2 in (2.5--6 cm) long and 1/2--1 in (13--25 mm) wide, thin, and slightly wavy.

Flower clusters (umbels) at leaf base 1--1 1/4 in (2.5--3 cm) long including flattened stalk of 1/2 in (13 mm). Flowers 5--12 on short stalk or none. Buds pear-shaped with blunt-pointed conical lid 3/8 in (10 mm) by 3/16 in (5 mm), usually with whitish waxy coating.

Seed capsules short-stalked, pear-shaped or conical, slightly narrowed at rim, thin, 5/16 in (8 mm) long and 1/4 in (6 mm) wide, with whitish waxy coating, narrow sunken disk, and 4-6 pointed thin valves slightly projecting and curved in, persisting on twigs back of leaves.

Wood pink to light reddish brown, moderately hard; very variable in density but averages moderately heavy (sp. gr. 0.57); of moderate strength and durability; straight grained with coarse texture. Outer wood in logs heavy and strong; inner wood of low density and frequently brash. Young plantation-grown timber has severe growth stress that causes end splitting, spring, and brittleheart in logs and lumber. Slightly lower in density than *E. robusta* and *E. saligna*. Subject to warping and other defects in seasoning, it has not been used in Hawaii. Elsewhere, it is used for poles, construction, mine props, and pulpwood.

Possibly a recent introduction to Hawaii. The first seed of true *E. grandis* was brought in 1957. Since then, several other races have been introduced. It is believed that most trees in Hawaii are hybrids with *E. saligna*. The Australian races that have been introduced have faster and more uniform growth than the local race of *saligna* X *grandis* and are also faster than races of pure *E. saligna* that have been compared (Skolmen 1986). Planted primarily on Hawaii and Kauai, but present on all larger islands.

Special areas--Keahua, Waiakea (best stand).

Champion--Height 110 ft (33.5 m), c.b.h. 4.3 ft (1.3 m), spread 35 ft (10.7 m). Honaunau Forest Reserve, Kailua-Kona, Hawaii (1968).

Range--Eastern Australia, coastal regions, in pure or almost pure stands. Equable summer rainfall climate. Extensively planted, for example, in South and East Africa, Brazil, and India.

Other common names--rose-gum; flooded-gum (Australia).

Eucalyptus microcorys F. Muell.*

103. Tallowwood eucalyptus

A large tree 100--150 ft (30--46 m) high. Trunks large, usually 2--3 ft (0.6--0.9 m) in diameter, sometimes much greater, straight, up to two-thirds of total height. Crown compact, narrow, with main branches horizontal and smaller branches erect. Bark reddish brown or gray, soft, fibrous, becoming rough and thick, furrowed into long shaggy plates with thin layers of cork. Inner bark light yellowish brown, fibrous, almost tasteless or slightly bitter. Twigs slender, angled, light green and pinkish.

Leaves alternate, with slender yellowish leafstalks of 1/2 in (13 mm) or less. Blades lance-shaped, 2 1/2--4 in (6--10 cm) long and 3/4--1 1/2 in (2--4 cm) wide, curved, tapering to long point, base short-pointed with sides mostly unequal, thin, above green and slightly shiny, beneath dull green, with moderately visible fine straight side veins, drooping. Juvenile leaves opposite for 4--5 pairs, stalkless or short-stalked, elliptical to broadly lance-shaped, 2--2 3/4 in (5--7 cm) long and 3/4--1 3/8 in (2--3.5 cm) wide, pale green.

Eucalyptus grandis W. Hill ex. Maid.*

102. Rosegum eucalyptus   *Eucalyptus grandis* W. Hill ex Maid.*
Leafy twig (left) and fruits (right), 1 X.

Flower clusters terminal (panicles) and branched or lateral (umbels) 1--2 in (2.5--5 cm) long. Flowers 4--7 at end of slender stalk of 1/2--1 in (13--25 mm), 5/8 in (15 mm) across the many spreading white stamens. Buds tapering to short stalk of 1/4 in (6 mm) or less, club-shaped, 1/4--5/16 in (6--8 mm) long and 3/16 in (5 mm) wide, with small half-round or conical lid.

Seed capsules narrowly conical, tapering to short stalk, 3/8 in (10 mm) long and 1/4 in (6 mm) wide, with flattened disk and 3--4 short valves often slightly protruding, light green.

Wood yellowish brown, shiny and rather greasy; very heavy (sp. gr. 0.8), hard, very strong and very durable. Moderately coarse texture, usually interlocked grain. Comparatively easy to work and easy to polish.

This strong, durable, easily worked timber has been in demand in Australia for many purposes including light and heavy construction, posts, poles, and railroad crossties. Very suitable for ballroom or dance floors because of the smooth greasy finish, indicated by the common name tallowwood. Rated as the best hardwood in New South Wales. It has not been used in Hawaii.

104. Gray-ironbark eucalyptus

A medium-sized tree 60--80 ft (18--24 m) high and 2--4 ft (0.6--1.2 m) in trunk diameter. Trunk of good form and to two-thirds of total height on better sites but shorter and crooked on poor sites. Crown dense and compact but more open and irregular with age. Bark dark or light gray, very rough, very thick, hard, deeply furrowed into long narrow ridges. Twigs very slender, slightly flattened or angled, yellow green to pinkish.

Leaves alternate, with slender yellow green leafstalks of 1/2--1 in (1.3--2.5 cm). Blades lance-shaped, 2 1/2--6 in (6--15 cm) long and 5/8--1 1/4 in (1.5--3 cm) wide, long-pointed at apex, base short-pointed often with sides unequal, thin, dull green on upper surface, slightly paler beneath, with many fine moderately visible straight side veins, drooping. Juvenile leaves opposite for 3--4 pairs, then alternate, short-stalked, broadly lance-shaped, 1--2 in (2.5--5 cm) long and 1/2--1 in (1.3--2.5 cm) wide, dark green.

Flower clusters (panicles) terminal, 1--3 in (2.5--7.5 cm) long. Flowers many, 3--9 in branched clusters (umbels) on short slender rounded stalks of 1/4 in (6 mm) or less, 3/4 in (2 cm) across the many spreading white stamens. Buds egg-shaped to diamond-shaped, 3/8 in (10 mm) long and 3/16 in (5 mm) wide, with conical base and rather short conical lid.

Seed capsules short-stalked, half-round, pear-shaped, or egg-shaped, 1/4--3/8 in (6--10 mm) long and wide, with disk sunken or flat and 3--4 (rarely 5) valves usually enclosed.

The Division of Forestry has planted more than 102,000 trees of this species, which was introduced in 1911. It grows exceptionally well at Kalopa State Park and in other areas where planted on all the larger islands. It is the most shade tolerant of the eucalypts in Hawaii and has often survived in a mixture with the taller, faster growing *E. saligna*. Not considered desirable for future planting because of the high density of its wood.

Special area--Kalopa.

Champion--Height 145 ft (44.2 m), 9.5 ft (3.0 m), spread 65 ft (20.0 m). Honaunau Forest Reserve, Kailua-Kona, Hawaii (1968).

Range--Coastal northern New South Wales and southern Queensland, Australia. Mixed with other eucalypts or as an overstory of rain forest. Equable summer rainfall climate. Planted in Africa, South America, and Asia.

Other common name--tallowwood (Hawaii, Australia).

The specific name, from Greek words meaning "little helmet," refers to the small lid of the flower.

Eucalyptus paniculata Sm.*

Wood brown or dark brown with a fine texture and interlocked grain. Extremely heavy (sp. gr. 1.0). Very hard, very strong, and very durable. Tough, not easy to work, but turns to a good finish. Tends to check in drying.

One of the hardest, heaviest, toughest, strongest, and most durable timbers in this large genus. Used largely for railroad crossties, bridge and wharf timbers, poles, and piles, also lumber for ships and construction. Has been used only for fenceposts in Hawaii.

There are about 1.8 million board feet of this species in Hawaii, mostly on Oahu. Most trees were planted in the late 1930's, and most are less than 20 in (51 cm) in diameter. The Division of Forestry has planted 137,000 trees. They may be seen at the Honouliuli Forest Reserve, on Tantalus Drive, and from the Aiea Loop Trial.

Champion--Height 107 ft (32.6 m), c.b.h. 13.2 ft (4.0 m), spread 111 ft (33.8 m). Ulupalakua Ranch, Ulupalakua, Maui (1968).

Range--Southeastern Australia along coast to 60 miles (97 km) inland. Scattered and not in pure stands. Equable summer rainfall climate. Planted in Australia as a shade and ornamental tree. Extensively introduced in East and South Africa, Brazil, and India.

Other common names--white ironbark (Hawaii); grey ironbark (Australia).

103. Tallowwood eucalyptus *Eucalyptus microcorys* F. Muell.*
Twig with flowers and fruits, 1 X (Mueller).

105. Blackbutt eucalyptus

Eucalyptus pilularis Sm.*

A large tree 120--180 ft (37--55 m) high. Trunk 2--4 ft (0.6--1.2 m) in diameter, straight and half to two-thirds of total height. Crown spreading and open. Bark gray brown or light gray, thick, rough, finely and deeply furrowed, fibrous, persistent on trunk only, partly shedding in strips on upper stem and branches and leaving a smooth white or yellow gray surface; branches smooth, whitish gray. Twigs slender, angled, pinkish.

Leaves alternate, with leafstalks of 1/2--1 in (13--25 mm). Blades lance-shaped, 3--5 in (7.5--13 cm) long and 3/4--1 1/2 in (2--4 cm) wide, often slightly curved, long-pointed at apex, base short-pointed and often with sides unequal, slightly thickened, shiny dark green on upper surface, paler beneath, with moderately distinct, fine straight side veins. Juvenile leaves opposite for several pairs, then alternate, stalkless or very short-stalked, oblong to narrowly lance-shaped, 1 1/2--3 in (4--7.5 cm) long and 1/2--1 1/2 in (1.3--4 cm) wide, green above, purplish beneath.

Flower clusters (umbels) at leaf base 1 in (2.5 cm) long, including angled or flattened stalk. Flowers 6--12, about 1/2 in (13 mm) across the many spreading white stamens. Buds on short stalk of 1/8 in (3 mm), half-round or egg-shaped, about 3/8 in (10 mm) long and 3/16 in (5 mm) wide, with conical base and sharply conical lid of equal length.

Seed capsules short-stalked, half-round or egg-shaped, 3/8--1/2 in (10--13 mm) in diameter, with flat or slightly sunken disk and 3--4 small valves enclosed or at rim.

Heartwood a pale brown similar to oak, hard, strong, very stiff and tough except when brittleheart, of moderate to good durability; moderately heavy (sp. gr. 0.66). Usually straight grained and readily worked. Has a large shrinkage in drying but less growth stress manufacturing problems than *E. robusta* or *E. saligna*.

One of the most important hardwoods of Australia. Used there for general and house construction and thought not highly durable, also for poles, posts, and railroad crossties. In Hawaii, used for flooring, truck beds, paneling, refuse flumes, irrigation canal stakes, and pallets.

The tree was introduced in 1911 and first planted in an experimental area in Nuuanu Valley, Oahu. Although it has been planted on all islands, Oahu has the most timber (2.5 million board feet). The Division of Forestry has planted more than 120,000 trees. Most of the blackbutt eucalyptus has been planted in drier areas, where it attains a height of about 100 ft (30 m). In areas with about 100 in (2,540 mm) rainfall, it grows to a great size, 180 ft (55 m), very rapidly. It has not grown well in very wet areas.

Special areas--Aiea, Pepeekeo.

Champion--Height 113 ft (34.4 m), c.b.h. 9.9 ft (3.0 m), spread 45 ft (13.7 m). Hilo Forest Reserve, Hawaii (1974).

Range--Coastal Northern New South Wales and southern Queensland, Australia. Frequently in pure stands or with other eucalypts on well-drained sites, commonly sandy loam soil.

Other common name--blackbutt (Hawaii, Australia).

The common name refers to the bases of the trunks commonly blackened by frequent fires in the native forest. The scientific name, "resembling a little pill," describes the rounded seed capsules.

106. Kinogum eucalyptus

Eucalyptus resinifera Sm.*

A large tree 150--200 ft (46--61 m) high and 2--3 ft (0.6--0.9 m) in trunk diameter. Crown dense, well branched, narrow to spreading. Bark reddish brown, rough, thick, persistent, very fibrous, becoming stringy; bark on branches light gray, smoothish. Inner bark reddish brown, fibrous, and tasteless. Bark resembles that of *E. microcorys* except that it is firmer and red, rather than yellowish brown on inner surfaces. Twigs long slender, pinkish tinged.

Leaves alternate, with pinkish tinged leafstalks of 1/2--1 in (13--25 mm). Blades lance-shaped, 4--7 in (10--18 cm) long and 3/4--1 1/4 in (2--3 cm) wide, slightly curved, long-pointed at apex, base short-pointed and often with sides unequal, thick, shiny or dull dark green above, dull and paler beneath, with many fine straight parallel side veins, horizontal or drooping. Juvenile leaves opposite for 3--4 pairs, then alternate, short-stalked, narrowly lance-shaped, 1 1/2--2 1/2 in (4--6 cm) long and 1/2--3/4 in (1.3--2 cm) wide, pale green.

Flower clusters (umbels) at leaf base about 1 in (2.5 cm) long, including angled stalk. Flowers 5--10 on rounded stalks of 1/4 in (6 mm), more than 3/4 in (2 cm) across the many spreading white stamens. Buds 1/2--5/8 in (13--15 mm) long and 1/4 in (6 mm) wide, with half-round base occasionally 2-ribbed and long conical lid abruptly long-pointed.

Seed capsules egg-shaped to half-round, 1/4 in (6 mm) in diameter, on stalk of 1/4 in (6 mm), with disk slightly convex and usually 4 prominent sharp-pointed raised valves.

Heartwood is reddish brown, pale when freshly cut and darkening on exposure to light. Grain usually straight, sometimes slightly interlocked. Moderately coarse texture. Wood is similar to *E. robusta* except that the grain is straighter. A heavy wood (sp. gr. 0.62) with

104. Gray-Ironbark eucalyptus *Eucalyptus paniculata* Sm.*
Twig with flowers and fruits, 1 X (Mueller).

105. Blackbutt eucalyptus　　　*Eucalyptus pilularis* Sm.*
Twig with flowers and fruits, twig with juvenile leaves (behind), 1 X (Mueller).

106. Kinogum eucalyptus   *Eucalyptus resinifera* Sm.*
Twig with flowers, fruits (lower left), twig with juvenile leaves (behind), 1 X (Maiden).

large shrinkage in drying, with manufacturing problems caused by growth stress. Wood grown in Hawaii has not been used. Since it has a similar density and appearance to E. robusta, it would probably be suitable for similar purposes.

Wood grown in Australia is much denser, and therefore stronger, than that from Hawaii. This durable, strong, easily worked wood is in demand for many uses, such as house construction, shipbuilding, general construction, railroad crossties, and as pulpwood.

The Division of Forestry has planted 90,000 trees on all islands. There are 1.3 million board feet of timber on the island of Hawaii, which has the largest volume of this species. The best stand of this tree is at Kalopa State Park, where there are many trees about 200 ft (61 m) tall.

Champion--Height 117 ft (35.7 m), c.b.h. 8 ft (2.4 m), spread 49 ft (14.9 m). Honaunau Forest Reserve, Kailua-Kona, Hawaii (1968).

Range--Eastern Australia, coastal regions. Usually scattered, occasionally in small almost pure stands, and more tolerant of shade than most eucalypts.

Other common names--red-mahogany eucalyptus (Hawaii); red-mahogany, red messmate (Australia).

This is one of the most promising of more than 30 species of Eucalyptus that have been tested experimentally in forest plantations in Puerto Rico. Widely grown in forest plantations elsewhere.

107. Robusta eucalyptus

Eucalyptus robusta Sm.*

The most common planted tree species and most common eucalypt in Hawaii. Recognized by the thick, soft reddish brown bark, the relatively broad lance-shaped leaves 1 1/4--2 1/2 in (3--6 cm) wide, the large flowers about 1 1/4 in (3 cm) across, the many spreading stamens, and the large bell-shaped seed capsules.

A moderately large tree 80--160 ft (24--48 m) high. Trunk relatively large, 3--4 ft (0.9--1.2 m) in diameter, straight and one-half to two-thirds the height of tree. Crown relatively dense, narrow to spreading, with long irregular branches. Bark reddish brown in wet sites, grayish brown where dry, very thick, deeply furrowed into long scaly ridges, fibrous and very soft, persistent. Inner bark whitish, fibrous, slightly bitter. Twigs stout, angled, yellowish green, becoming reddish brown.

Leaves alternate, with yellowish or pinkish tinged leafstalks of 1/2--1 in (13--25 mm). Blades broadly lance-shaped, 4--8 in (10--20 cm) long and 1 1/4--2 1/2 in (3--6 cm) wide, long-pointed at apex, base short-pointed and often with sides unequal, thick, shiny or dull dark green above, dull light green beneath, with fine regular almost parallel side veins and vein at margin. Juvenile leaves opposite for 3--4 pairs, then alternate, lance-shaped or narrowly ovate, 3--5 in (7.5--13 cm) long and 2--2 1/2 in (5--6 cm) wide, light green, thick.

Flower clusters (umbels) at leaf base, to 2 1/2 in (6 cm) long including flattened stalk of 3/4--1 1/4 in (2--3 cm). Flowers 5--10, short-stalked, 1 1/4 in (3 cm) across the many spreading white or cream-colored stamens. Buds 1/2--3/4 in (13--19 mm) long and 3/8 in (10 mm) wide, with long conical or bell-shaped base and long narrow conical pointed lid.

Seed capsules stalked, bell-shaped or cylindrical, 1/2--3/4 in (13--19 mm) long and 3/8--1/2 in (10--13 mm) wide, with flattened or sunken disk and 3--4 short-pointed valves not protruding.

Because robusta eucalyptus is Hawaii's most plentiful exotic timber species, its wood properties have been studied intensively. Sapwood is pale brown and heartwood reddish brown, pale when fresh and darkening to a rich mahogany color on exposure. Grain is interlocked and produces a ribbon figure on quarter-sawn faces. Texture is coarse. Moderately heavy (sp. gr. 0.6) but quite variable in density, as in other eucalypts. Wood from the interior of logs is often lightweight (sp. gr. 0.35), while that near the log surface may be heavy (sp. gr. 0.8). The wood compares with white oak (*Quercus alba* L.) in most strength properties but is unstable and shrinks and swells markedly with humidity changes.

It is a difficult and expensive wood to process in sawmilling, as are most eucalypts, because of a peculiarity called growth stress. Growth stresses are longitudinal forces in logs that cause log ends to split open, interior wood near the pith to be crushed into the brittleheart with numerous compression failures, and logs and boards to spring, crook, split, and jam when being sawed. Robusta eucalyptus and saligna eucalyptus grown in Hawaii have severe growth stress.

The wood can be seasoned with little degrade if done with care. It is resistant to decay and termites, more so than western redcedar, but less so than redwood. About 1 million board feet of lumber have been produced each year in Hawaii (none currently) and the wood has been used for many purposes including house siding, framing, and flooring. It has been used in furniture mostly for upholstered frames but is attractive as finish as well. It has performed well as boat framing and conveyor slats used in sugar mills. It has also been used extensively in pallets and irrigation canal stakes. It has also been chipped and sent to Japan for kraft pulp manufacture.

Robusta eucalyptus is the most commonly planted tree in Hawaii. Although it is not currently planted, having been replaced by *E. saligna* as the preferred species, it has a demonstrated ability to thrive on almost any site with a 500--3,500 ft (152--1,067 m) altitude that has more than 40 in (1,016 mm) rainfall. The Division of Forestry planted more than 2.3 million trees before 1960, and nearly equal numbers were planted by private landowners. On the islands of Hawaii and Maui, there

107. Robusta eucalyptus *Eucalyptus robusta* Sm.*
Twig with flowers and fruits, 1 X (Mueller).

are more than 150 million board feet of sawtimber. The trees may be seen in many places, such as along the road to Koloa, Kauai. The stand near the Nuuanu Pali, Oahu, was planted about 1900. The 1,300--acre (526--ha) stand at Opana, Maui, is said to be the world's largest single block of robusta eucalyptus. There are stands along the highways near Honokaa, Hawaii, and the nearby windbreaks from Mud Lane to Waimea. Recorded from Niihau.

Apparently, this is the best adapted of more than 30 species of Eucalyptus that have been tested for forestry in Puerto Rico. It is also planted in Florida. As a street tree in California, this species has suffered from broken tops in strong winds. In Hawaii, it withstands continuous battering by trade winds quite well and is an important windbreak tree.

108. Saligna eucalyptus

Eucalyptus saligna Sm.*

A very large tree 130--200+ ft (39--61+ m) high. Trunk 2--4 ft (0.6--1.2 m) in diameter, straight, excellent form, and two-thirds of total height of tree. Crown open, spreading, irregular. Bark dull greenish gray, smooth, peeling off slightly and exposing yellow layer; on large trunks usually 4--8 ft (1.2--2.4 m), but sometimes up to 30 ft (9 m) from base, becoming gray brown, thick, rough, furrowed into short narrow ridges or plates, persistent. Inner bark with green cap, then whitish, fibrous, bitter. Twigs slender, angled, yellow green to pink. Presently, the principal species being used for forestation in Hawaii. This species and No. 102, rosegum eucalyptus, *Eucalyptus grandis* W. Hill ex Maid., are compared in the key to species.

Leaves alternate, on slender slightly flattened yellow green to pink leafstalk of 1/2--1 in (13--25 mm). Blades lance-shaped, 4--8 in (10--20 cm) long, 5/8--1 1/4 in (1.5--3 cm) wide, often curved, tapering to long point, short-pointed at base, dull green or dark green above, dull light green beneath, with fine straight parallel side veins scarcely visible, hanging obliquely or horizontal. Juvenile leaves opposite for 3--4 pairs, then alternate, short-stalked, lance-shaped, 1 1/4--2 1/2 in (3--6 cm) long and 3/4--1 1/2 in (2--3 cm) wide, thin, slightly wavy, pale green.

Flower clusters (umbels) at leaf bases and along twigs about 3/4 in (2 cm) long including flattened or angled stalk. Flowers 4--9, usually 7, short-stalked or stalkless, 1/2--3/4 in (13--19 mm) across the many spreading white stamens. Buds 3/8 in (10 mm) long and 3/16 in (5 mm) wide, with conical to bell-shaped base, lid conical and bluntly or sharply pointed.

Seed capsules on short stalk of 1/4 in (6 mm) or none, conical or slightly bell-shaped, about 1/4 in (6 mm) long and wide, dark brown, with narrow sunken disk and 3--5 narrow pointed valves slightly projecting and spreading.

Suitable for shade and ornament and also a honey plant.

Special areas--Wahiawa, Aiea, Kalopa, Tantalus, Pepeekeo.

Champion--Height 97 ft (29.6 m), c.b.h. 16.6 ft (5.1 m), spread 72 ft (21.9 m). Kaupakuea, Hilo, Hawaii (1968).

Range--Southeastern Australia in a very narrow coastal strip from sea level to a few hundred feet (about 100 m) altitude; it is frost free or nearly so. Grows mainly in coastal swamps and on edges of saltwater estuaries, usually scattered or in narrow belts. Summer rainfall climate. Extensively planted worldwide in tropics.

Other common names--swamp-mahogany eucalyptus (Hawaii); swamp-mahogany, swamp messmate (Australia); beakpod eucalyptus.

Botanical synonym--*Eucalyptus multiflora* Poir.

Sapwood pale brown and heartwood light reddish brown. Grain varies from straight to strongly interlocked, the latter producing a pronounced ribbon-stripe figure on quartered faces. Generally paler in color and finer in texture than wood of *E. robusta*. Moderately heavy (sp. gr. 0.61), but variable between inner and outer wood from logs. Slightly stronger than robusta in most properties because of its slightly straighter grain. Wood is very stiff and is highly subject to brittleheart formation. Compression failures are very common in interior wood of logs. Wood of trees less than 12 years old is of low density but free of brittleheart and, though of low grade, easy to work and suitable for general construction (Skolmen 1974). Normal wood from older trees is moderately resistant to decay but not resistant to termites.

Saligna eucalyptus wood in Hawaii used mostly for flooring and pallets. Otherwise its uses have been identical with those of *E. robusta*, but on a lesser scale. It is a good pulpwood and has been chipped and shipped to Japan. Because of serious growth stress problems, pulpwood is probably its best use. It is a major pulpwood species in Brazil and South Africa, although the trees there are often hybrids with *E. grandis*. Now, this is the main species in biomass fuelwood plantations in Hawaii.

Introduced to Hawaii about 1880 and planted fairly extensively on all islands, particularly after 1960. Before 1960, the Division of Forestry planted 437,000 trees. Since then, more than 1 million have been planted. There are presently about 10 million board feet on Hawaii and 8 million on Oahu. The tree has remarkable growth on good sites. One tree was measured at 106 ft (32 m) at 5 years of age. The tallest hardwood tree in the United States is a saligna eucalyptus growing in North Kona that was 276 ft (84 m) high in 1980. Recently, a tree only 44 years old was felled in a stand near Umikoa, Hawaii, and measured on the ground as 236 ft (72 m).

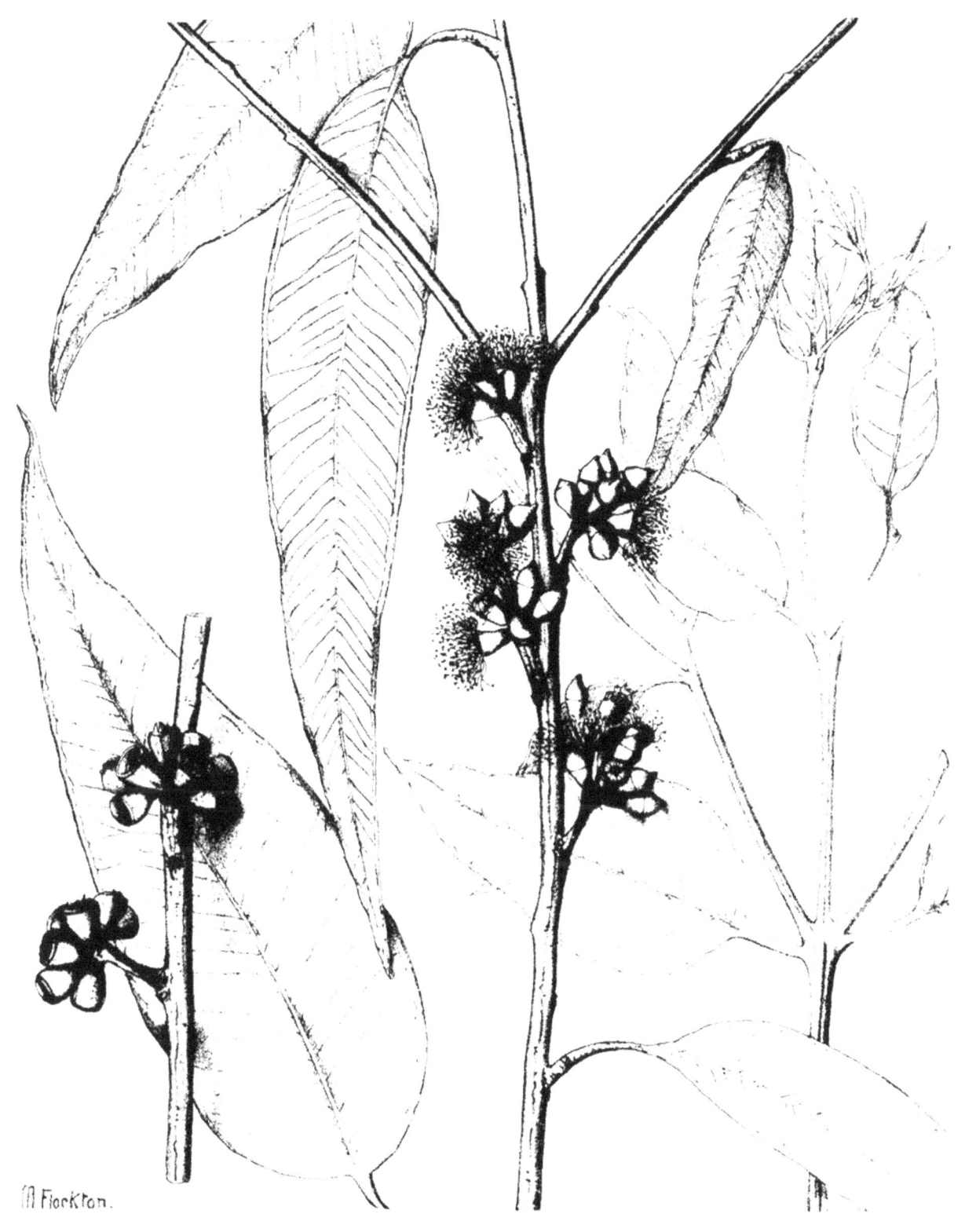

108. Saligna eucalyptus *Eucalyptus saligna* Sm.*
Twig with flowers, fruits (lower left), twig with juvenile leaves (right), fruits (bottom), 1 X (Maiden).

Four other trees from the same stand were 194--208 ft (59--63 m) measured on the ground.

The local race of saligna eucalyptus planted most extensively in Hawaii is believed to be a hybrid with E. grandis and possibly other species. Plantings made with third or fourth generation seed show considerable variation because of inbreeding depression. Recent plantings have been made with trees with provenances from Australia and Brazil to offset this problem.

It is classed as a honey plant, and is too large for ornamental planting.

Trees may be seen on the right side of the Volcano Highway at Olaa Forest Park at Mountain View, Hawaii, along the Hana Road near Puaakaa Park on Maui, along the Aiea Loop Trail on Oahu, and in the Puu Ka Pele Forest Reserve near Kokee Park, Kauai.

Special areas--Keahua, Waiakea, Aiea, Kalopa.

Range--Southeastern Australia within 100 miles (161 km) of the coast, commonly mixed with other eucalypts, occasionally in pure stands. It grows on slopes while E. grandis grows on bottom land in moist coastal forests.

Other common names--Sydney bluegum eucalyptus, flooded-gum (Hawaii); Sydney bluegum, bluegum (Australia).

Champion--Apparently the tree mentioned above as 276 ft (84 m) high in 1980. This is the tallest hardwood tree in the United States.

109. Red-Ironbark eucalyptus

Eucalyptus sideroxylon A. Cunn. ex Woolls*

Characterized by blackish or dark brown bark, very thick, deeply furrowed and ridged, and by the leaves dull gray green on both surfaces. Flowers clustered and bending down on curved stalks at leaf base, sometimes pink or reddish instead of white. Crushed foliage very aromatic, with odor like camphor.

Where native, a large tree 70--100 ft (21--30 m) high. Trunk 2 1/2--4 ft (0.8--1.2 m) in diameter, straight, usually less than half of total height. Crown irregular. Bark blackish or dark brown, very thick, hard, deeply furrowed into long narrow irregular rough ridges, containing eucalyptus gum or kino. Outer dead bark light brown, slightly corky, inner bark reddish. Branches smooth and light gray. Twigs very slender, yellow green, pink tinged.

Leaves alternate, drooping on very slender leafstalks of 1/2--3/4 in (13--19 mm), yellow green, pink tinged. Blades lance-shaped, 2 1/2--5 in (6--13 cm) long and 3/8--3/4 in (1--2 cm) wide, long-pointed at both ends, thin, dull gray green on both surfaces, with narrow midvein and side veins faint or scarcely visible, with gland dots visible under lens. Juvenile leaves opposite for 3--4 pairs, then alternate, short-stalked, very narrow (linear) to oblong, 1 1/2--3 in (4--7.5 cm) long and 1/4--5/8 in (6--15 mm) wide, dull green.

Flower clusters (umbels) at leaf base, 1--1 1/2 in (2.5--4 cm) long, with 3--7 long-stalked flowers from curved stalk. Flowers bending down on curved stalks, 3/4--1 in (2--2.5 cm) across the many spreading white or sometimes pink or reddish stamens. Buds egg-shaped, 3/8--5/8 in (10--15 mm) long and 1/4--3/8 in (6--10 mm) wide, with conical lid shorter than conical base, covered with gland dots.

Seed capsules on stalk less than 1/4 in (6 mm) long, egg-shaped and narrowed at base, about 3/8 in (10 mm) long and 5/16 in (8 mm) wide, with narrow opening, wide sunken disk, and 5 enclosed valves.

Wood dark reddish brown, with interlocked grain and fine texture, very heavy (sp. gr. 0.85), very hard, very strong, and very durable, but difficult to work.

The wood has been used in Hawaii only for fenceposts. In Australia it is used for railroad crossties and general construction, occasionally for large heavy beams.

In Hawaii this species has been planted mainly in arid to semi-arid areas, where it has been the only tree tried that survived. The Division of Forestry planted about 150,000 trees before 1960, mostly on Oahu at the Honouliuli and Mokuleia Forest Reserves, both very dry sites. It was also used on the dry ridges of Molokai. Parker Ranch planted it in several large blocks, some of which may be seen along the highway between Waimea and Puu Waawaa and lower Saddle Road near Waikii.

Champion--Height 83 ft (25.3 m), c.b.h. 7.9 ft (2.4 m), spread 51 ft (15.5 m). Haleakala Ranch, Makawao, Maui (1968).

Range--Southeastern Australia, sometimes in almost pure open stands or scattered with other eucalypts. Native in dry areas and adapted to winter rainfall climate. Widely planted, for example, in North Africa.

Other common names--red ironbark (Hawaii); red ironbark, ironbark, mugga (Australia).

110. Java-plum

Eugenia cumini (L.) Druce*

Small to medium-sized evergreen tree commonly naturalized in windward and some leeward lowlands, recognized by the clusters of oblong blackish edible fruits on old twigs back of leaves. To 50 ft (15 m) in height and 2 ft (0.6 m) in trunk diameter, with rounded crown. Bark light gray, smoothish. Inner bark with thin green outer layer, mottled light brown, astringent and

109. Red-Ironbark eucalyptus *Eucalyptus sideroxylon* A. Cunn. ex. Woolls.*
Twig with flowers, fruits (bottom), 1 X (Maiden).

bitter. Twigs light green, becoming light gray, slightly flattened, hairless.

Leaves paired (opposite), hairless, with slender light yellow stalks. Blades narrowly elliptical, 2 3/4--5 in (7--13 cm) long and 1 1/2--3 in (5--7.5 cm) wide, abruptly short-pointed at apex, short-pointed at base, slightly thickened and leathery, dull light green, paler beneath, with light yellow midvein, many straight fine parallel side veins, with many tiny gland-dots visible under a lens.

Flower clusters (cymes) on old twigs back of leaves, 2--2 1/2 in (5--6 cm) long and wide, with many paired stout branches at nearly right angles, end flower opening first. Flowers nearly stalkless, with cuplike conical light green base (hypanthium) about 1/8 in (3 mm) long and broad, with 4 tiny rounded calyx lobes on rim. Petals 4, white, rounded concave, more than 1/167 in (2 mm) long. Stamens many, white threadlike, 3/16 in (5 mm) long. Pistil with inferior ovary, many minute ovules, and stout style.

Fruits (berries) many, crowded and almost stalkless along twigs back of leaves, oblong to elliptical, 5/8--1 in (1.5--2.5 cm) long, dark purple to black, juicy, sour, edible, 1-seeded.

The wood is white to yellowish white, moderately coarse-textured, with interlocked grain. It has been used on a small scale in Hawaii as decorative veneer for manufacture of marquetry designs on place mats.

Common to abundant, thoroughly naturalized in both moist and dry lowlands of Hawaii. Classed as a weed in waste places, pastures, and rangelands. Very common in gulches and flats along windward sides of all islands, reaching fairly large size in wetter areas such as near Kaneohe, Oahu, and Hanalei, Kauai. In dry sites it is confined to moist gulches. Recorded by Hillebrand (1888) as cultivated (before 1871).

Special areas--Waimea Arboretum, Foster.
Champion--Height 91 ft (27.7 m), c.b.h. 9.9 ft (3.0 m), spread 54 ft (16.5 m). Ruddy Tong, Kapaa, Kauai (1968).
Range--Native of Indo-Malaysian region. Introduced in many tropical countries.
Other common names--jambolan, jambolan-plum, palama; mesegerak (Palau).
Botanical synonyms--*Eugenia jambolana* Lam., *Syzygium cumini* (L.) Skeels.

This genus commemorates Prince Eugene of Savoy (1663-1736), patron of botany and horticulture.

111. Rose-apple, 'ōhi'a loke

Eugenia jambos L.*

Naturalized tree with handsome foliage and edible fruits, distinguished by paired shiny dark green lance-shaped leaves, few large yellowish white 4-petaled flowers 3--4 in (7.5--10 cm) across numerous threadlike stamens, and pale yellowish or pinkish tinged, rounded or elliptical fruits 1 1/4--1 1/2 in (3--4 cm) long, with odor and flavor like rose perfume.

Small evergreen tree 15--30 ft (4.6--9 m) tall, often with several crooked trunks 4--8 in (0.1--0.2 m) in diameter, and spreading dense opaque dark green crown of many branches. Bark brown, smoothish with many small fissures. Inner bark whitish or light brown, astringent. Twigs green when young, becoming dark brown, hairless.

Leaves opposite, hairless, with short leafstalks of 3/16--3/8 in (5--10 mm). Blades lance-shaped, 3 1/2--8 in (9--20 cm) long and 5/8--1 3/4 in (1.5--4.5 cm) broad, long-pointed at apex, short-pointed at base, not toothed on edges, leathery, shiny dark green on upper surface, dull green beneath, and with tiny gland-dots visible under a lens.

Flower clusters (corymbs) terminal, commonly with 4--5 large flowers. Conical pinkish green tubular base (hypanthium) about 1/2 in (13 mm) high and wide, enclosing ovary and bearing other parts; calyx of 4 rounded broad lobes 1/2 in (13 mm) long, persistent on fruit; 4 rounded concave whitish petals about 5/8 in (15 mm) long, faintly tinged with green, coarsely gland-dotted; numerous threadlike stamens; and pistil consisting of inferior 2-celled ovary and persistent whitish slender style 1 3/4 in (4.5 cm) long.

Fruits (berries) have 4 calyx lobes at apex, pale yellow firm flesh with little juice. Seed 1 (sometimes 2) rounded brown, 3/8 in (1 cm) diameter in large cavity.

Wood dull brown, hard, and heavy (sp. gr. 0.7). Not durable in soil and very susceptible to attack by dry-wood termites. Seldom used. Elsewhere, coarse baskets and barrel hoops have been made from young branches and poles from larger limbs.

Planted in the tropics for ornament, primarily for the showy flowers and handsome foliage. Sometimes used for windbreaks and shade. Occasionally, the insipid fruits are made into jellies, preserves, and salads; it is a good honey plant. Elsewhere seeds and roots utilized in home remedies. Trees reproduce naturally from seeds and sprout vigorously when cut. Shade beneath pure thickets generally kills out all vegetation.

Planted and sparsely naturalized through the Hawaiian Islands in moist areas such as pastures, waste places, and stream banks, from sea level to 1,600 ft (488 m), rarely to 4,000 ft (1,219 m) altitude. Usually found as an understory tree in mixture with species No. 118, guava (*Psidium guajava*). Classed as a weed.

Special areas--Waimea Arboretum, Tantalus.
Range--Native of southeastern tropical Asia but now widely cultivated and naturalized through the tropics, including Puerto Rico and Virgin Islands. Planted also in Florida and southern California.

110. Java-plum *Eugenia cumini* (L.) Druce*
Flowering twig, fruit (lower right), 1 X.

Introduced into Hawaii about 1825, apparently for the edible though insipid aromatic fruit.

Other common name—pomarrosa (Puerto Rico, Spanish); youenwai (Pohnpei).

Botanical synonyms—*Jambosa jambos* (L.) Millsp., *Syzygium jambos* (L.) Alston.

112. 'Ōhi'a ai, mountain-apple *(color plate, p. 39)*

Eugenia malaccensis L.**

This species with its red, sometimes pink or white, pear-shaped fruits of applelike color undoubtedly was introduced by the early Hawaiians. It has large pretty, purplish red or rose purple flowers (sometimes whitish) composed of a mass of many threadlike stamens 2 1/2--3 in (6--7.5 cm) across, spreading like pins in a pin cushion, which drop to form a purplish red carpet on the ground beneath.

A small to medium-sized evergreen tree 20--50 ft (6--15 m) tall with erect trunk 4--8 in (0.1--0.2 m) or more in diameter and with dense conical or columnar crown of dark green foliage. Bark light brown, smoothish to slightly fissured. Inner bark brownish streaked and slightly astringent. Twigs light brown, green when young, hairless with slightly raised leaf-scars.

Leaves opposite, hairless, with stout green to brown leafstalks of 1/2--3/4 in (1.3--2 cm). Blades large oblong, 7--12 in (18--30 cm) long and 3--5 in (7.5--13 cm) broad, long-pointed at apex, short-pointed at base, leathery, slightly curved upward on both sides of midvein, side veins slightly sunken and connected near straight margins. Upper surface dark green or green and slightly shiny, lower surface dull light green, with scattered tiny gland-dots visible under a lens.

Flower clusters (cymes or panicles) 4--5 in (10--13 cm) across several to many almost stalkless flowers at end of short branched green lateral axis along branches back of leaves and trunk. Flowers odorless, with funnel-shaped light purplish green base (hypanthium) 3/4 in (2 cm) long and more than 3/8 in (1 cm) wide at top, enclosing ovary and extending as broad tube 3/16 in (5 mm) beyond. Calyx of 4 broad rounded thickened persistent sepals 1/8--3/16 in (3--5 mm) long; 4 spreading rounded concave purplish red petals 1/2 in (13 mm) long; many spreading stamens 1--1 1/4 in (2.5--3 cm) long, purplish red with yellow dot anthers; and pistil composed of inferior 2-celled ovary and persistent purplish red straight style about 1 1/4 in (3 cm) long.

Fruits (berries) pear-shaped, 2--3 in (5--7.5 cm) long and 1--2 in (2.5--5 cm) in diameter, red (sometimes pink or white), with 4 calyx lobes at apex, thin soft skin, and white crisp juicy pulp with pleasant slightly sour or sweet flavor suggesting apple. Seed 1, large rounded, 3/4 in (2 cm) in diameter, light brown. Flowering mainly in March and April or irregularly and maturing fruits in summer and autumn.

Sapwood light brown, heartwood reddish brown, straight-grained, and fine-textured. Wood described as hard, tough, very heavy, tending to warp and difficult to work. Hawaiians hewed the trunks into posts, rafters for houses, and enclosures for temples. Religious images were carved from the wood, which was regarded as sacred. A tea brewed from bark was used as a sore throat remedy.

Widely cultivated through the tropics. Fruits are eaten raw, cooked, or preserved. It is said that the slightly sour stamens are good in salads. Elsewhere astringent bark, flowers, and leaves have been used in folk remedies.

An attractive ornamental and shade tree, planted elsewhere for windbreaks. One author places it among the most beautiful flowering trees of the tropics. Easily propagated from seeds, it has moderately rapid growth. Fruits are produced in 7--8 years.

Mountain-apple was introduced into Jamaica in 1793 from Tahiti by Captain William Bligh of the British ship *Providence*. This, along with breadfruit, was one of several trees brought in to provide inexpensive food for slaves.

Naturalized in moist lowland forests through the Hawaiian Islands, mainly on windward sides in valleys and gorges. It forms almost pure stands to about 1,800 ft (549 m) altitude. Accessible groves on Oahu are in Waimano Gulch and Sacred Falls Valley (Kaliuwaa), also along Trail Number 1 in upper Makiki Valley and Kahana Valley. The form with white flowers and fruits, rare on Oahu, was named 'ōhi'a'ai hua keokeo or 'ōhi'a kea.

Special areas—Waimea Arboretum, Foster.

Champion—Height 35 ft (10.7 m), c.b.h. 4.8 ft (1.5 m), spread 20 ft (6.1 m). Kaelakehe, Kailua-Kona, Hawaii (1968).

In Puerto Rico and Virgin islands, trees are grown around buildings as ornamentals, also occasionally for windbreaks. Uncommon in southern Florida.

Range—Native probably of Malay Peninsula or Malay Archipelago. Widely planted through tropics.

Other common names—Malay-apple; manzana malaya, pomarrosa malaya (Puerto Rico); macupa (Guam); kidel (Palau); faliap (Yap); faniap (Truk); paniap (Pohnpei).

Botanical synonyms—*Jambosa malaccensis* (L.) DC., *Syzygium malaccense* (L.) Merr. & Perry.

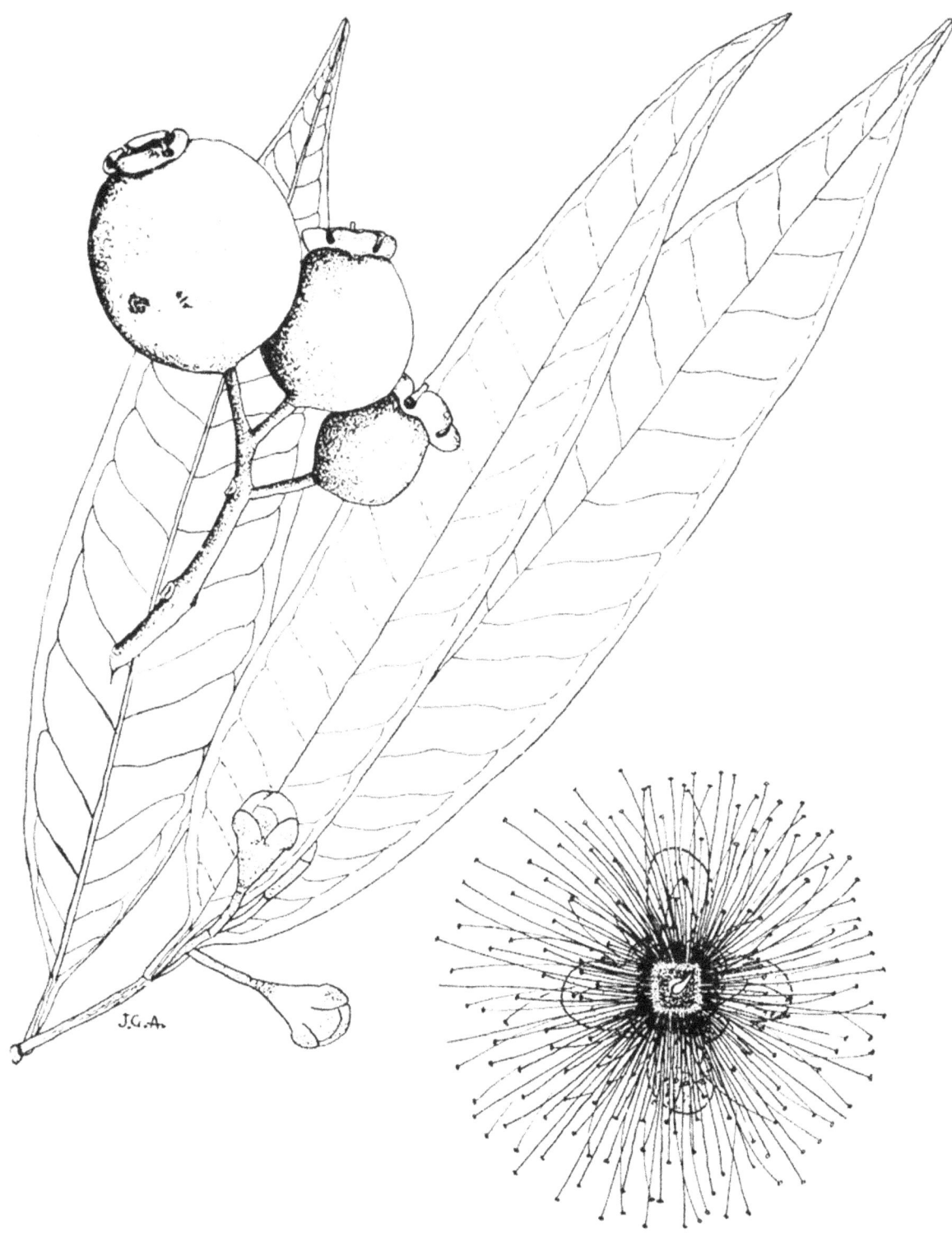

111. Rose-apple, 'ōhi'a loke *Eugenia jambos* L.*
Fruits, twig with flower buds, flower (lower right), 1 X (P. R. v. 1).

113. 'Ōhi'a ha

Eugenia sandwicensis Gray

Evergreen native tree of wet forests (except on the Island of Hawaii), characterized by 4-angled twigs, paired elliptical to oblong leaves, small white or pinkish flowers with many tiny stamens and small rounded shiny red edible fruits. A large forest tree to 60 ft (18 m) high and 3 ft (0.9 m) in trunk diameter or shrubby on exposed ridges. Bark gray to reddish brown, smoothish to slightly fissured; inner bark light brown, slightly astringent. Twigs 4-angled and slightly winged, slightly enlarged at ringed nodes, hairless, greenish when young, turning brown.

Leaves opposite, hairless, with short leafstalks of 1/8--1/2 in (3--13 mm). Blades variable in shape, elliptical to oblong, 1--4 in (2.5--10 cm) long and 3/4--2 in (2--5 cm) wide, rounded and usually notched at apex, blunt at base, often widest beyond middle and turned under at edges, slightly thick and leathery, curved up on sides, upper surface shiny green with side veins inconspicuous, beneath light green, with gland-dots visible under lens. Crushed leaves emit a distinctive odor.

Flower clusters (cymes) 1 1/2--3 in (4--7.5 cm) long at bases of upper leaves, with 4-angled branches. Flowers several to many on stalks of 1/8 in (3 mm), about 5/16 in (8 mm) long and wide, composed of funnel-shaped greenish base (hypanthium), 4 pinkish rounded calyx lobes on rim, 4 rounded fringed white or pinkish petals less than 1/8 in (3 mm) long, many tiny white stamens, and pistil with inferior 2-celled ovary and short style.

Fruits (berries) rounded and slightly flattened, 5/16--3/8 in (8--10 mm) in diameter, with calyx at top, shiny red, with slightly sour edible white pulp. Seeds 1--2, 1/8 in (3 mm) long. Fruits often abundant in late summer.

The wood is described as reddish brown, hard, and durable; it was used as fuel and for house construction by the Hawaiians.

The bark furnished a black dye for tapa or bark cloth.

Common and widespread in lower and middle wet forests to 4,000 ft (1,219 m) altitude.

Special area--Kokee.
Range--Kauai, Oahu, Molokai, Lanai, and Maui.
Other common names--hā, pā'ihi (Maui), Hawaiian syzygium.
Botanical synonym--*Syzygium sandwicense* (Gray) Ndz.

114. Manuka

Leptospermum scoparium J. R. & G. Forst.*

Introduced aromatic ornamental evergreen shrub or small tree with needlelike leaves on very slender erect twigs, whitish or pinkish flowers borne singly on short twigs, and shreddy bark on angled trunks.

Shrub or small tree to 20 ft (6 m) high and 6 in (15 cm) in trunk diameter, with irregular crown of many nearly erect wiry branches. Bark gray brown, fissured and shreddy, with thick outer layer. Inner bark light pink, fibrous, slightly bitter or astringent. Twigs very slender, erect or nearly so, gray and finely hairy when young, becoming fissured and rough.

Leaves borne singly (alternate), stalkless, needle-like and very narrow (linear), 5/16--5/8 in (8--15 mm) long and about 1/16 in (1.5 mm) wide, sharp-pointed, narrowed to stalkless base, thin, slightly stiff, with whitish pressed hairs when young, dull green above, paler beneath, with tiny gland-dots visible under lens.

Flowers single and almost stalkless at end of short side twigs, about 5/8 in (15 mm) across. The green conical base (hypanthium) 1/8 in (3 mm) long bears on rim 5 whitish pointed sepals, 5 rounded petals whitish or turning pinkish, and many threadlike white stamens, and has pistil with inferior ovary and slender style.

Fruits (seed capsules) nearly stalkless, bell-shaped, 1/4 in (6 mm) long and broad, gray, hard, and becoming very wrinkled, 5-celled, opening at flattened top along 5 spreading starlike lines, remaining attached on older twigs. Seeds many, tiny, reddish brown, 1/16 in (1.5 mm) long.

Wood light brown, hard, used elsewhere for tool handles and in small boats. Not used in Hawaii.

Planted as an ornamental shrub in Hawaii and southern California, where several horticultural varieties are distinguished. Propagated by seeds and cuttings. It is naturalized and has become a pest on Lanaihale, Lanai. Also naturalized on Kauai and Oahu at 1,100--3,900 ft (335--1,189 m).

Range--Native to Australia and New Zealand.
Other common names--teatree, broom teatree, New-Zealand-tea.

The common name teatree refers to the use of the aromatic foliage as a substitute for tea by early settlers in Australia.

115. Paperbark, cajeput-tree

Melaleuca quinquenervia (Cav.) S. T. Blake*

Paperbark, introduced for ornament and watershed cover, is easily recognized by its odd whitish bark, which splits and peels in many papery layers, by the lance-shaped or narrowly elliptical leaves with mostly 5 veins

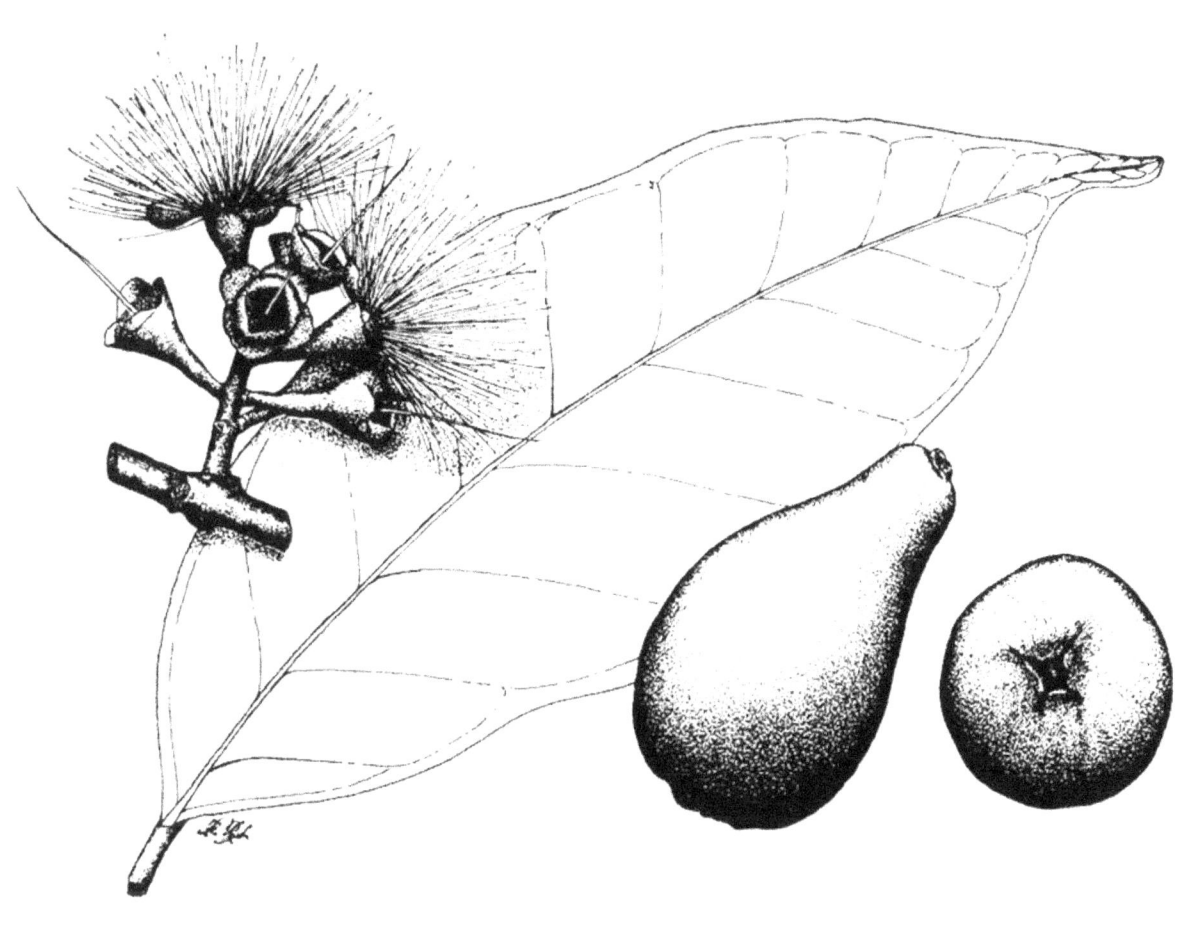

112. 'Ōhi'a ai, mountain-apple   *Eugenia malaccensis* L.**
Flowers, leaf, fruits, 1 X (P. R. v. 1).

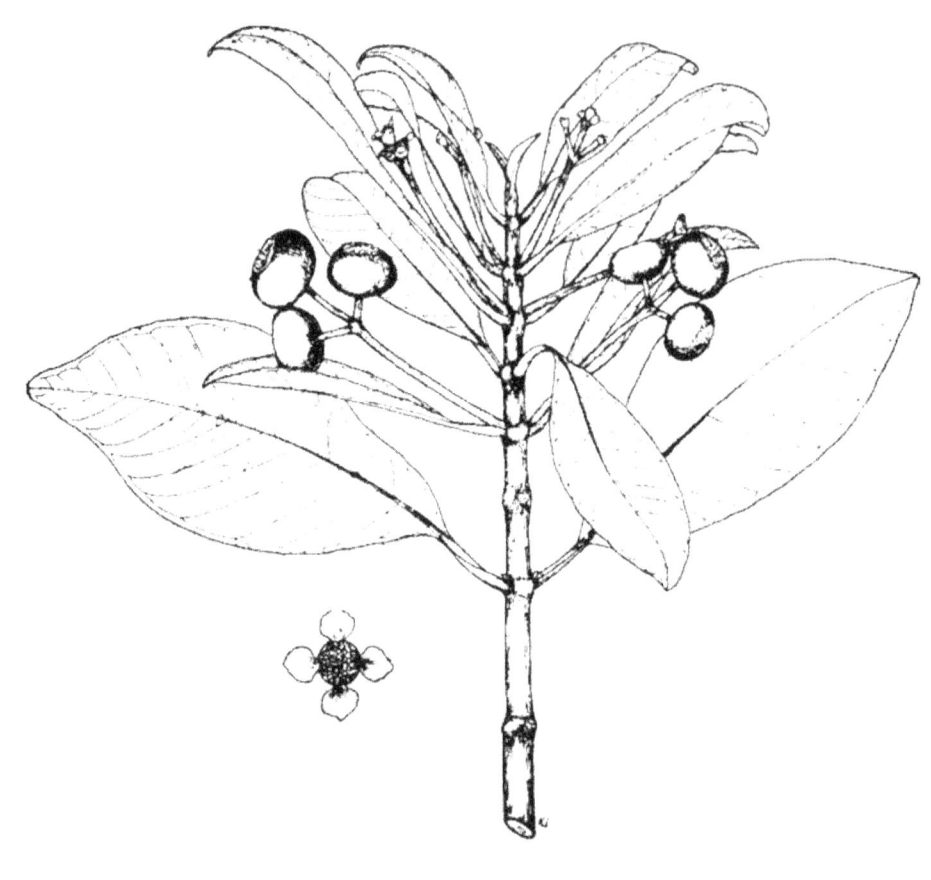

113. 'Ōhi'a ha   *Eugenia sandwicensis* Gray
Twig with flowers and fruits, 1 X; flower (lower left), 3 X (Degener).

114. Manuka *Leptospermum scoparium* J. R. & G. Forst.*
Twig with flowers and fruits, 1 X.

from base to apex, and many white flowers with many threadlike stamens, crowded and stalkless, suggesting a bottlebrush. Crushed leaves have a resinous odor and taste somewhat like those of eucalyptus.

An evergreen resinous ornamental and forest tree 20--50 ft (6--16 m) high, the trunk 1 ft (0.3 m) in diameter, slightly angled and grooved, with main axis and irregular branches forming a narrow or open crown. In age, a large tree to 75 ft (23 m) and 2--3 ft (0.6--0.9 m). Bark of trunk and branches whitish, very thick, corky or spongy, composed of many light pink fibrous papery layers. Inner bark light brown, slightly sour. Twigs long and slender, often drooping, light brown and finely hairy when young, turning gray. End and side buds round to cylindrical, 1/8--1/4 in (3--6 mm) long, greenish brown, composed of many rounded overlapping scales.

Leaves alternate, with finely hairy light green leafstalks of 1/8 in (3 mm). Blades 1 1/2--3 1/2 in (4--9 cm) long and 1/4--3/4 in (6--19 mm) wide, long-pointed at both ends, not toothed on edges, slightly thickened and stiff, upper surface gray green and hairless, with 5 (sometimes 7) veins (as specific name indicates), faint and nearly parallel, lower surface paler and often slightly hairy.

Flower clusters (spikes) 1--3 in (2.5--7.5 cm) long and 1 1/2 in (4 cm) across, at end of twig, which elongates and forms new leaves beyond, composed of many crowded stalkless flowers 5/8 in (15 mm) long. Flowers have base (hypanthium) 1/16 in (1.5 mm) long; calyx of 5 half-round lobes less than 1.5 mm long; 5 concave whitish petals nearly 1/8 in (3 mm) long; and about 30 threadlike white stamens nearly 5/8 in (15 mm) long, slightly united in 5 groups at base and falling together early; and pistil composed of inferior 2--4-celled ovary with many ovules, long threadlike white style, and brown dot stigma.

Seed capsules, many crowded and stalkless in groups 1--3 in (2.5--7.5 cm) long on gray twigs back of leaves or between groups of leaves, short cylindrical, 1/8 in (3 mm) long and 3/16 in (5 mm) wide, gray brown, hard and persistent, opening at flattened apex by 3--4 blunt hairy valves or teeth. Seeds many, minute, less than 1/16 in (1.5 mm) long, very narrow, brown.

Sapwood is pale brown and heartwood light pinkish brown. Wood heavy (sp. gr. 0.58) with fine texture.

116. 'Ōhi'a lehua (color plate, p. 40)

'Ōhi'a lehua, the most common and most widespread large tree of Hawaii's wet forests, is well known by its showy clusters of large red flowers formed by a mass of threadlike stamens and by its crowded paired small leaves. Extremely variable and divided into numerous varieties based on leaf shape, hairiness, and flower color, varying from scarlet to pink, salmon, and yellow. The red-flower form is the official county flower of the Island of Hawaii.

Wood grown in Australia is reputed to be resistant to decay and termites, but these properties have not been evaluated for Hawaii-grown wood. Australian wood (sp. gr. 0.67) is quite strong in bending, tough, and hard, comparable to somewhat denser eucalypts in these properties. Tests of wood from Florida indicate that it is similar in properties to dogwood (*Cornus* spp.). The wood has not been used in Hawaii but should be suitable for fenceposts. The problem with utilization is that the inside bark diameter of the stems is generally quite small.

A fast growing hardy tree resistant to wind, drought, fires, and salt water and suitable for windbreaks and beach planting. Propagated from seeds.

The thick papery bark has served elsewhere as packing material for fruits, caulking for boats, and as torches. Cajeput oil of medicine is obtained from the leaves and twigs of this and related species by steam distillation.

In Hawaii, the paperbark is grown extensively in pure and mixed forest stands for watershed cover, windbreaks, and as an ornamental. Naturalized, but not a pest as in Florida. Forest fires, which cause the seed pods to open and disseminate the seeds, are rare in Hawaii while common in Florida. The first recorded planting by the Division of Forestry was in 1917 on the land of Luaalaea in Manoa Valley, Oahu. Since then, the paperbark has become the third most commonly planted tree in Hawaii, with more than 1.7 million trees planted in the Forest Reserves, because it thrives on very harsh eroded and wet sites where others do not.

Special areas--Tantalus, Kalopa.

Champion--Height 75 ft (22.9 m), c.b.h. 10.2 ft (3.1 m), spread 66 ft (20.1 m). State Forestry Arboretum, Hilo, Hawaii (1968).

Range--Native from eastern Australia to New Caledonia and Papua, where it grows on coastal flats. Planted and naturalized in tropical regions. Naturalized and very common in southern Florida. Planted also in southern Texas, southern California, and Puerto Rico.

Other common names--paperbark-tree, bottlebrush, punk-tree; cayeputi, cayeput (Puerto Rico).

Formerly referred to *Melaleuca leucadendron* (L.) L., a related species of northern and northeastern Australia, southern New Guinea, and Amboina.

Metrosideros polymorpha Gaud.

Large evergreen tree becoming 80 ft (24 m) tall and 3 ft (0.9 m) in trunk diameter, sometimes larger, with compact to irregular crown, or only a shrub in exposed places. Giants to 100 ft (30 m) have been recorded. Trunks straight to twisted and crooked, large ones often composed of compacted stiltlike roots. Bark light gray, becoming rough and thick, fissured and scaly, sometimes shaggy. Inner bark light brown, slightly astringent.

115. Paperbark, cajeput-tree *Melaleuca quinquenervia* (Cav.) S. T. Blake*
Twig with flowers and fruits, 1 X (P. R. v. 2).

Twigs slender, from hairless to covered with dense coat of white hairs.

Leaves crowded, opposite, short-stalked to nearly stalkless, from hairless to white hairy, the young leaves often reddish. Blades small, very variable, elliptical, ovate, or rounded, mostly 3/4–3 in (2–7.5 cm) long and 1/2–1 1/2 in (1.3–4 cm) wide, short-pointed to rounded at both ends, sometimes notched at base, not toothed on edges, slightly thick and leathery, upper surface commonly shiny green or dull and hairless, with many fine side veins usually inconspicuous or not visible, and lower surface paler and often hairy, sometimes with dense coat of white hairs.

Flower clusters (cymose corymbs) terminal, branched, of many short-stalked flowers in groups of 3 forming mass of threadlike stamens to 3 in (7.5 cm) in diameter, varying in color from scarlet to pink, salmon, and yellow. Each flower is composed of a bell-shaped base (hypanthium) to 1/4 in (6 mm) long, which bears 5 blunt calyx lobes 1/8 in (3 mm) long, 4 petals 1/8–1/4 in (3–6 mm) long, and many spreading threadlike stamens 1–1 1/4 in (2.5–3 cm) long with dot anthers; pistil with inferior 3-celled ovary and long threadlike style.

Fruit (capsule) bell-shaped, 1/4–3/8 in (6–10 mm) long and broad, brown, often hairy, slightly 3-lobed, 3-celled, with calyx remaining. Seeds many, minute.

Sapwood is pale brown, grading gradually into reddish-brown heartwood. The wood is heavy (sp. gr. 70), very hard, strong, and has a large shrinkage in drying. It is not resistant to decay but is moderately resistant to subterranean termites. Although variable in this characteristic, most 'ōhi'a wood is readily pressure treated with preservatives.

The Hawaiians used the wood for construction, carved images, household implements, and wear-strips along the gunwales of canoes.

Principal modern uses include flooring, ship blocking, marine construction, irrigation canal stakes, pallets, fenceposts, and decorative poles. Objects made from the wood have ranged from ukulele keys to railroad crossties. The Santa Fe Railroad Company used about 5 million 'ōhi'a ties on the West Coast in the early part of the century until the wood was found not durable. Drawbacks to wider use of 'ōhi'a are its density and shrinkage, which limit its usefulness, coupled with expensive logging because it occurs only in low-volume stands.

The flowers are used in weaving beautiful garlands or leis and are considered sacred to Pélé, goddess of volcanoes, and thus the subject of songs and legends. Native birds feed upon the secreted nectar. The small bird called 'i'iwi has scarlet plumage, which matches the flowers. Another, 'apapane, is dark red like the darker red-flowered lehua 'apane. However, green birds such as 'amakihi also feed on the flowers.

Young red leaves called liko lehua are used in leis and folk remedies and make a pleasant tea. Because of the abundant nectar, this species is classed as a honey plant.

The seeds commonly germinate on trunks of treeferns in the forest understory. Then the seedlings send roots down to the ground. These roots may enclose the fern trunk. Thus, the larger tree is often supported on its roots, having shaded and killed its host. Because of this habit, early Hawaiians regarded the fern as the parent. Mature trees sometimes produce large masses of aerial roots that hang down from branches and may take up some moisture from the air. 'Ōhi'a lehua seedlings also get an early start on fresh aa lava flows. The pioneer fern *Sadleria* serves as a nurse plant for the germination and establishment of an 'ōhi'a lehua tree. But most 'ōhi'a lehua trees originate from seed that has germinated in the soil. Seedlings are commonly found along the edges of roads. As with koa and other trees, soil disturbance forms a good seed bed. Intolerant of shade as a seedling, this species requires an opening to get started.

This species is distributed from sea level to timber line at 8,500 ft (2,591 m) through wet to dry forests. It is the dominant or most common native tree through the islands. It may occur as a shrubby tree on exposed lava flows or even as a low creeping or prostrate shrub exposed on mountain ridges and in bogs.

Special areas--Kokee, Keahua, Waimea Arboretum, Aiea, Haleakala, Kalopa, Volcanoes, Kipuka Puaulu.

Champion--Height 84 ft (25.6 m), c.b.h. 17.8 ft (5.4 m), spread 78 ft (23.8 m). Waipunalei, Hilo, Hawaii (1968).

Range--Six largest Hawaiian Islands.

Other common name--lehua 'ōhi'a.

Botanical synonym--*Metrosideros collina* (J. R. & G. Forst.) Gray subsp. *polymorpha* (Gaud.) Rock.

The common name *lehua* meaning hair, may have been suggested by the numerous threadlike stamens. Many other names were applied to variations in flowers and leaves, according to Rock. Lehua mamo had orange yellow flowers; lehua pua kea, white flowers; lehua kū makua, stalkless heart-shaped leaves; lehua lau li'i, very small leaves.

The Hawaiian trees have been classed also as a subspecies of *Metrosideros collina* (J. R. & G. Forst.) Gray, which occurs on high South Pacific islands to New Zealand (Rock 1917b). However, that species, as defined in the monograph by Dawson (1970) and supported by St. John (1979), is restricted to Tahiti and Rarotonga. The shorter name with rank of species, *Metrosideros polymorpha* Gaud., is simpler and appropriate for the numerous Hawaiian variations distinguished further as about 8 varieties. Also, 4 other local species within Hawaii have been accepted.

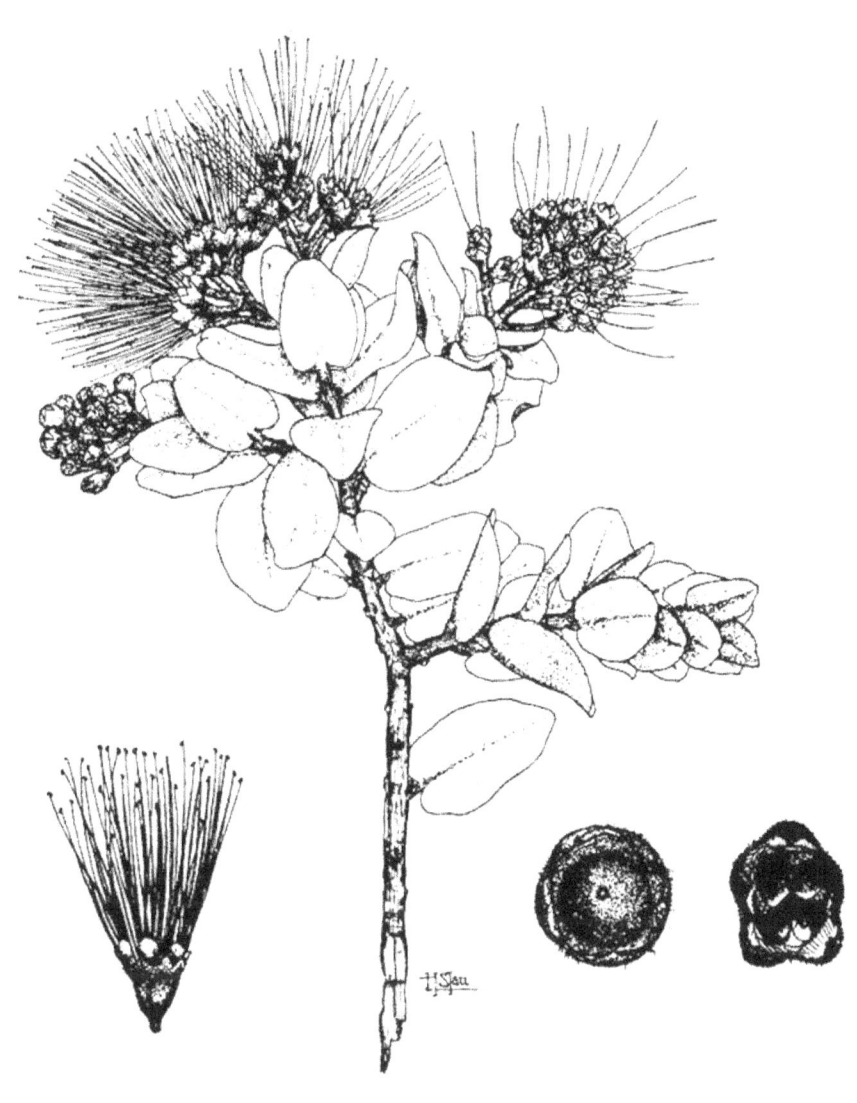

116. ʻŌhiʻa lehua *Metrosideros polymorpha* Gaud.
Flowering twig, 1/2 X; flower (lower left), 1 X; fruits (lower right), 2 X (Degener).

117. Strawberry guava
Psidium cattleianum Sabine*

Small evergreen tree or shrub planted for its round or elliptical dark reddish purple or yellow edible fruits and widely naturalized and forming thickets in lowland areas. Becoming 20--50 ft (6--15 m) high, with slightly angled trunk 4--12 in (0.1--0.3 m) in diameter. Bark gray or light brown, smooth, peeling off and exposing light greenish brown inner layers. Inner bark light pink, bitter and astringent. Twigs brown, hairless.

Leaves paired (opposite), hairless, leafstalks green, 1/4--1/2 in (6--13 mm) long. Blades elliptical, 1 1/2--3 1/2 in (4--9 cm) long, and 1--1 3/4 in (2.5--4.5 cm) wide, short-pointed at apex, broadest beyond middle, and tapering to long-pointed base, slightly thick and leathery and turned under at edges, above shiny dark green with inconspicuous veins, beneath dull light green with tiny gland-dots.

Flowers 1--2 on short stalks at leaf bases, white, less than 1 in (2.5 cm) across, composed of greenish conical base (hypanthium) less than 1/4 in (6 mm) long, 4--5 rounded green calyx lobes that remain at top of fruit, 4--5 elliptical white petals, many threadlike white stamens, and pistil with inferior ovary and slender style.

Fruits (berries) 1--2 on slender stalks at leaf bases or back of leaves, round or elliptical, 1--1 1/2 in (2.5--4 cm) long, dark reddish purple or sometimes yellow, with 4 rounded thick calyx lobes 1/4 in (6 mm) long at apex, thick-walled, pinkish or whitish, juicy, slightly sour to sweet edible pulp, aromatic. Seeds many, rounded or elliptical, 3/16 in (5 mm) long, hard, light brown or yellow.

Sapwood is yellowish white and heartwood pale reddish brown. A fine-textured, moderately heavy wood used only for fuelwood in Hawaii.

Fruits are eaten raw or made into jam or jelly with strawberry flavor. One variety of larger size with large yellow fruits is called yellow strawberry guava.

Introduced into Hawaii in 1825 for the edible fruit, but now thoroughly naturalized and established. It forms thickets in moist lowland areas up to about 2,500 ft (762 m) elevation, rarely to 4,000 ft (1,220 m). Classed as a weed in pastures, rangelands, and waste places. Grown also in southern California and Florida.

Special areas--Waimea Arboretum, Wahiawa, Tantalus, Aiea, Kalopa, Pepeekeo.

Range--Native of Brazil.

Other common names--purple strawberry guava, Cattley guava, waiawī.

Botanical synonym--*Psidium littorale* Raddi.

118. Guava, kuawa
Psidium guajava L.*

Guava is an evergreen shrub or small tree cultivated for its rounded yellow edible fruits and widely naturalized as a weed tree in lowland thickets. It differs from No. 117, strawberry guava (*Psidium cattleianum*), in the larger flowers, larger yellow fruits, and in thinner leaves with many sunken parallel side veins.

Shrub or small tree to 30 ft (9 m) high, with trunk to 8 in (0.2 m) in diameter and widely spreading crown. Bark brown, smooth, thin, peeling off in thin sheets, exposing greenish brown inner layers. Inner bark brown, slightly bitter. Twigs 4-angled and slightly winged when young, hairy and green, becoming brown.

Leaves opposite, with short broad leafstalks of 1/8--1/4 in (3--6 mm). Blades oblong or elliptical, 2--4 in (5--10 cm) long and 1--2 in (2.5--5 cm) wide, short-pointed or rounded at both ends, slightly thickened and leathery, with edges a little turned under. Upper surface green or yellow green, dull or slightly shiny, almost hairless at maturity, with many sunken parallel side veins, and lower surface paler, finely hairy, with side veins raised, and with tiny gland-dots visible under a lens.

Flowers mostly 1, sometimes 2--4, scattered at leaf bases on stalks of 3/4--1 in (2--25 cm), white, fragrant, about 1 1/2 in (4 cm) across. The green finely hairy tubular base (hypanthium) 3/8 in (10 mm) long and broad encloses the ovary and bears other parts; calyx of 4--5 yellow green rounded, slightly thickened, finely hairy lobes 3/8--5/8 in (10--15 mm) long, which remain at top of fruit; 4--5 elliptical rounded white petals 5/8--3/4 in (15--19 mm) long; very many spreading threadlike white stamens; and pistil with inferior 4--5-celled ovary and slender white style.

Fruits (berries) rounded or sometimes pear-shaped, yellow, 1 1/4--2 in (3--5 cm) in diameter and as much as 3 in (7.5 cm) long, smooth or slightly rough, with 4--5 calyx lobes at apex, with strong mellow odor at maturity, edible. Outer layer thin, yellow, slightly sour or sweet, and juicy pinkish or yellow pulp. Seeds many, elliptical, more than 1/8 in (3 mm) long, yellow.

Sapwood light brown, heartwood reddish brown. The hard, strong, heavy wood (sp. gr. 0.8) has been used for tool handles, implements, and charcoal.

Elsewhere, the bark has been employed in tanning. Extracts from leaves, bark, roots, and buds have served in folk medicine. Hawaiians made a medicinal tea from leaf buds.

Commonly cultivated through the tropics for the fruits, which are unusually rich in vitamin C. Fruits can be eaten raw, although the pulp is many seeded. Guava paste, jelly, preserves, and juice are prepared from the fresh fruit and guava powder from dehydrated fruits.

Several horticultural varieties have been named. Hawaiians distinguished a few by their fruits, according to Neal: kuawa-lemi or lemon guava with pink pulp; kuawa-ke'oke'o with whitish pulp; and kuawea-momona with larger seeds, sweet pink pulp, and thicker skin.

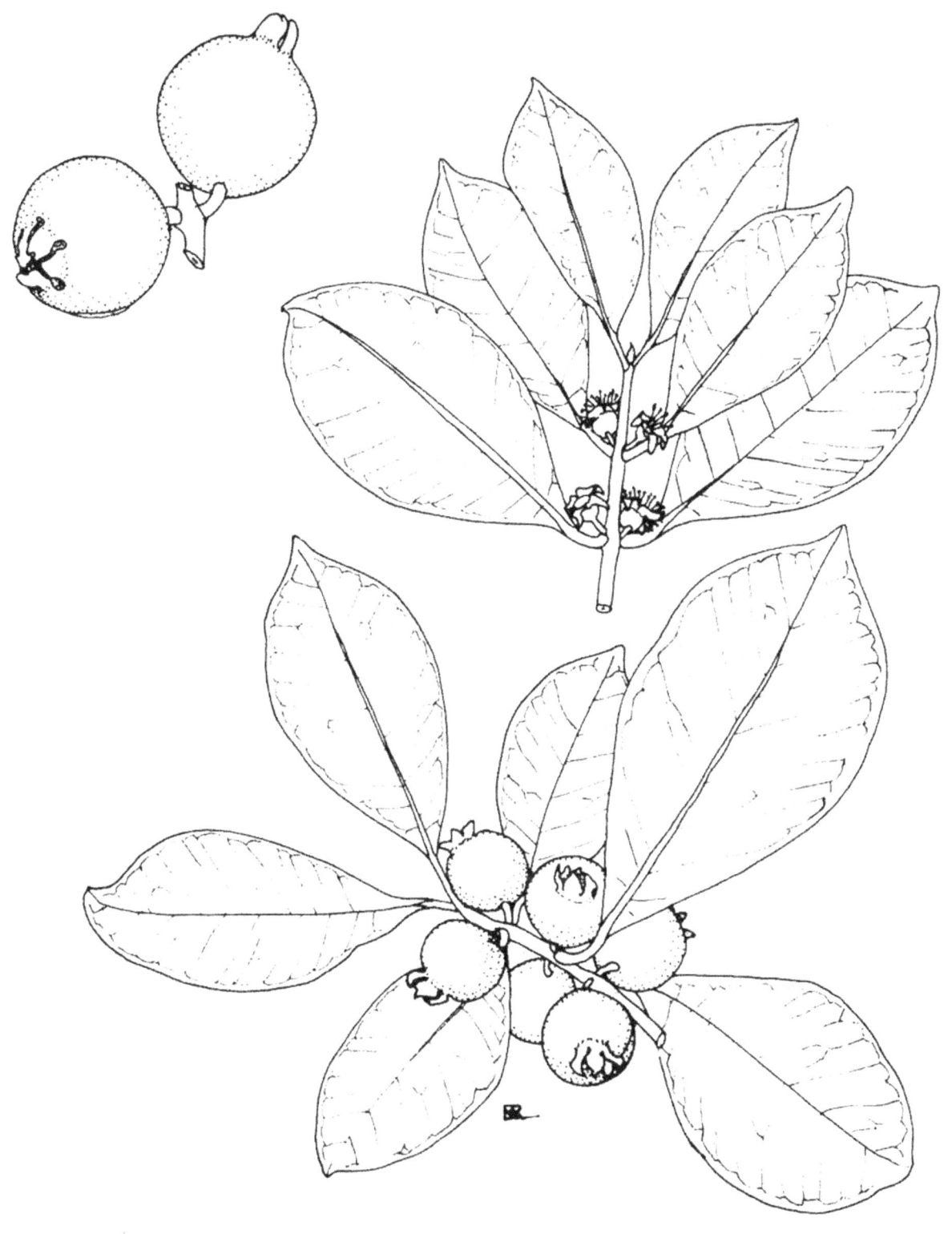

117. Strawberry guava *Psidium cattleianum* Sabine*
Fruits (upper left), flowering twig (upper right), and fruiting twig (below), 1 X.

Guava plants are propagated by root and stem cuttings. They begin to bear at 3 to 4 years, are best at 15 to 25, and die at about 50.

Guava was introduced into Hawaii early in the 19th century, apparently by Don Marín. The oriental and Mediterranean fruit flies and melon flies, which infest about half of the fruits, came later.

In Hawaii, guava is planted and abundantly naturalized through the islands. Classed as an undesirable weed in pastures, rangelands, and waste places. More than a century ago, Hillebrand (1888, p. 130), who left Hawaii in 1871, noted that guava had "spread over many parts of the islands, in some valleys forming close thickets, to the exclusion of every other shrub or tree."

The seeds are scattered by cattle, wild hogs, and birds. Guava thickets in pastures destroy forage plants.

119. Turpentine-tree

Large resinous evergreen tree related to *Eucalyptus*, introduced in forest plantations and for shade, differing in the leaves 2--4 at a node and the small flowers and seed capsules united in round balls. To 100 ft (30 m) high and 3 ft (0.9 m) in trunk diameter, slightly enlarged at base, reported to become much larger where native. Bark gray or reddish brown, very thick, very deeply furrowed into long irregular ridges, becoming shreddy or shaggy and peeling off. Inner bark fibrous, reddish brown, the innermost with blackish resin. Twigs slightly 4-angled, light green, turning brown, with tiny pressed hairs.

Leaves 2--4 at a node, leafstalks 1/2--1 in (13--25 mm) long, light yellow to brown, grooved above, with tiny hairs. Blades narrowly elliptical, 2 1/2--5 in (6--13 cm) long and 1--1 3/4 in (2.5--4.5 cm) wide, short- to long-pointed at both ends, slightly turned under at edges, slightly thickened, above dull green and almost hairless, beneath dull whitish green with tiny hairs and gland-dots visible under a lens.

Flower clusters (heads) round like a ball, single on stalks of about 1 in (2.5 cm) at base of new leaves. Flowers 6--11 in a ball, small, white, united at bell-shaped base (hypanthium). Calyx 4-lobed on rim, remaining on fruit; petals usually 4, white, rounded, 1/8 in (3 mm) long; many stamens on rim; and pistil with inferior 3-celled ovary and long slender persistent style.

Fruit (multiple) rounded, 1/2--3/4 in (13--19 mm) in diameter, gray with tiny hairs, hard, composed of 6--11

120. Brushbox

Large evergreen tree related to *Eucalyptus*, introduced in forest plantations. Planted as an ornamental and for shade. Differs from *Eucalyptus* by the leaves, which are crowded at ends of twigs, and the very numerous stamens united in 5 groups.

Eradication is difficult because of root sprouts. In the late 1800's firewood gatherers on the windward side of Oahu were stopped from cutting guava, the principal fuelwood, because the cutting caused extensive root sprouting and spread of the plants in pastures.

Special areas--Kokee, Waimea Arboretum, Wahiawa, Aiea, Tantalus.

Range--Native of tropical America, probably from southern Mexico to South America, the range greatly extended through cultivation in tropical and subtropical regions of the world. Naturalized in southern Florida and Puerto Rico and Virgin Islands, and planted in California.

Other common names--common guava; guayaba (Puerto Rico, Spanish); abas (Guam); apas (N. Marianas); guabang (Palau); abas (Yap); guahva (Pohnpei).

Syncarpia glomulifera (Sm.) Niedz.*

united seed capsules like bells 1/4 in (6 mm) across rim, each opening at top. Seeds many, tiny, brown, 1/16 in (1.5 mm) long, very narrow.

Sapwood light yellow, heartwood reddish brown. The wood is described as hard, heavy, fine-textured. Australian wood (sp. gr. 0.7) is very durable, and resistant to decay, marine borers, and termites. It is difficult to burn and almost fireproof, thus not suitable for fuel. In Australia, used for marine piling, railway crossties, fenceposts, construction, and flooring. The bark yields turpentine.

Of rapid growth and suitable as a shade tree. This tree is reported to attain a height of 200 ft (61 m) and a trunk diameter of 9.5 ft (2.9 m). Planted also in southern California.

The Division of Forestry has planted more than 83,000 trees in the forest reserves on all islands, the largest number on Oahu in the Honolulu, Honoululi, and Mokuleia Forest Reserves. Trees may be seen at Waahila Ridge State Park (St. Louis Heights), Keaiwa Heiau State Park (Aiea Heights) on Oahu, at Waihou Spring Forest Reserve on Maui, and at Pepeekeo Arboretum (Puukauku) near Pepeekeo, Hawaii.

Special areas--Wahiawa, Aiea, Kalopa.

Champion--Height 75 ft (22.9 m), c.b.h. 14.5 ft (4.4 m), spread 38 ft (11.6 m). Iole, Hawi, Hawaii (1968).

Range--Native to Australia.

Other common name--turpentine-myrtle.

Botanical synonym--*Syncarpia laurifolia* Ten.

Tristania conferta R. Br.*

To 60 ft (18 m) high with straight trunk 1 1/2 ft (0.5 m) in diameter or larger, with narrow rounded crown of dense foliage. Bark gray brown, becoming rough, thick, slightly scaly with long fissures. Inner bark light brown, fibrous, slightly bitter. Twigs light green and with tiny pressed hairs when young, turning brown and shedding

118. Guava, kuawa *Psidium guajava* L.*
Flowering twig, fruit (below), 2/3 X (P. R. v. 1).

119. Turpentine-tree *Syncarpia glomulifera* (Sm.) Niedz.*
Twig with flowers, fruits (lower right), 1 X (Maiden).

120. Brushbox *Tristania conferta* R. Br.*
Twig with flowers and fruiits, 1 X.

outer layer, with rings of raised half-round leaf scars. End buds light green, with rounded overlapping scales.

Leaves mostly 4--5 at enlarged nodes though borne singly (alternate), gray hairy when young, becoming hairless. Leafstalks slender, light green, 1/2--1 in (13--25 mm) long, flattened. Blades elliptical or narrowly ovate, 2 1/2--6 in (6--15 cm) long and 1--2 1/2 in (2.5--6 cm) wide, long-pointed at apex, short-pointed at base, slightly thick and leathery, above dull green with light yellow midvein and very fine side veins, beneath dull light green.

Flowers 3--7 clustered at end of short flattened unbranched stalk 1/2 -1 in (13--25 mm) long at leaf bases and back of leaves, white, fragrant, about 1 in (2.5 cm) across. Basal cup (hypanthium) conical, 1/4 in (6 mm) long and broad, light green, hairy, bearing 5 pointed green calyx lobes, 5 rounded white petals about 1/2 in (13 mm) long, short-stalked and hairy, and very numerous short threadlike white stamens united in 5 columns 3/8 in (10 mm) long; pistil with half inferior 3-celled ovary and threadlike style.

Fruits (seed capsules) 1--7 clustered at end of flattened stalk on twig back of leaves, cup-shaped, 3/8--5/8 in (1--1.5 cm) in diameter, light green to brown, opening at flattened apex, hard, 3-celled. Seeds many, light brown, less than 1/8 n (3 mm) long, narrow.

Sapwood pale brown and heartwood pinkish to grayish brown. Wood heavy (sp. gr. 0.61), with fine texture and mildly interlocked grain but little figure. It has a relatively large shrinkage in drying and tends to warp in seasoning. It is not subject to severe growth stress problems in manufacturing as are certain eucalypts. Wood grown in Hawaii is moderately resistant to decay and termites. Wood from Australia is classed as very resistant to both. In Hawaii, it has been used for pallets, flooring, and pulp chips, generally mixed indiscriminately with the wood of Eucalyptus saligna. Elsewhere, used for construction, shipbuilding, bridges, railway crossties, and mallets. A sizable amount of flooring of this species has been imported to Hawaii from Australia.

Planted in moist areas of Hawaii in forest plantations and as a handsome shade tree. The Division of Forestry has planted more than 396,000 trees in the forest reserves on all islands, but mostly on Oahu and Hawaii. Oahu has 1.1 million board feet of timber and Hawaii 1.3 million. Trees may be seen at Waahila Ridge State Park (St. Louis Heights) and Keaiwa Heiau State Park (Aiea Heights). Also planted for shade in the Fort Street Mall in downtown Honolulu. Grown also in southern California and Florida.

Special areas--Wahiawa, Aiea, Tantalus.

Champion--Height 60 ft (18.3 m), c.b.h. 12.9 ft (3.9 m), spread 47 ft (14.3 m). Ulupalakua Ranch Co., Ulupalakua, Maui (1968).

Range--Native of east coast of Australia.

Other common names--Brisbane-box, vinegartree.

Botanical synonym--Lophostemon confertus (R. Br.) P. G. Wilson & J. J. Waterhouse.

Wilson and Waterhouse (1982) concluded that Tristania was a heterogeneous assemblage and divided it into 5 genera.

MELASTOME FAMILY (MELASTOMATACEAE*)

121. Glorybush

Tibouchina urvilleana (DC.) Cogn.*

This introduced ornamental has escaped from gardens and become naturalized as a weed, forming impenetrable thickets. It has large showy flowers 3--4 in (7.5--10 cm) across the 5 spreading violet or purple petals. Distinguished also by the paired elliptical, velvety hairy leaves with 5 main veins from base.

A tall evergreen shrub commonly 15 ft (4.6 m) high or becoming a small tree to 40 ft (12 m) high and 4 in (10 cm) in trunk diameter, with few slender erect branches. Bark light gray, smoothish, thin. Twigs 4-angled and slightly winged, stout, light green, with long greenish or pinkish spreading hairs, ringed at nodes; older twigs shedding hairy bark and becoming round.

Leaves opposite, elliptical, with very hairy leafstalk 1/4--3/4 in (6--19 mm) long. Blades elliptical, 2 1/4--5 in (3--13 cm) long and 1--2 1/2 in (2.5--6 cm) wide, long-pointed at apex, blunt or rounded at base, edges straight, with 5 (sometimes 7) main veins from base; these and curved smaller veins sunken on upper surface and raised beneath. Upper surface yellow green, covered with pressed hairs; lower surface silvery or light green, velvety hairy. Dying leaves turning red above, silvery orange beneath.

Flower clusters (panicles) terminal, erect, large, branched, 3--5 in (7.5--13 cm) long. Flowers several but not opening together, very large and showy, on short hairy stalks, composed of 2 large hairy pointed pinkish bracts or scales 1 in (2.5 cm) long and rose-red buds; densely hairy calyx with narrow tube 5/8 in (1.5 cm) long, 5 narrow spreading hairy, reddish-tinged lobes almost as long, lobes shedding; 5 violet or purple petals about 1 1/2 in (4 cm) long and nearly as broad, oblong, broad and straight at apex, widely spreading and falling early; 10 long threadlike purple stamens of 2 sizes, bent in middle, with narrow curved anthers; and pistil with hairy 5-celled ovary, many tiny ovules, and long threadlike curved purple style.

121. Glorybush *Tibouchina urvilleana* (DC.) Cogn.*
Flowering twig and undeveloped fruits (lower right), 3/5 X (Degener).

Fruit an egg-shaped pale brownish capsule 5/16 in (8 mm) long, 5-celled, with many small round seeds.

The wood is reddish brown, hard, strong, and difficult to cut. Not used.

This species has become naturalized and is classed as a weed in pastures and wastelands. It forms dense thickets by spreading vegetatively from roots, and cut branches also bear roots and grow.

Scattered, forming impenetrable tangles in moist open forests and roadsides at 1,500--4,000 ft (457--1,219 m) altitude, especially in the Volcano area of the island of Hawaii.

Special areas--Wahiawa, Volcanoes.

Range--Naturalized through the Hawaiian Islands from Kauai and Oahu to Hawaii. Native of Brazil (Rio de Janeiro and Rio Grande do Sul) and widely cultivated as an ornamental.

Other common names--Urville glorybush, Hawaiian glorybush, tibuchina, lasiandra.

Botanical synonym--*Lasiandra urvilleana* DC. Formerly identified as *Tibouchina semidecandra* Cogn., a related species also of Brazil.

Degener (1930) stated that it was introduced in about 1910 from South America, traced to an estate near Kurtistown, Hawaii. The first herbarium specimen was collected by Rock in August 1917 at Kalanilehua, Kilauea, Hawaii. It was spread by amateur horticulturists who took cuttings to their gardens. By 1930 this weed was observed in a garden near Honolulu. At that time Degener predicted the spread and also the replacement of native vegetation, which has since occurred. As he observed in 1970, this species exemplifies the difficulty of eradicating ornamentals cultivated in the islands that become weeds.

Two additional shrubby naturalized species in this family have become noxious weeds. Koster's curse, *Clidemia hirta* (L.) D. Don, has small white flowers and small blackish or purplish edible berries. *Melastoma candidum* D. Don (*M. melabathricum* auth., not L.) has large pink to purple flowers and small bristly berrylike fruits. They are of no forage value and may be found in wastelands, pastures, and lowland forests.

GINSENG OR ARALIA FAMILY (ARALIACEAE)

122. Lapalapa

Cheirodendron platyphyllum (Hook. & Arn.) Seem.

Lapalapa is an aromatic small evergreen tree scattered in wet forests of mountains of Oahu and Kauai. It is distinguished by paired large compound leaves with 3 broad half-round leaflets wider than long, sometimes with few teeth on edges, continuously trembling on their long slender stalks. Crushed foliage and bark have an odor like that of carrot or oil and a spicy or turpentine taste.

A small tree to 26 ft (8 m) high and 8 in (20 cm) in trunk diameter, with rounded open crown, hairless throughout. Bark gray, smoothish; inner bark greenish, slightly spicy, aromatic. Twigs stout, enlarged and ringed at nodes, with raised half-round leaf-scars, purplish, becoming brownish, weak, and brittle.

Leaves opposite, 4--8 in (10--20 cm) long, with very slender purplish or greenish leafstalks 2--4 in (5--10 cm) long, slightly flattened, enlarged and slightly clasping at base. Leaflets 3, spreading on slender flattened stalks of 1 1/4--2 in (3--5 cm). Blades half-round or broadly ovate, wider than long, 1 1/2--3 1/4 in (4--8 cm) long and 2--4 in (5--10 cm) wide, rounded with abrupt narrow point at apex, nearly straight at base, edges often with few small teeth, slightly thickened, upper surface shiny green with many fine straight side veins, lower surface dull light green.

Flower clusters (panicles) terminal, 4--6 in (10--15 cm) long, with many slender forking purplish branches and many flowers spreading on short equal stalks (umbels). Flowers 1/4 in (6 mm) long, purplish, composed of cuplike base (hypanthium) 1/8 in (3 mm) long, calyx of 5 tiny teeth, 5 narrow spreading petals 1/8 in (3 mm) long and shedding early, 5 short stamens, and pistil with inferior 5-celled ovary and 5-dotlike stigmas (parts sometimes in 4's).

Fruits (berries) round, about 1/4 in (6 mm) in diameter, shiny purplish black, with ring of calyx and stigmas at apex, with spicy flesh, 5-angled when dry. Seeds (nutlets) 5 or fewer, more than 1/8 in (3 mm) long, brown.

Soft whitish wood will burn when freshly cut. No other uses are reported. It is quite likely that the tree had uses similar to those of No. 123, 'ōlapa (*C. trigynum*). The leaves and bark were probably used to make a bluish dye, and poles were probably cut from the tree because it is soft and easily cut.

Common in wet forests and swamps at middle altitudes of 2,200--5,000 ft (671--1,524 m) on Kauai and Oahu. On Oahu confined to summit ridges of Koolau Range and swamp on top of Mt. Kaala.

Range--Oahu and Kauai only.

Botanical synonym (or variety)--*Cheirodendron kauaiense* Krajina.

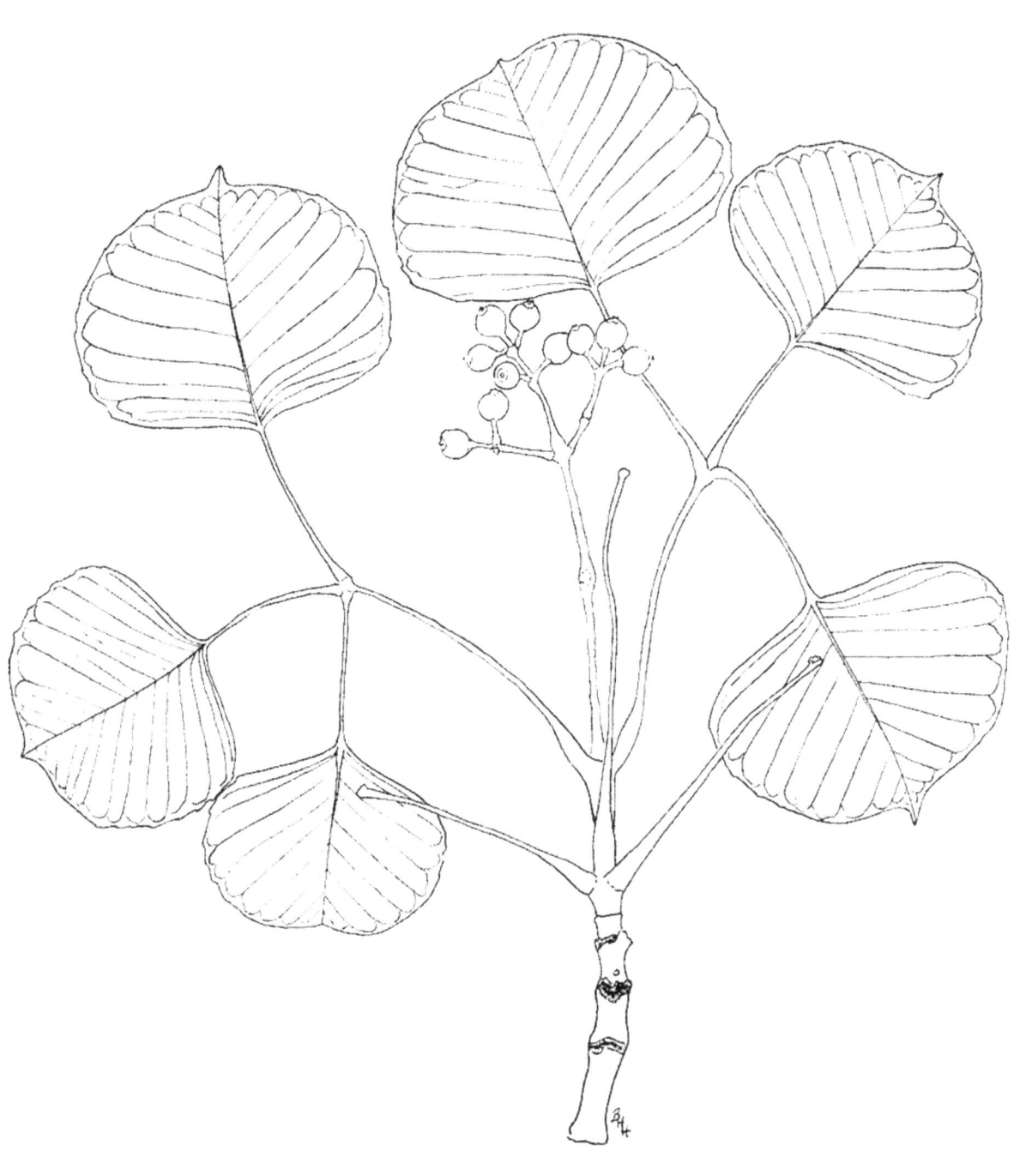

122. Lapalapa *Cheirodendron platyphyllum* (Hook. & Arn.) Seem.
Twig with leaves and fruits, 2/3 X.

123. 'Ōlapa, common cheirodendron

Cheirodendron trigynum (Gaud.) Heller

'Ōlapa, an aromatic medium-sized evergreen tree common in wet forests through the Hawaiian Islands. It is characterized by paired large palmate leaves with 3--5 finely toothed elliptical or ovate leaflets, constantly moving on their long slender stalks. The crushed foliage and bark have a strong odor like that of carrot or oil and a spicy or turpentine taste.

A tree to 40--50 ft (12--15 m) high and 2 ft (0.6 m) in trunk diameter, with rounded open crown, hairless throughout. Bark gray, smoothish or sometimes rough and scaly. Inner bark greenish white, slightly spicy, aromatic. Twigs stout, enlarged and ringed at nodes, green or purplish, becoming brownish, weak and brittle.

Leaves opposite, 4--8 in (10--20 cm) long, palmate or digitate, with very slender green, purplish or brownish leafstalks of 2--4 in (5--10 cm), slightly flattened, enlarged and slightly clasping at base. Leaflets 3--5 (7), spreading on slender slightly flattened stalks of 3/8--1 1/2 in (1--4 cm) long. Blades elliptical or ovate, 2--5 in (5--13 cm) long and 1 1/4--2 1/4 in (3--6 cm) wide, rounded or blunt with narrow curved point at apex, straight to blunt at base, edges mostly with small curved teeth or sometimes none, thin or slightly thickened, upper surface shiny green with fine side veins, lower surface dull light green.

Flower clusters (panicles) terminal, 3--6 in (7.5--15 cm) long, with many slender forking greenish or purplish branches and many flowers spreading on equal stalks (umbels). Flowers 3/16 in (5 mm) long and wide, greenish, composed of cuplike base (hypanthium) 1/8 in (3 mm) long, calyx of 5 tiny teeth, 5 narrow spreading petals 1/8 in (3 mm) long, 5 short stamens, and pistil with inferior ovary commonly 3--4-celled and 3--4 dotlike stigmas.

Fruits (berries) round, about 1/4 in (6 mm) in diameter, blackish, with ring of calyx and stigmas at apex, purplish flesh juice or slightly bitter. Seeds (nutlets) 3--4, more than 1/8 in (3 mm) long, brown.

The wood is pale yellow without distinct heartwood, moderately heavy, and moderately hard. It burns when green.

A bluish dye for coloring tapa, or bark cloth, was obtained from the fruits, leaves, and bark. Fruits eaten by native birds such as the rare 'ō'o, 'ō'u, and 'ōma'o. The wood was used for bird hunting spears. Distinctive leis can be made by binding or tying together the compound leaves.

The Hawaiian term for graceful dancers is also 'ōlapa. Performers of the native hula dance were divided into two groups, the 'ōlapa and ho'opa'a. The 'ōlapa were the dancers, perhaps because their movements were like the fluttering movement of the tree leaves. The ho'opa'a stayed in one place, chanting and playing musical instruments.

Common and widespread through the Hawaiian Islands, mostly in wet forests at 2,000--7,000 ft (610--2,134 m) altitude. It is the most prevalent understory tree in the forest on Hawaii in the zone within 1,000--4,500 ft (305--1,372 m), where koa and 'ōhi'a commonly intermix.

Special areas--Haleakala, Volcanoes, Kipuka Puaulu.

Champion--Height 37 ft (11.3 m), c.b.h. 4.5 ft (1.4 m), spread 28 ft (8.5 m). Hawaii Volcanoes National Park, Hawaii (1968).

Range--Hawaiian Islands only.

Other common names--'ōlapalapa, māhu, kauila māhu.

Botanical synonym--*Cheirodendron gaudichaudii* (DC.) Seem.

Many varieties of this species have been named but currently are not accepted. The generic name *Cheirodendron*, from Greek hand and tree, refers to the palmate or digitate leaves with 5 leaflets like fingers in a hand.

124. 'Ohe makai, Hawaiian reynoldsia

Reynoldsia sandwicensis Gray

This distinctive deciduous tree of dry forests is easily recognized by its stout smooth, light gray trunk with spreading crown of stout crooked branches, clear yellow brown tasteless resin, and large pinnate leaves with 7--11 ovate wavy margined leaflets heart-shaped at base. Also, many small greenish flowers in large branched clusters are conspicuous mostly in early spring while the tree is leafless.

A tree 15--50 ft (4.6--15 m) high with stout trunk 1 1/2--2 ft (0.5--0.6 m) in diameter, leafless from late fall until after spring rains. Bark light gray, smooth, becoming slightly warty and fissured. Inner bark light brown or whitish streaked, slightly bitter, yielding yellow brown tasteless gum. Branches stout but easily broken. Twigs few, stout, 5/8--3/4 in (1.5--2 cm) thick, light gray, with broad crescent-shaped leaf-scars, powdery hairy when young. End bud conical, less than 1/4 in (6 mm) long, covered by overlapping scales.

Leaves alternate, crowded at end of twig, 10--12 in (25--30 cm) long, powdery hairy when young, becoming hairless, with slender light green axis enlarged and almost clasping at base and ringed at base of leaflets. Leaflets mostly 7--11, paired except at end, hairless, on slender stalks of 3/8--1 in (10--25 mm), broadly ovate, 2--4 in (5--10 cm) long and 1 1/4--3 1/4 in (3--8 cm) wide, blunt at apex, heart-shaped at base, wavy margined,

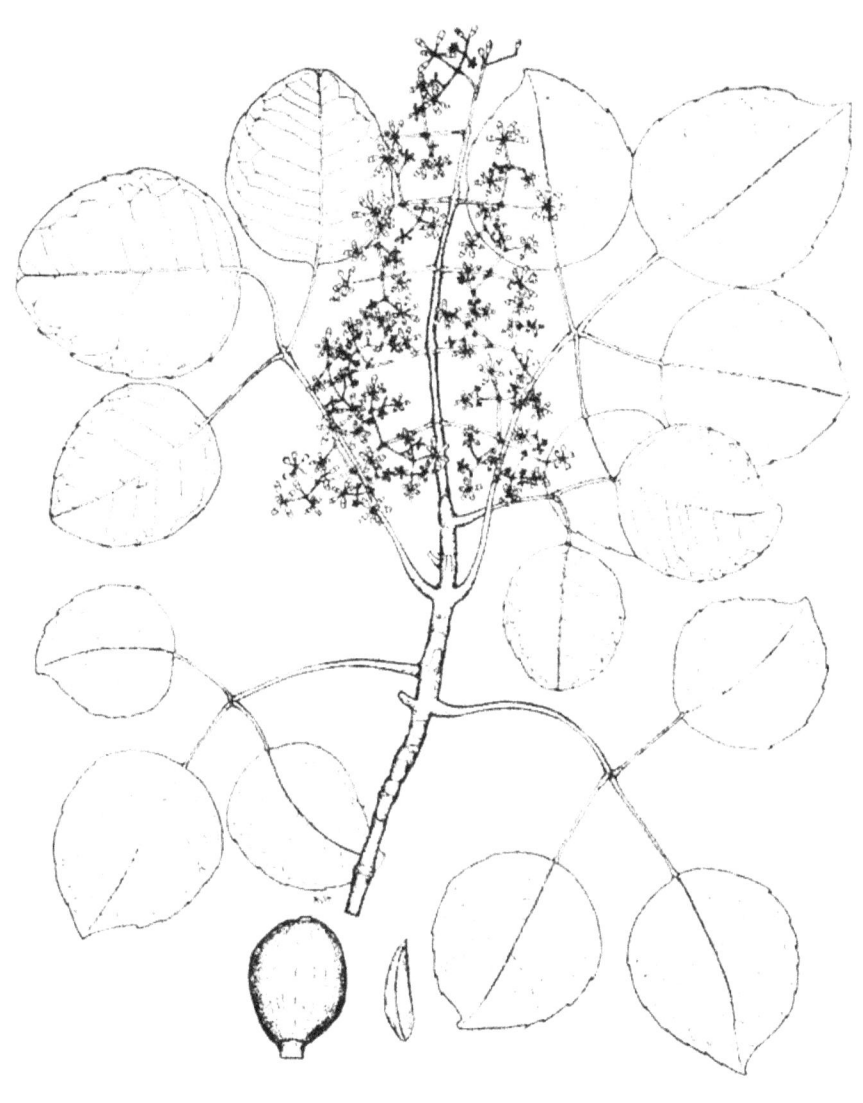

123. 'Ōlapa, common cheirodendron   *Cheirodendron trigynum* (Gaud.) Heller
Flowering twig, 1/3 X; fruit and nutlet (below), 3 X (Degener).

slightly thickened and fleshy, dull light green above, paler beneath.

Flower clusters (panicles) 1--3, terminal, 3--8 in (7.5--20 cm) long, with stout axis and many slender spreading branches, resinous. Flowers many, greenish, on short slender stalks, from elliptical green resinous buds more than 1/4 in (6 mm) long, composed of conical base (hypanthium) 1/8 in (3 mm) high, 5 tiny calyx teeth, 8--10 narrow petals 1/4 in (6 mm) long, 8--10 spreading stamens alternate with petals, and pistil with inferior ovary of 8--10 cells, very short style in thick disk, and 8--10 dot stigmas in conical mass.

Fruits (berries) rounded about 5/16 in (8 mm) in diameter, with stigmas at apex, purplish, juicy, bitter, becoming dry with ridges. Seeds (nutlets) 10 or fewer, elliptical, flattened, 3/16 in (5 mm) long.

The soft whitish wood is not presently used. The Hawaiians made stilts from it for a game. The resin or gum was also used.

Scattered in dry lowland areas, especially aa or rough lava fields, to 2,600 ft (792 m) altitude; becoming rare.

Special areas--Waimea Arboretum, Wahiawa, Volcanoes.

Champion--Height 48 ft (14.6 m), c.b.h. 6.7 ft (2 m), spread 23 ft (7 m). Puuwaawaa, Kailua-Kona, Hawaii (1968).

Range--This genus is found on Niihau, Oahu, Molokai, Lanai, Maui, and Hawaii, but is absent from Kauai.

Other common names--'ohe-kukuluae'o, 'ohe.

Botanical synonyms--*Reynoldsia degeneri* Sherff, *R. hillebrandii* Sherff, *R. hosakana* Sherff, *R. huehuensis* Sherff, *R. mauiensis* Sherff, *R. oblonga* Sherff, *R. venusta* Sherff.

In the broad sense this genus in Hawaii has a single variable species, though once divided into 8. A few others are found in Polynesia.

The genus *Reynoldsia*, discovered by the United States Exploring Expedition under Capt. Charles Wilkes, was named by Asa Gray in 1854 with this explanation: "I dedicate the genus to J. N. Reynolds, Esq., who merits this commemoration for the unflagging zeal with which he urged upon our Government the project of the South Sea Exploring Expedition, and also for having made, under trying circumstances, an interesting collection of dried plants in Southern Chili, many years ago [1837]."

125. Octopus-tree

Schefflera actinophylla (Endl.) Harms*

This distinctive ornamental is easily recognized by the several trunks mostly unbranched, the few very large palmately compound leaves with 7--12 leaflets slightly drooping in a circle at end of long leafstalk as in an umbrella, and the large showy clusters of many dark red or crimson flowers on 10--20 widely spreading dark purple axes suggesting arms of an octopus.

Small evergreen introduced tree 20--40 ft (6--12 m) high, with several trunks from base 4--12 in (0.1--0.3 m) or more in diameter, unbranched or with few stout branches, and with flattened or rounded open crown, hairless throughout. Bark light gray, smoothish or becoming slightly fissured. Twigs few, very stout, 3/4--2 in (2--5 cm) in diameter, green, with light brown lines (lenticels).

Leaves alternate, palmately compound (digitate), about 2--3 ft (0.6--0.9 m) long. Leafstalks very long, 1--2 ft (0.3--0.6 m), relatively slender, enlarged at both ends, round, light green. In angle above leafstalk, also forming bud at end of twig, is a light green persistent stipule or very narrow long-pointed scale 1 1/2--2 in (4--5 cm) long. Leaflets mostly 7--12 (5--18), spreading in circle at end of leafstalk on slender spokelike stalks of 2--4 1/2 in (5--11 cm). Leaflet blades oblong or elliptical, mostly 6--12 in (15--30 cm) long and 3--5 in (7.5--13 cm) broad, rounded and abruptly short-pointed at apex, rounded or short-pointed at base, slightly turned under at edges, slightly thickened and leathery. Upper surface shiny dark green with grooved light green midvein and inconspicuous side veins, and lower surface dull light green with slightly raised veins.

Flower clusters (panicles) large terminal, composed of 10--20 widely spreading stout axes. Flowers are borne 10--12 crowded stalkless in rounded heads 3/4 in (2 cm) across on dark purple stalks of 3/8--1/2 in (10--13 mm) along axis. Top half of the rounded dark red bud nearly 1/4 in (6 mm) in diameter is composed of 10--12 narrow pointed thick petals 3/16 in (5 mm) long, dark red on outer surface and whitish on inner surface, shedding early as half-round cap. Other flower parts are of same number as petals, calyx represented by narrow rim with minute teeth. Stamens 10--12, 3/16 in (5 mm) long, red, with stout filaments and large anthers, erect and slightly spreading 3/8--1/2 in (10--13 mm) across, soon shedding. Pistil half-round, turning from light to dark red, consists of inferior ovary whitish within and slightly resinous and aromatic, with 10--12 narrow cells each containing 1 ovule, and same number of dot stigmas in ring.

In fruit, head is composed of 10--12 berries, each bordered by 4 brownish black scales in form of a cup, which is persistent after shedding. Berry round or top-shaped, blackish, 1/4 in (6 mm) in diameter, with ring of stigmas at apex, ring slightly above middle, also vertical ridges corresponding to the 10--12 cells. Seed (nutlet) 1 in each cell, elliptical flattened, brown, 1/8 in (3 mm) long. Flowering from April to October in Hawaii.

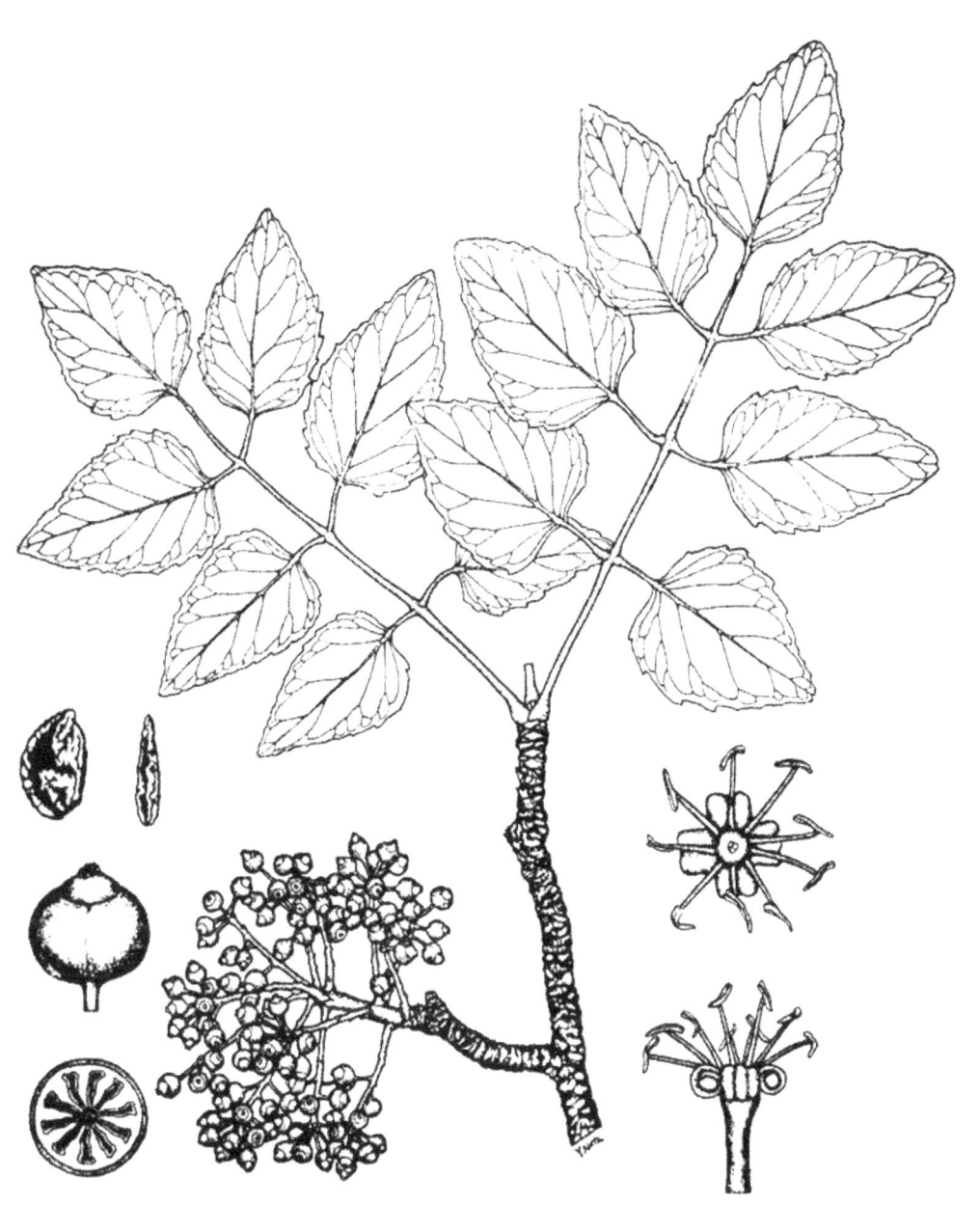

124. 'Ohe makai, Hawaiian reynoldsia   *Reynoldsia sandwicensis* Gray
Twig with fruits, 1/3 X; seeds, 3 X, and fruits, 2 X (lower left); flowers (lower right), 2 X (Degener).

Wood soft, not durable, not used.

Propagated from cuttings and seeds. The plants will grow in poor sand and can be pruned and topped. The seeds sometimes germinate on other trees and send roots down to the ground.

Common as an ornamental in lowlands of Hawaii and escaping from cultivation, becoming a real tree weed. Introduced about 1900.

In southern and central Florida, this popular small tree is recommended for tropical effect in confined areas of office buildings, parking lots, patios, and homes. However, the berries stain sidewalks. This species withstands a few degrees of freezing temperature and flowers in about 10 years where located in the sun. Northward, it is grown indoors as a potted plant for the handsome foliage. Cultivated also in Puerto Rico.

Special areas--Waimea Arboretum, Foster, Tantalus, Iolani.

Champion--Height 50 ft (15.3 m), c.b.h. 17.3 ft (5.3 m), spread 58 ft (17.7 m). State Forestry Arboretum, Hilo, Hawaii (1968).

Range--Native to Queensland, Australia. Introduced as an ornamental through the tropics and as a potted plant northward in temperate regions.

Other common names--umbrella-tree, brassaia, schefflera.

Botanical synonym--*Brassaia actinophylla* Endl.

126. 'Ohe'ohe *Tetraplasandra hawaiiensis* Gray

The genus *Tetraplasandra*, known as 'ohe'ohe, has about 6 species in Hawaii. They are small evergreen trees with few branches, recognized by alternate pinnate leaves turned under at edges, many small greenish flowers spreading on slender equal stalks, and many small rounded blackish juicy fruits. The example described here is easily distinguished further by the leaflets densely whitish or grayish hairy beneath and by the flower clusters branching along an axis.

Small to large tree recorded to 40--80 ft (12--24 m) high and 1--2 ft (0.3--0.6 m) in trunk diameter, with broad flat crown. Bark gray, smoothish, becoming fissured, rough, and scaly. Inner bark light brown and whitish streaked, slightly spicy and gritty. Twigs very stout, 1 1/4 in (3 cm) in diameter, gray, finely hairy, with very large half-round leaf-scars.

Leaves several, alternate, crowded at ends of twigs, very large, pinnate, 12--18 in (30--46 cm) long, with very stout gray hairy axis enlarged at base. Leaflets 5 or 7, paired except largest at end, on stout gray hairy stalks of about 1/2 in (1.2 cm), oblong or narrowly elliptical, 3--7 in (7.5--18 cm) long and 1 1/2--3 in (4--7.5 cm) wide, blunt at apex, short-pointed and unequal-sided at base, turned under at edges, thick, leathery, and stiff, upper surface slightly shiny green and hairless, hairy when young, and lower surface densely whitish or grayish hairy.

Flower clusters (umbellate panicles) terminal large, 9--12 in (23--30 cm) long and 6--10 in (15--25 cm) wide, with spreading stout gray hairy branches, bearing many flowers, mostly 2--3 together, erect on stout stalks, from gray hairy buds. Flowers about 3/8 in (1 cm) long, gray hairy. Cup-shaped base (hypanthium) 3/16 in (5 mm) high, bears calyx of 5--8 tiny teeth, 5--8 gray hairy pointed petals more than 1/4 in (6 mm) long, and 20--32 stamens. Pistil has inferior rounded ovary with 7--13 cells and ovules and rounded stigmas.

Fruits (berries or drupes) rounded, nearly 5/16 in (8 mm) in diameter, with ring of calyx and pointed stigma, blackish, juicy, aromatic, with ridges when dry. Seeds (nutlets) 7--13, elliptical, more than 1/8 in (3 mm) long, flattened, brown.

The wood in this genus is white with a silvery luster, without distinct heartwood. It is lightweight, fine-textured, straight-grained, and of moderate hardness.

Scattered in wet and dry forests, mostly in rain forests at 500--2,600 ft (150--792 m).

Special areas--Foster, Wahiawa, Volcanoes.

Champion--Height 67 ft (20.4 m), c.b.h. 14.2 ft (4.3 m), spread 69 ft (21 m). Hoomau Ranch, Honomalino, Hawaii (1968).

Range--This species with varieties is recorded from Hawaii, Maui, Lanai, and Molokai.

Other common name--'ohe.

A related endemic tree, *Munroidendron racemosum* (Forbes) Sherff, is rare and found only on Kauai, in lowland dry forests, but can be seen also in cultivation at Wahiawa Botanic Garden and Waimea Arboretum on Oahu. It is usually 12--15 ft (3.7--4.6 m) tall, to 40 ft (12 m). It has a straight trunk, smooth gray bark, and spreading branches. Leaves, similar to *Tetraplasandra*, are hairy. The long flower clusters (racemes) bear many drooping cream-colored flowers with deep red stigmas and ovary tops.

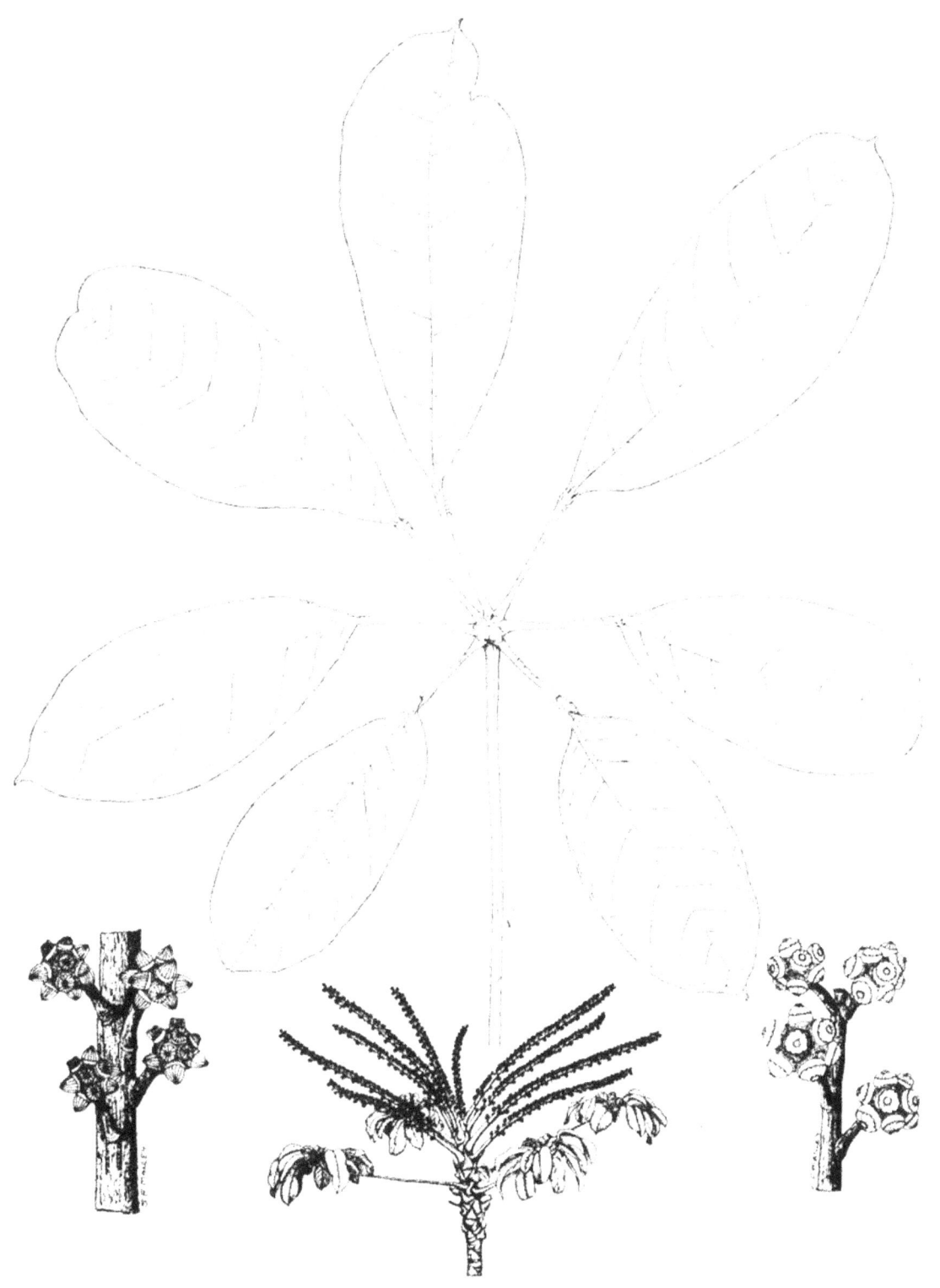

125. Octopus-tree   *Schlefflera actinophylla* (Endl.) Harms*
Leaf, 1/3 X; flowers (lower left) and fruits (lower right), 2/3 X (P. R. v. 2). Flowering twig (below), 1/18 X (Degener).

126. 'Ohe'ohe *Tetraplasandra hawaiiensis* Gray
Twig with mature and young leaves and flowers, and fruits (lower right), 2/3 X.

127. Pūkiawe *Styphelia tameiameiae* (Cham. & Schlecht.) F. Muell.
Twig with fruits (above), twig with flowers (below), 1 X.

EPACRIS FAMILY (EPACRIDACEAE)

127. Pūkiawe

Styphelia tameiameiae (Cham. & Schlecht.) F. Muell.

Pūkiawe is a large evergreen shrub or sometimes small tree when in understory of wet forests. Common through the islands from low to high altitudes. Recognized by the small needlelike leaves whitish beneath, very small whitish flowers, and small round red, white, or pink fruits.

A shrub of 3--10 ft (0.9--3 m), and in forests a small tree to 15 ft (4.6 m) high with twisted trunk to 5 in (13 cm) in diameter, with many spreading irregular slender branches. Bark gray, finely fissured, becoming scaly and shaggy; inner bark thin, greenish, fibrous and slightly bitter. Twigs very slender and wiry, finely hairy, pinkish when young, becoming brown, with tiny rounded raised leaf scars.

Leaves many, alternate, scattered and spreading along twig on tiny hairy yellowish leafstalks. Blades bent at right angle to twig, very narrow (linear to oblong), 1/4--1/2 in (6--13 mm) long and 1/8 in (3 mm) or less in width, sharp-pointed, rounded at base, turned under at edges, slightly thickened and stiff, hairless, upper surface dull green without visible veins, lower surface whitish, with many long fine nearly parallel veins.

Flowers few, single and almost stalkless at leaf bases, bell-shaped, 1/8 in (3 mm) long and and wide, composed of several overlapping scales at base; calyx with 5 overlapping hairy sepals green and pinkish tinged; white bell-shaped tubular corolla with 5 spreading narrow pointed lobes; 5 tiny stamens in notches of corolla; and pistil with rounded ovary, short style, and dot stigma.

Fruits (berries) several at leaf bases, round, 3/16--1/4 in (5--6 mm) in diameter, red, pink, or white, slightly shiny, with calyx at base and style at apex, slightly fleshy, mealy, tasteless or slightly astringent, becoming brown and dry. Seed or stone 1, elliptical, brown, more than 1/8 in (3 mm) long.

Sapwood is light reddish brown and heartwood dark reddish brown. A fine-textured heavy wood of moderate hardness.

The wood was used for cremating bodies of outlaws. Also, according to Hawaiian historian David Malo, when a high ranking chief wanted to mingle with commoners, he would enter a smoke house and be smudged with smoke of pūkiawe wood while a priest chanted a prayer for dispensation.

The bright beadlike fruits served in Hawaiian garlands or leis.

Common and widespread from understory of wet forests to border of dry forest and exposed ridges and waste places, from near sea level to 10,600 ft (3,231 m) or above. Of largest size in forests at 5,000--6,000 ft (1,524--1,829 m) and almost creeping in bogs. It is very common in Hawaii Volcanoes and Haleakala National Parks. In pastures and rangelands at higher altitudes classed as a weed of no forage value (Hasselwood and Motter 1966).

Special areas--Kokee, Haleakala, Volcanoes.

Range--Through the 6 large Hawaiian Islands and in Marquesas Islands.

Other common names--maiele, kāwa'u, 'a'ali'imahu, kawai, Kamehameha styphelia, Hawaiian-heather, kānehoa, pūpūkiawe, puakiawe.

Botanical synonym--*Styphelia douglasii* (Gray) Skottsb.

The scientific name honors King Kamehameha I (1758--1819), who united the Hawaiian Islands into one kingdom. The spelling shows the difficulty in transcribing the spoken letters k and t, which were almost interchangeable before the language was written.

MYRSINE FAMILY (MYRSINACEAE)

128. Shoebutton ardisia

Ardisia elliptica Thunb.*

Introduced evergreen shrub or small tree with elliptical leathery evergreen leaves and many small pinkish starlike flowers, ornamental and becoming naturalized in moist lowland areas. To 20 ft (6 m) high and 3 in (7.5 m) in trunk diameter. Twigs slightly stout, light green, purplish tinged, hairless.

Leaves alternate, elliptical, 3--4 1/2 in (7.5--11 cm) long, 1--1 1/2 in (2.5--4 cm) wide, short-pointed at apex, broadest beyond middle, tapering toward long-pointed base, edges straight, slightly thickened and leathery with side veins not visible, hairless, dull light green on both surfaces, beneath with gland dots sometimes blackish. Leafstalk about 1/4 in (6 mm) long, purplish, flattened above.

Flower clusters (panicles) at base of upper leaves, 1 1/2--3 in (4--7.5 cm) long, branched. Flowers many on slender green talks of about 1/2 in (13 mm), from pointed pink bud 3/8 in (1 cm) long, spreading starlike, composed of cuplike calyx with 5 rounded lobes, whitish with brown dots; corolla with 5 lobes from short tube spreading starlike about 1/2 in (13 mm) wide, pink with brown dots; 5 stamens inserted near base of corolla opposite lobes, with large long-pointed brown anthers

128. Shoebutton ardisia   *Ardisia elliptica* Thunb.*
Flowering twig (above), fruiting twig (below), 1 X.

united; and pistil with round greenish ovary containing many ovules and with slender whitish style.

Fruits (drupes) many, rounded, becoming 3/8 in (1 cm) in diameter, slightly flattened, turning from pink to black, with gland-dots, calyx and base of style remaining, with purplish juice. Seed 1, round.

Heartwood is pale brown with large conspicuous darker colored rays showing prominently on all surfaces. Of moderate density and hardness, with a fine texture; not used.

Planted as an ornamental for the pretty flowers and glossy leaves but escaping and becoming naturalized as an weed in moist lowland areas, especially Oahu, Maui, and Hawaii.

129. Kōlea

Two common and widely distributed species (Nos. 129 and 130) will serve as examples of the genus *Myrsine*, of small to medium-sized native evergreen trees in which as many as 20 Hawaiian species have been distinguished. In this genus the leaves generally are many, crowded, narrowly elliptical or lance-shaped and broadest beyond middle, tapering to base and short leafstalk, with gland-dots visible under a hand lens; many small short-stalked greenish 5-parted flowers crowded along twig back of leaves; and many small round, blackish 1-seeded fruits (berries).

This species characterized by larger leaves than others is a small or medium-sized tree to 60 ft (18 m) tall and 1--2 ft (0.3--0.6 m) in trunk diameter. Bark light gray, smooth to finely fissured, thick. Inner bark with brown outer layer, pink or red with brown streaks, bitter. Twigs mostly stout, hairless, green when young, turning gray and becoming warty, with raised half-round leaf-scars, ending in narrow pointed bud of young leaves.

Leaves many, alternate but crowded, hairless, narrowly elliptical, 3 1/4--5 1/2 in (8--14 cm) long and 1--1 1/2 in (2.5--4 cm) wide, slightly thick and leathery or slightly fleshy, apex blunt to rounded or pointed, base short- to long-pointed, pink when young, tapering at base to short winged greenish or pinkish leafstalk of 1/4 in (6 mm) or almost stalkless, upper surface slightly shiny green with inconspicuous veins, lower surface dull light green with many blackish gland-dots visible under a lens.

Flowers many, small, about 1/8 in (3 mm) long and broad, in groups of 3--7 on short slender stalks of 1/4 in (6 mm) along twigs and short spurs back of leaves and at base of oldest leaves. Calyx light green, of 5 pointed lobes; corolla of 5 elliptical lobes yellowish with red dots;

130. Kōlea lau-li'i

This handsome much branched, spreading evergreen shrub or small tree, with small many crowded dark green spoon-shaped leaves whitish beneath, is

Special area--Waimea Arboretum.
Range--Native of southeastern Asia. Widely cultivated through the tropics and becoming naturalized. Introduced and naturalized locally in south Florida.
Other common name--elliptical-leaf ardisia.
Botanical synonym--*Ardisia solanacea* Roxb., *A. humilis* auth., not Vahl.

The common name apparently refers to the resemblance of the round blackish fruits to buttons formerly worn on ladies' high-topped shoes.

Myrsine lessertiana A. DC.

5 short stalkless stamens attached on corolla and opposite lobes; and pistil with conical ovary and almost stalkless stigma.

Fruits (berries) round or elliptical, about 1/4 in (6 mm) in diameter, turning from green to reddish or black, with calyx at base and pointed stigma at apex. Seed 1, round.

The wood is pinkish yellow with prominent reddish brown rays providing a prominent figure on all surfaces. It is moderately hard, easily worked and polished, and suitable for cabinetwork but not used at present. Used by the Hawaiians as timber for houses and anvils on which to beat tapa. A black dye for tapa was made from the charcoal.

This very variable species is widespread through the islands, especially in wet forests and open areas at 700--4,000 ft (213--1,219 m) altitude.

Special areas--Haleakala, Volcanoes, Kipuka Puaulu.
Champion--Height 67 ft (20.4 m), c.b.h. 6.8 ft (20.7 m), spread 25 ft (7.6 m). Hawaii Volcanoes National Park, Hawaii (1968).
Range--Kauai, Oahu, Molokai, Lanai, Maui, and Hawaii only.
Other common name--kōlea lau nui (meaning large-leaf kolea).
Botanical synonyms--*Myrsine meziana* (Lévl.) Wilbur, *Suttonia lessertiana* (A. DC.) Mez, *Rapanea lessertiana* (A. DC.) Deg. & Hosaka.

Rock noted that this species is one of the most variable in the genus and that two trees are hardly alike. He reported that red sap exudes from cut trunks and formerly served for dyeing the tapa or bark cloth.

Myrsine sandwicensis A. DC.

found in wet forests at middle to moderately high altitudes through the islands though absent from Kauai. A shrub or sometimes small tree 13--25 ft (4--7 m) high.

129. Kōlea *Myrsine lessertiana* A. DC.
Twig with fruits, 2/3 X.

Twigs slender, light green tinged with pink when young, finely hairy, with 2 lines below each leaf, turning gray, with raised half-round leaf-scars.

Leaves alternate, very numerous and crowded, spoon-shaped (obovate), 5/8--1 in (15--25 mm) long and 1/4--3/8 in (6--10 mm) wide, thick and leathery, hairless, rounded and slightly notched at apex, widest beyond middle and tapering to base and short leafstalk of 1/8 in (3 mm), edges turned under, with gland-dots visible under a lens, upper surface shiny green or dark green without visible side veins, lower surface whitish green. Youngest leaves pinkish tinged. Leaves on small plants larger, to 1 1/2 in (40 mm) long and 5/8 in (15 mm) wide.

Flowers scattered at leaf bases, 2--8 together on slender stalks of 1/8--1/4 in (3--6 mm), 1/8 in (3 mm) long and wide, composed of calyx of 5--7 minute lobes, 5--7 narrow yellowish or reddish petals with reddish gland dots, 5--7 short stamens inserted on corolla opposite the lobes, and pistil with round ovary and dot stigma.

Fruits (drupes) round, 3/16--5/16 in (5--8 mm) in diameter, purplish black with gland-dots, with calyx at base and stigma at apex. Seed 1, round.

The wood of *M. sandwicensis* is similar to that of *M. lessertiana* A. DC.

Common and widespread through the islands except Kauai, in wet forests at middle to high altitudes of 1,000--4,800 ft (305--1,463 m).

Special area--Volcanoes.

Range--Oahu, Molokai, Lanai, Maui, and Hawaii only.

Other common names--kolea, Hawaiian rapanea.

Botanical synonyms--*Suttonia sandwicensis* (A. DC.) Mez, *Rapanea sandwicensis* (A. DC.) Deg. & Hosaka.

The pinkish young leaf shoots are sometimes used effectively in leis.

SAPODILLA FAMILY (SAPOTACEAE)

131. Keahi, Hawaiian nesoluma

Nesoluma polynesicum (Hillebr.) Baill.

This evergreen small tree of dry lowland forests, with milky sap and many dark purple or brown fruits like olives, is recognized also by the rusty brown hairs of young foliage, lower leaf surfaces, twigs, and flowers. A small tree to 33 ft (10 m) high and 1 ft (0.3 m) in trunk diameter, with rounded crown of spreading and drooping branches, or a shrub. Bark gray, rough, thick, becoming furrowed into rectangular plates. Twigs long, with raised half-round leaf-scars and round hairy buds and their scars.

Leaves alternate, scattered along twig, with slender leafstalks mostly 3/4--1 1/4 in (2--3 cm) long. Blades elliptical, 1 1/4--3 1/4 in (3--8 cm) long and 3/4--2 in (2--5 cm) wide, blunt or rounded at both ends, turned under at edges, thick and leathery, upper surface shiny green and hairless, lower surface with pressed rusty brown hairs and often becoming nearly hairless in age.

Flowers many, 1--5 at leaf base or on twig back of leaves on stalks of 1/4 in (6 mm), rusty hairy, fragrant, 3/16 in (5 mm) long and wide, composed of hairy calyx with usually 4--5 (3--6) overlapping pointed lobes; urn-shaped greenish white hairless corolla with short tube usually 7--9 (5--12) pointed, rounded, or toothed lobes; usually 7--9 (5--12) small stamens inserted on base of corolla and opposite lobes; and pistil with elliptical hairy ovary of usually 3--5 (2--6) cells, 1 ovule in each cell, short stout style, and dot stigma.

Fruits (berries) elliptical, many along twigs 1/4--5/8 in (12--15 mm) long and 3/8 in (10 mm) wide, shiny dark purple or brown, with calyx and base and pointed style at apex, with thin flesh, not edible. Seed 1, elliptical or rounded, 5/16 in (8 mm) long, shiny brown, with large rounded scar at base. Fruits many, from May to September.

The hard durable wood apparently has not been used.

Scattered in dry forests through the Hawaiian Islands at 400--2,000 ft (122--610 m) altitude. Formerly common, now rare.

Special area--Wahiawa.

Range--Oahu, Molokai, Maui, Hawaii, also Raivavae, Rapa in South Pacific.

Botanical synonym--*Chrysophyllum polynesicum* Hillebr.

This variable species has several named forms generally not distinguished.

132. 'Āla'a

Pouteria sandwicensis (Gray) Baehni & Deg.

Medium-sized native evergreen tree of dry forests mainly, with milky sap, recognized by the bronze or reddish brown lower surfaces of the leathery oblong or elliptical leaves. To 50 ft (15 m) high and 16 in (0.4 m) in trunk diameter. Bark gray, thick, fissured to furrowed; inner bark orange with dark red outer layer, gritty and slightly bitter, with milky sap. Twigs with minute pressed brown hairs when young, with raised half-round leaf scars.

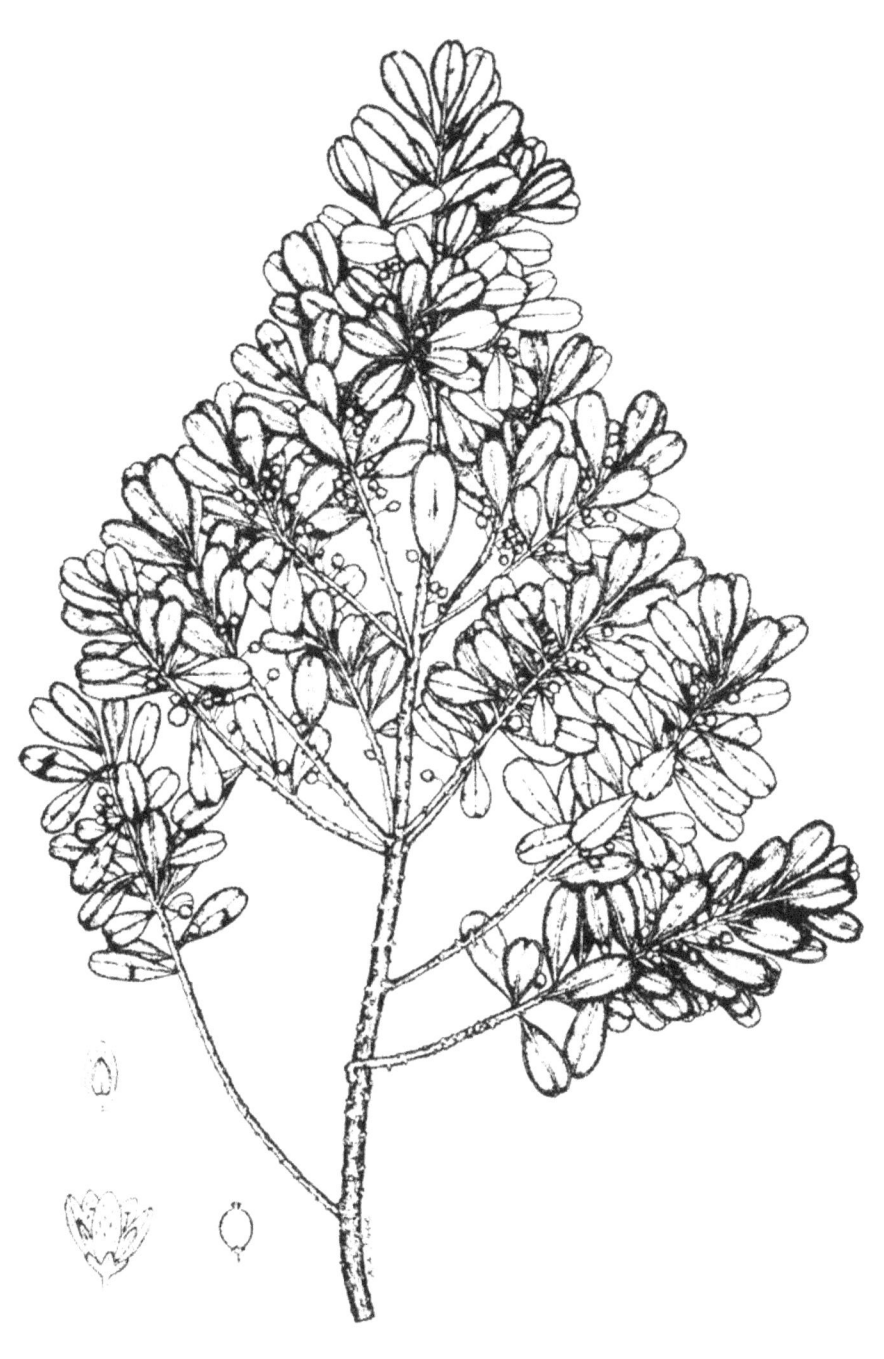

130. Kōlea lau-li'i   *Myrsine sandwicensis* A. DC.
Twig with fruits, 1/2 X; flower (lower left), 3 X (Degener).

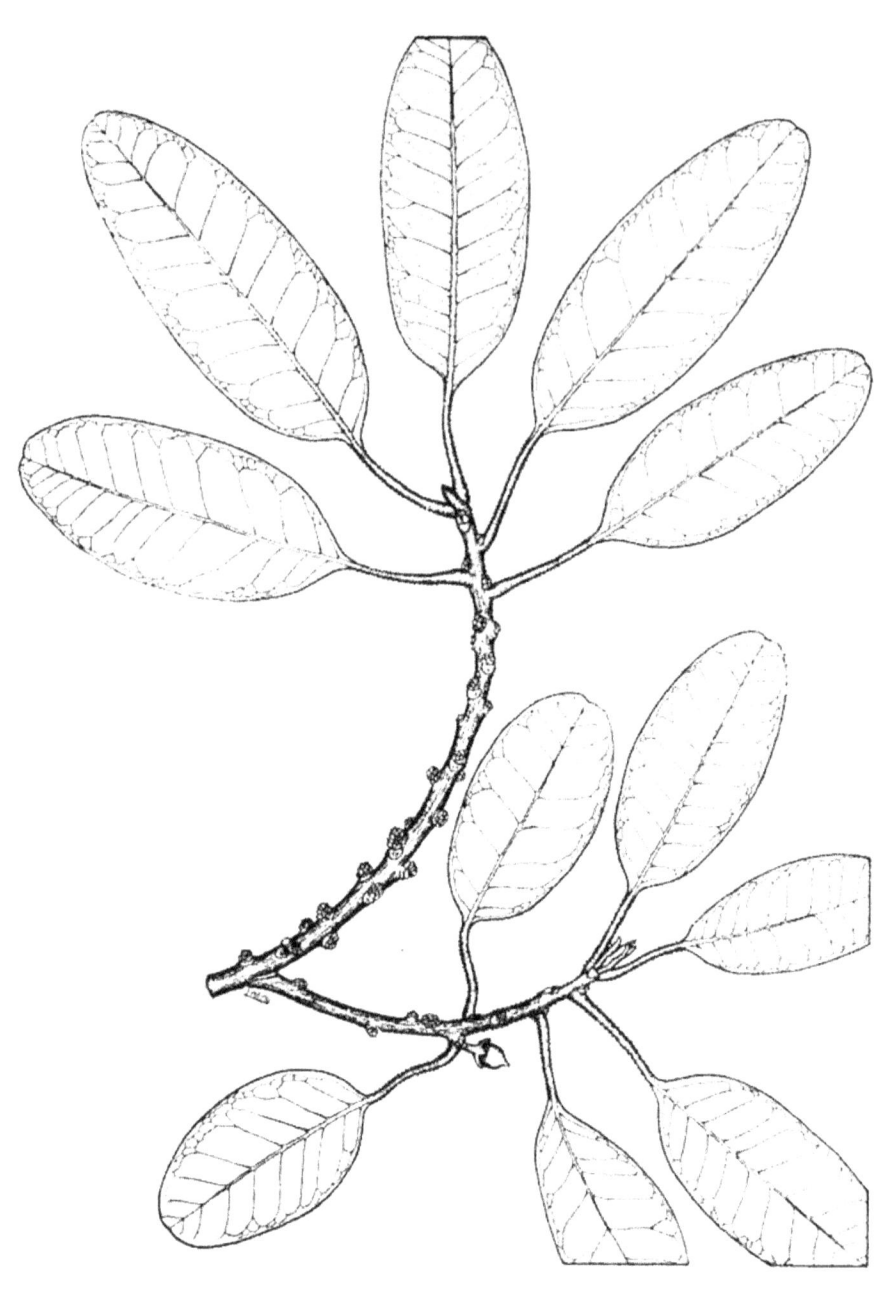

131. Keahi, Hawaiian nesoluma *Nesoluma polynesicum* (Hillebr.) Baill.
Twig with flower buds and immature fruit, 1/2 X (Degener).

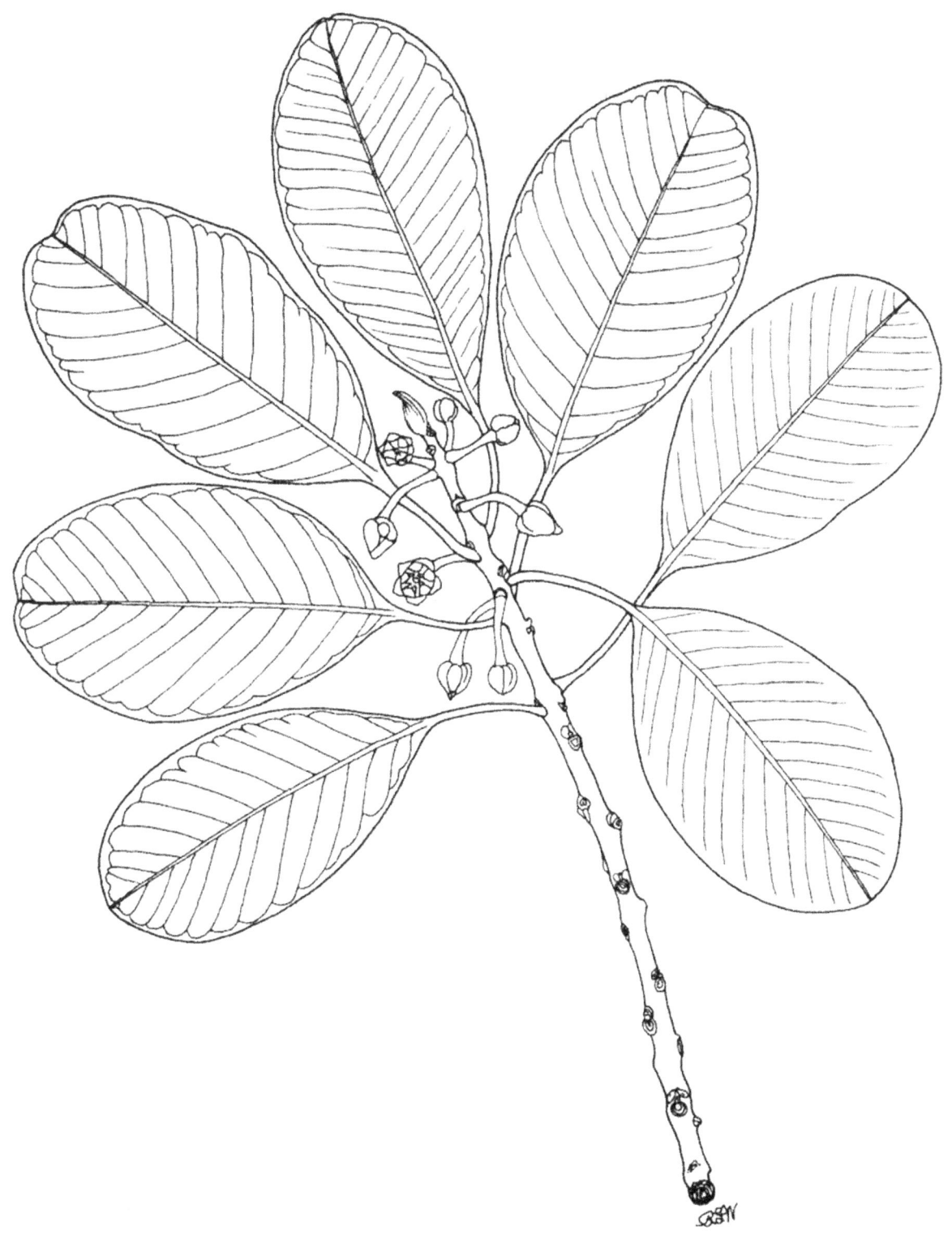

132. 'Āla'a　　*Pouteria sandwicensis* (Gray) Baehni & Deg.
Flowering twig, 1 X.

Leaves alternate, with leafstalks 3/4–1 1/4 in (2–4 cm) long, finely brown hairy. Blades oblong or elliptical, 2 1/2–5 1/2 in (6–14 cm) long and 1 1/4–2 1/2 in (3–6 cm) wide, thick and leathery, rounded at apex, short-pointed at base, not toothed on edges, with many fine straight parallel side veins almost at right angle with slightly sunken midvein, upper surface shiny green and nearly hairless, lower surface dull with bronze or reddish brown pressed hairs or becoming nearly hairless.

Flowers 1–4 at base of leaf on straight or curved brown hairy stalks of about 3/4 in (2 cm), bell-shaped, about 1/4 in (6 mm) long. Calyx of 5 broad rounded light brown hairy sepals; corolla with short tube and 5 broad rounded lobes, light green and hairless; 5 short stamens attached within tube and opposite lobes and 5 small sterile stamens (staminodia) in notches; and pistil with hairy conical 5-celled ovary and short style.

Fruit (berry) elliptical, rounded, or pear-shaped, 1 1/4–1 1/2 in (3–4 cm) long, yellow, orange, or purplish black, dry. Seeds 1–5, about 3/4 in (2 cm) long, elliptical and flattened, with long scar, shiny yellow brown.

Wood yellow with black streaks, hard, straight-grained, and with faint growth rings. Formerly used for house construction, o'o (digging sticks), and spears.

The milky sap was used by the Hawaiians as birdlime for catching small birds.

Dry forests mainly, in dry gulches to moist ridges at 600–4,000 ft (182–1,219 m) altitude, through the islands.

Special areas--Kokee, Wahiawa.
Champion--Height 38 ft (11.6 m), c.b.h. 5.3 ft (1.6 m), spread 29 ft (8.8 m). Puuwaawaa, Kailua-Kona, Hawaii (1968).
Range--Hawaiian Islands only.
Other common names--kaulu, āulu.
Botanical synonyms--*Sideroxylon sandwicense* (Gray) Benth. & Hook. f., *Planchonella sandwicensis* (Gray) Pierre, *P. puulupensis* Baehni & Deg., *Pouteria auahiense* Rock, *P. ceresolii* (Rock) Fosberg, *P. rhynchosperma* (Rock) Fosberg, *P. spathulata* (Hillebr.) Fosberg.

This species, treated here as the only Hawaiian representative of its genus, includes 5 or more variations formerly accepted as species.

EBONY FAMILY (EBENACEAE)

133. Lama, Hillebrand persimmon *Diospyros hillebrandii* (Seem.) Fosberg

This species of lama is confined to Oahu and Kauai. It has large dark green leaves which have a network of fine veins on upper surface and also the calyx lobes short-pointed. The pink flushes of new foliage on red twigs over a tree are showy.

A small evergreen tree to 30 ft (9 m) high and 5 in (13 cm) in trunk diameter, with horizontal branches. Bark blackish gray, smoothish; inner bark light brown, slightly bitter. Twigs gray, hairless, with raised dots. Buds about 1/4 in (6 mm) long, of spreading pointed nearly hairless scales.

Leaves alternate in 2 rows, hairless, with short leafstalks less than 1/4 in (6 mm) long. Blades oblong, 3–6 in (7.5–15 cm) long and 1 1/4–2 1/2 in (3–6 cm) wide, blunt at apex, rounded at base, not toothed on edges, slightly thickened, the upper surface shiny dark green with a prominent network of fine veins when dry, the lower surface dull light green.

Flowers male and female on different plants (dioecious), single and stalkless at leaf bases, 3/8 in (10 mm) long with overlapping scales at base. Male flowers have green narrow tubular calyx of 1/4 in (6 mm) with 3 short-pointed lobes hairy at end, tubular bell-shaped pink hairy corolla with 3 spreading lobes, and 9 short stamens. Female flowers have calyx, corolla, and pistil with hairy ovary and 3-forked style.

Fruits (berries) elliptical, 3/4–1 in (20–25 mm) long, slightly curved on 1 side and widest beyond middle, hairy toward blunt apex with point from style, orange, and at base the enlarged calyx 1/4 in (6 mm) long with 3 short-pointed lobes, nearly dry, edible but insipid. Seeds elliptical, 5/8 in (15 mm) long, shiny brown black.

Sapwood whitish yellow (heartwood color unknown), hard, similar in appearance and uses to *D. ferrea*.

Scattered in wet forests of Koolau and Waianae Ranges of Oahu to 2,000 ft (610 m) altitude. In mountains near Honolulu and not uncommon on Kauai at Kokee and in Kipu Range.

Special area--Wahiawa.
Range--Oahu and Kauai only.
Other common name--ēlama.
Botanical synonym--*Maba hillebrandii* Seem.

This species honors William Hillebrand (1831-86), German-born physician and botanist, who lived in Honolulu 20 years. His classic "Flora of the Hawaiian Islands" (1888) remains a very useful reference a century later.

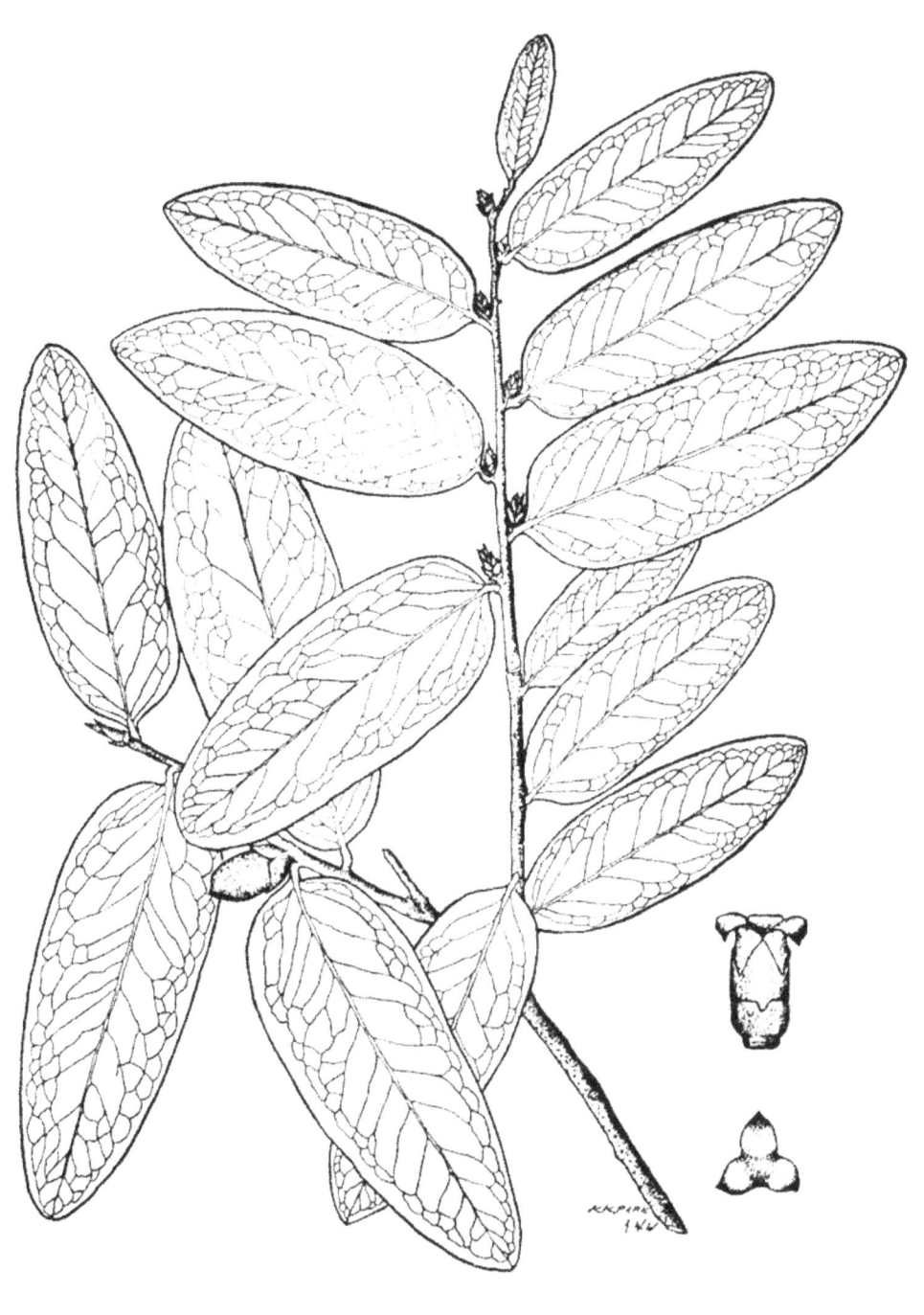

133. Lama, Hillebrand persimmon *Diospyros hillebrandii* (Seem.) Fosberg
Twig with flowers, twig with fruit (lower left), 1/2 X; flower (lower right), 2 X (Degener).

134. Lama
Diospyros sandwicensis (A. DC.) Fosberg

Lama is widespread through the islands, especially in dry forests, and one of the most common trees. It has thick oblong or elliptical dull green leaves spreading alternate in 2 rows on nearly horizontal twigs, small whitish flowers mostly single and almost stalkless at leaf bases, and yellow to orange egg-shaped or elliptical fruits 5/8--3/4 in (15--20 mm) long.

A handsome evergreen small to medium-sized tree of 20--40 ft (6--12 m) in height and 1 ft (0.3 m) in trunk diameter. Bark blackish or dark gray, thick, from smoothish becoming rough and furrowed into irregular squares. Inner bark pink streaked, astringent and bitter. Twigs gray, finely hairy when young, with raised dots, becoming cracked and rough. Buds less than 5/16 in (5 mm) long, covered by finely hairy overlapping scales.

Leaves alternate in 2 rows, with short brown hairy leafstalks of 1/4 in (6 mm). Blades oblong or elliptical, 1 1/4--2 1/2 in (3--6 cm) long and 5/8--1 1/4 in (1.5--3 cm) wide, blunt or short-pointed at apex, rounded at base, slightly turned under at edges, thick and stiff, slightly curved up on sides from midvein, with side veins fine and not visible, red with pink flushes and hairy when young, above dull green, beneath dull light green.

Flowers male and female on different plants (dioecious), single and almost stalkless at leaf bases, about 1/4 in (6 mm) long. Male flowers have hairy greenish tubular calyx with 3--4 blunt lobes, tubular bell-shaped whitish corolla with 3 spreading lobes, and 12--18 stamens. Female flowers have calyx, corolla, and pistil with hairy 3-celled ovary and 3-forked style.

Fruits (berries) egg-shaped or elliptical, 5/8--3/4 in (15--20 mm) long, slightly curved on one side to blunt apex with point from style; at base is the enlarged cup-shaped finely hairy light green calyx 1/4 in (6 mm) long with 3--4 blunt or rounded lobes; yellow flesh or dry. Seeds 1 or sometimes 2, elliptical and flattened, 5/8 in (15 mm) long, brown.

Fruits often are abundant at maturity in late winter, especially February. Mature fruits are sweetish and eaten by birds and people. Immature fruits, however, are slightly astringent, like those of their close relatives on the continental United States, persimmons (*Diospyros virginiana* L.).

The sapwood is wide and white. The heartwood is described as rich reddish-brown with redder and yellower zones, very hard, fine-textured, and straight-grained. Hawaiians used the white wood as blocks in their altars to symbolize the goddess Laka, and to fence sacred enclosures. The place name Kapalama in Honolulu means lama fence and referred to a former fence around the school for the chief's children there. The wood is not presently used.

Widespread and common in dry and wet regions through the islands down to sea level.

Special areas--Kokee, Waimea Arboretum, Wahiawa, Volcanoes.

Champion--Height 35 ft (10.7 m), c.b.h. 2.8 ft (0.9 m), spread 17 ft (5.2 m). Puuwaawaa Ranch, Kailua-Kona, Hawaii (1968).

Range--All main Hawaiian Islands.

Botanical synonyms--*Maba sandwicensis* A. DC., *Diospyros ferrea* var. *sandwicensis* (A. DC.) Bakh.

This variable species has many varieties. The common name lama means enlightenment.

OLIVE FAMILY (OLEACEAE)

135. Tropical ash
Fraxinus uhdei (Wenzig) Lingelsh.*

Tropical ash was originally introduced from Mexico as a shade tree but was found suitable for forest plantations. This large deciduous tree is identified by the paired, pinnately compound leaves with 5--9 lance-shaped finely saw-toothed leaflets and the many brown key fruits with long narrow wing.

A large forest tree 80 ft (24 m) or more in height and 3 ft (0.9 m) in trunk diameter. Bark gray or brown, rough, thick, furrowed into ridges. Inner bark whitish and bitter. Twigs green, turning brown, hairless except when young, with paired raised half-round leaf-scars. Buds paired, 3/16 in (5 mm) long, blunt, covered by few finely hairy brown scales.

Leaves opposite, pinnate, 6--11 in (15--28 cm) long, composed of slender green hairless axis and 5-9 leaflets paired except at end, on slender stalks of 1/8--1/2 in (3--13 mm). Leaflet blades mostly 2--4 in (5--10 cm) long and 3/4--2 in (2--5 cm) wide, long-pointed at apex, short-pointed or blunt at base, finely saw-toothed on edges, slightly thickened, upper surface dull green and hairless, lower surface light green with small hairs along midvein.

Flower clusters (panicles) at sides of twigs, 5--8 in (13--20 cm) long, much branched, with many slender-stalked small greenish flowers without petals, male and female on different trees (dioecious). Male flowers have tiny 4-toothed calyx and 2 stamens 1/8 in (3 mm) long. Female flowers have 4-toothed calyx and pistil 3/16 in (5 mm) long with ovary and 2-forked style.

Key fruits (samaras) 3/4--1 1/2 in (2--4 cm) long, composed of small nearly cylindrical dark brown body 1/4--3/8 in (6--10 mm) long at base and long light brown

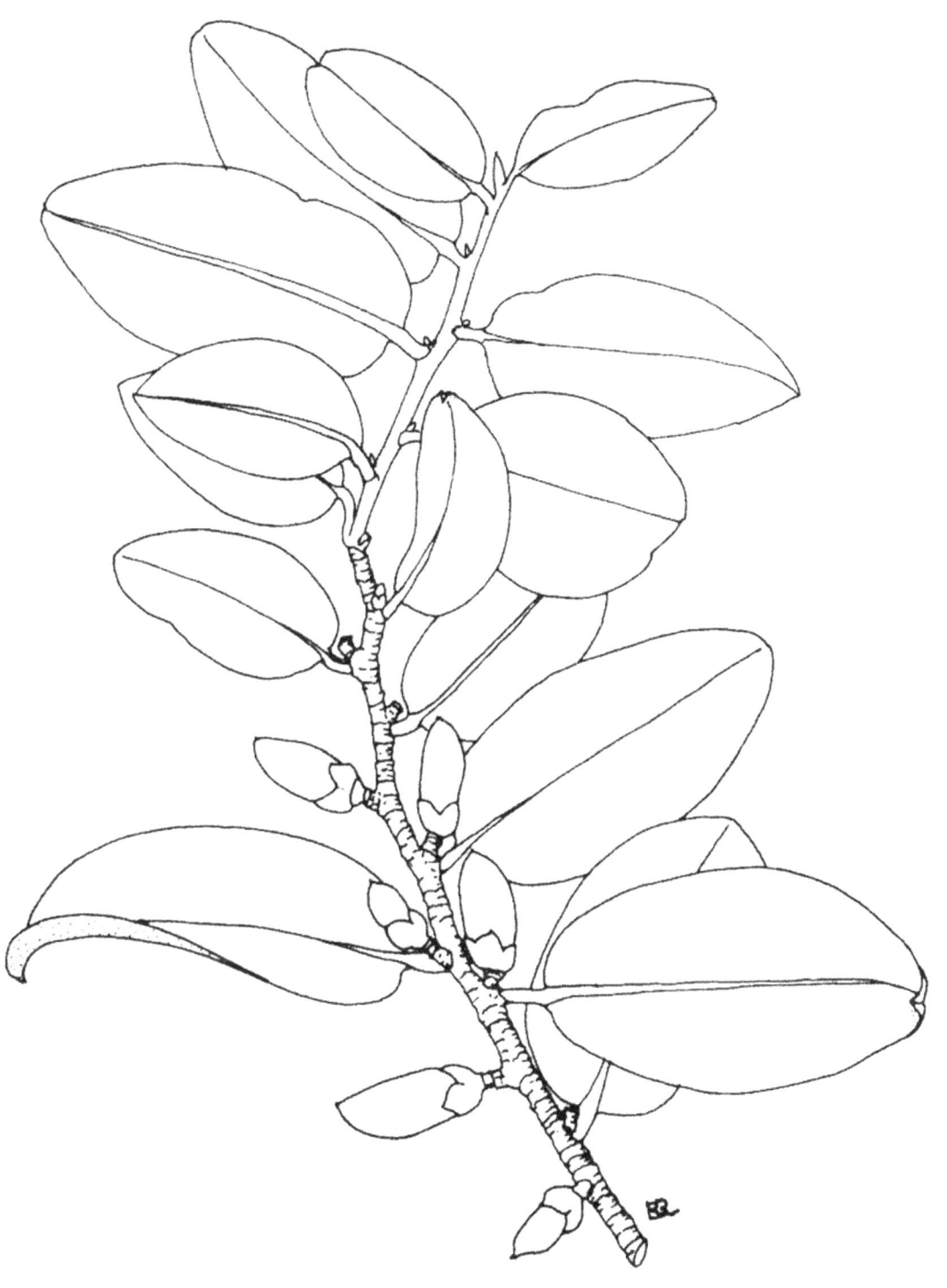

134. Lama *Diospyros sandwicensis* (A. DC.) Fosberg
Twig with fruits, 1 X.

wing to 1/4 in (6 mm) wide, extending down to middle of body and slightly notched at end.

The wood is blond in color without a differentiated sapwood. It is lightweight (sp. gr. 0.47), ring porous, straight-grained, moderately fine-textured, and almost indistinguishable from white ash, except in its lower density and hardness. It is more stable than white ash and easier to work, so is better suited for furniture. It is not as tough as white ash and not well suited for handle stock as its wide-ringed characteristic would suggest. Easy to season and to work, but not resistant to decay or insect attack. It has been used in Hawaii for furniture and paneling. Tests have shown it to be a good veneer species as well.

In the late 1800's two trees were planted on Oahu, one in Kalihi Valley and the other in Nuuanu Valley. According to Lester W. "Bill" Bryan, these trees were the seed source for the tropical ash planted in Hawaii. The Division of Forestry began planting this species for watershed cover about 1920 and since then has planted over 700,000 trees on all islands. For a time in the early 1960's, the species was extensively planted as a potential timber. It was found to have such a poor form, however, that it has now been dropped from consideration. The poor form may result from inbreeding depression, since the entire population originates from one or two parent trees. Tropical ash is a shade-tolerant tree when young and regenerates prolifically in moist sites where planted.

Special areas--Tantalus, Kula, Waihou, Waiakea.

Champion--Height 92 ft (28.0 m), c.b.h. 11.7 ft (3.6 m), spread 95 ft (29.0 m). Kohala Forest Reserve, Muliwai, Hawaii (1968).

In Mexico City and elsewhere in subtropical parts of Mexico where native, this species is a popular street and shade tree. Introduced in southern Arizona.

Range--Native of western and southern Mexico from Sinaloa to San Luis Potosí and Oaxaca south to Guatemala.

Other common names--Hawaiian ash, Shamel ash; fresno (Spanish).

Related to white ash, *Fraxinus americana* L., of continental United States.

136. Olopua, pua

Osmanthus sandwicensis (Gray) Knobl.

This medium-sized evergreen tree is common in dry forests through the islands. It has paired elliptical or lance-shaped leaves, small yellowish green flowers clustered at leaf bases, and small green turning to bluish black fruits. Usually a small tree of 20 ft (6 m) in height with short stout trunk to 8 in (0.2 m) in diameter and rounded crown, sometimes medium-sized to 66 ft (20 m) and 3 ft (0.9 m). Bark gray, rough, furrowed into ridges and plates; inner bark brown, dry, bitter. Twigs light gray or light brown with raised dots and paired raised half-round leaf-scars. Buds scaly pointed gray, 1/8 in (3 mm) long, finely hairy.

Leaves opposite, hairless, with light yellow leafstalks of 1/4--1/2 in (6--13 mm). Blades elliptical or lance-shaped, 2 1/2--6 in (6--15 cm) long and 3/4--2 1/2 in (2--6 cm) wide, long- or short-pointed at apex, short-pointed at base, straight or wavy at edges, slightly stiff and leathery, curved up on sides, above dull green or dark green with midvein and few fine side veins yellowish, beneath dull light green.

Flower clusters (racemes) 1--2 in (2.5--5 cm) long at leaf bases, unbranched. Flowers several in pairs, short-stalked, less than 1/4 in (6 mm) long and broad, composed of 4-toothed minute green calyx, yellowish or whitish green corolla more than 1/8 in (3 mm) long, deeply 4-lobed, 4 stamens attached on base of corolla and alternate with lobes, and pistil with conical greenish 2-celled ovary, short style, and 2-lobed stigma.

Fruits (drupes) egg-shaped, 1/2--7/8 in (13--22 mm) long, blunt at apex, green turning to bluish black, slightly fleshy but becoming dry, the large stone 1-seeded.

Sapwood is yellow and heartwood yellowish brown or dark brown with blackish streaks. Wood fine-textured, very heavy, and very hard, with indistinct growth rings. It is very durable and takes a fine polish. Formerly used by the Hawaiians for tool handles such as adzes and as a rasp for fishhook manufacture. It was a preferred fuelwood because it gave a hot fire even when green. Not presently used.

This species is a favorite host for Hawaiian land snails.

Widespread through the islands, especially in dry forests, at 1,000--4,200 ft (305--1,280 m) altitude.

Special areas--Kokee, Wahiawa, Volcanoes, Kipuka Puaulu.

Champion--Height 54 ft (16.5 m), c.b.h. 10.2 ft (3.1 m), spread 31 ft (9.4 m). Hoomau Ranch, Honomolino, Hawaii (1968).

Range--Through the Hawaiian Islands, not known elsewhere.

Other common names--Hawaiian-olive, ulupua.

Botanical synonym--*Nestegis sandwicensis* (Gray) O. Deg., I. Deg., & L. Johnson.

This is the only native species of the olive family, Oleaceae. Several others are introduced.

Eleven tree species in this handbook commemorate the early English name for Hawaii, the Sandwich Islands. Six, including the olopua were named by Asa Gray, U.S. botanist of Harvard University. Captain James Cook of the British Navy, named the islands in 1778 in honor of John Montagu, Earl of Sandwich, who had sponsored Cook's expedition.

135. Tropical ash　　　*Fraxinus uhdei* (Wenzig) Lingelsh.
Female flowers (upper right), twig with leaf, and fruits (lower left), 1 X (P. R. v. 2).

DOGBANE FAMILY (APOCYNACEAE)

137. Hao, Hawaiian rauvolfia *Rauvolfia sandwicensis* A. DC.

Hao is a variable species of small evergreen native trees or shrubs widespread in mainly dry areas, characterized by milky sap, elliptical or oblong light green leaves mostly 4--5 at ringed nodes, small greenish yellow or whitish tubular flowers, and 2-lobed black fruits. A small tree 20 ft (6 m) high, sometimes to 40 ft (12 m), and 1 ft (0.3 m) in trunk diameter, or a shrub. Bark light gray, smooth or lightly fissured. Inner bark yellowish or brown streaked with pink, bitter, when cut exuding white, slightly bitter sap or latex. Twigs stout, hairless, light green, turning gray, with raised half-round leaf-scars at ringed nodes.

Leaves mostly 4--5 at node (whorled), hairless, with slender light green leafstalk 5/8--1 1/4 in (1.5--3 cm) long, slightly flattened and winged, and in angle at leaf base 5--12 tiny stalked glands. Blades elliptical or oblong, 2 1/2--5 in (6--13 cm) long and 1--2 in (2.5--5 cm) wide, thin or slightly thickened, blunt to long-pointed at both ends and straight on edges, upper surface dull light green with light yellow midvein and many fine parallel side veins, lower surface paler.

Flower clusters (cymes) mostly terminal, branched. Flowers many, crowded, nearly stalkless, fragrant, 3/8--1/2 in (10--13 mm) long, composed of green calyx of 5 pointed overlapping lobes, greenish yellow or whitish narrow tubular corolla with 5 rounded overlapping spreading lobes, 5 small stamens inside tube, and pistil with 2 ovaries, threadlike style, and long stigma.

Fruits (drupes) 2-lobed, about 1/2 in (13 mm) long and broad, slightly flattened and heart-shaped, with calyx at base, black, fleshy.

Sapwood light yellow and heartwood deep reddish brown. Moderately heavy, fine-textured, moderately hard, durable, straight-grained and without growth rings. Not used by the Hawaiians for fuel because the smoke was thought to be poisonous, nor for charcoal because it burned completely to ashes. It was however, considered a good wood for construction.

Scattered mostly in dry forests on leeward slopes to about 2,000 ft (610 m) altitude. A stunted shrub on aa rough lava flows.

Special area--Volcanoes.
Range--Hawaiian Islands only.
Other common name--devilpepper.
Botanical synonyms--*Rauvolfia degeneri* Sherff, *R. forbesii* Sherff, *R. helleri* Sherff, *R. mauiensis* Sherff, *R. molokaiensis* Sherff, *R. remotiflora* Deg. & Sherff.

This variable species was divided into 7 differing mainly in sizes of leaves and flowers (Sherff 1947). The 6 segregates were questioned in a monograph by Rao (1956) and are treated here as synonyms.

The genus *Rauvolfia* (formerly also spelled *Rauwolfia*) commemorates Leonhard Rauwolf (1535--96), German botanist and physician. He collected plants and made an early herbarium on his travels through the Near East in 1573--75.

The root of a shrubby species of snakeroot of this genus in India (and another in Africa) has yielded the drug reserpine for the treatment of high blood pressure and certain mental illnesses. Local use by traditional healers in India for snakebite and lunacy has been traced back many centuries. In recent years chemical tests of other species in this tropical genus have been made in screening searches for additional drugs.

BORAGE FAMILY (BORAGINACEAE)

138. Kou *(color plate, p. 41)* *Cordia subcordata* Lam.**

Small evergreen tree with broad crown, uncommon along shores, apparently introduced by the early Hawaiians. Characterized by large broadly ovate to elliptical blunt-pointed leaves and showy large funnel-shaped orange flowers. To about 26 ft (7 m) high and 15 in (38 cm) in trunk diameter, reported to reach somewhat larger size formerly. Bark gray, thick rough, furrowed into narrow ridges. Twigs stout, light gray, sometimes hairy, with raised half-round leaf-scars.

Leaves alternate, with long stout leafstalks 1 1/2--4 in (4--10 cm) long. Blades broadly ovate to elliptical, 3--7 in (7.5--18 cm) long and 2--5 in (5--13 cm) wide, blunt at apex, rounded or unequally short-pointed at base, often slightly wavy on edges, thin, above shiny or dull green with few long yellowish veins and hairless, beneath dull light green with raised veins and fine whitish hairs mostly along veins and in vein angles. Leaves turn bright yellow and dark brown before falling.

Flower clusters (cymes or panicles) terminal and lateral, short, about 2 in (5 cm) long. Flowers several, short-stalked, showy, about 1 1/4 in (4 cm) long and broad. Calyx cylindrical, fleshy, yellow green, 5/8 in (15 mm) long, with 3--5 short broad teeth; corolla orange, funnel-shaped with 5--7 rounded slightly wrinkled lobes; stamens 5--7, threadlike, opposite, inserted in tube and slightly longer; and pistil with conical 4-celled

136. Olopua, pua   *Osmanthus sandwicensis* (Gray) Knobl.
Twigs with fruits and flowers, 1 X.

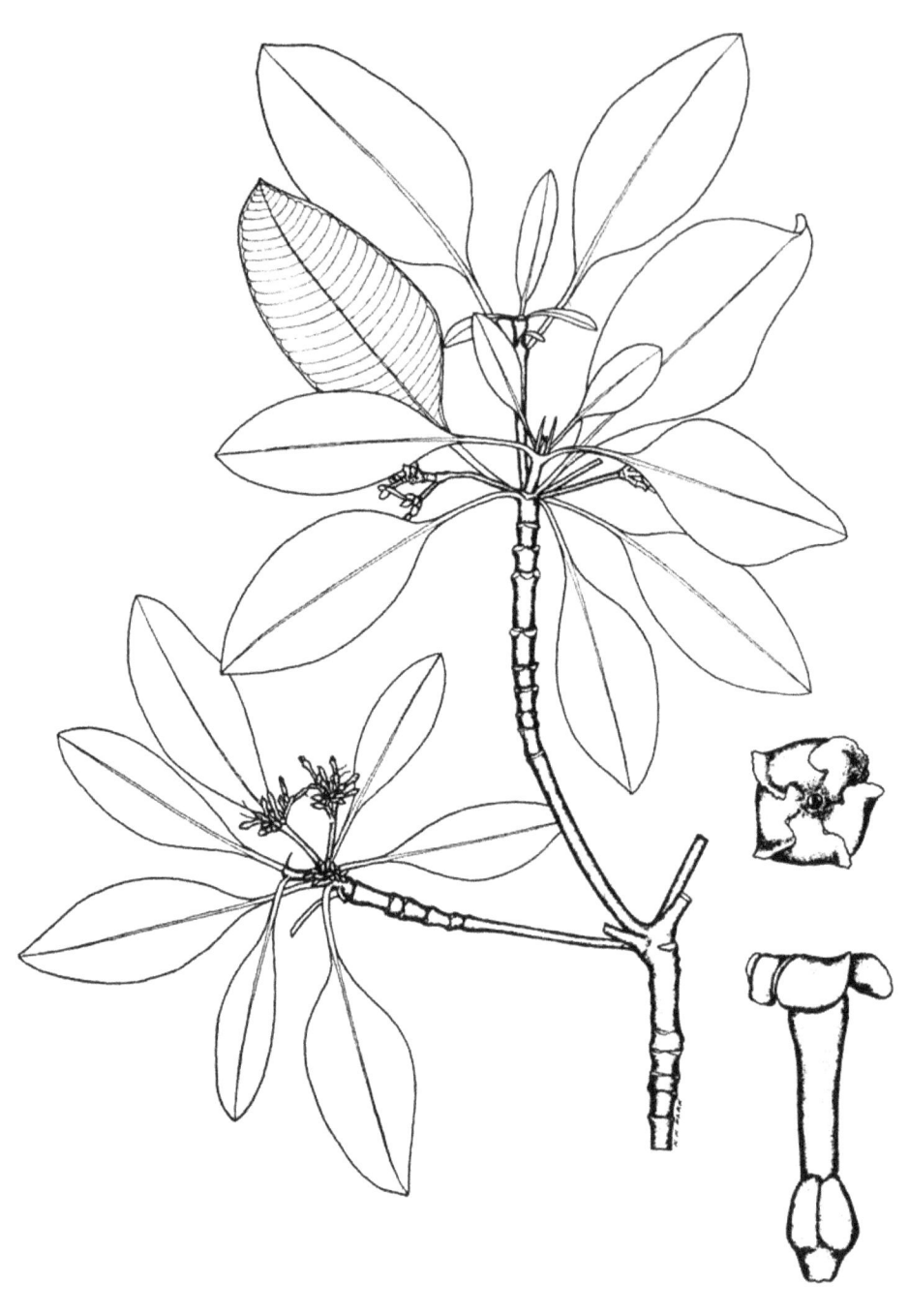

137. Hao, Hawaiian rauvolfia — *Rauvolfia sandwicensis* A. DC.
Flowering twig, 1/2 X; flowers (lower right), 5 X (Degener).

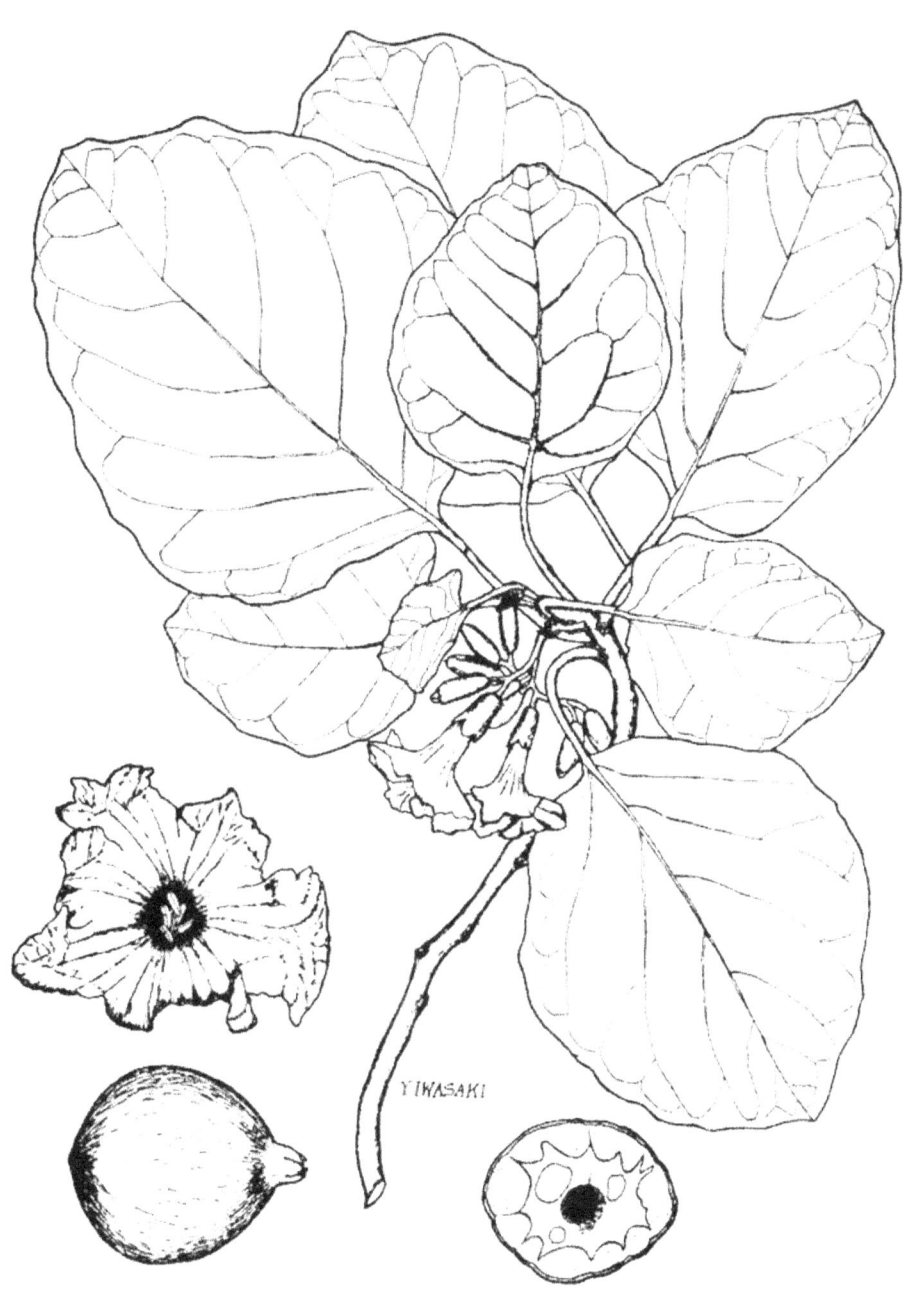

138. Kou   *Cordia subcordata* Lam.**
Flowering twig, 1/2 X; flower and fruit (lower left), 1 X (Degener).

ovary and slender orange style branched twice near apex.

Fruits (drupes) several, egg-shaped, 1 in (2.5 cm) long, enclosed by the brown calyx with teeth and base of style at apex, green, becoming dry and brown, with large hard stone. Seeds 4 or fewer, white, about 1/2 in (13 mm) long, narrow.

Sapwood is pale yellowish brown and heartwood is light brown prominently marked by dark brown or black streaks in the growth rings. Figure resembles that of "Circassian" walnut. Lightweight (sp. gr. 0.45), soft, easily worked, durable, and takes a fine polish. It was used by the Hawaiians for their handsome bowls, cups, and dishes. The wood was favored for utensils because it did not impart a flavor to foods as do koa and other native woods. It is still used occasionally for craftwood but is in very short supply.

Formerly, this very useful tree was more common as a shade tree of rapid growth around houses and along the shore. The flowers were made into necklaces or leis, and the seeds were eaten. Many trees were destroyed by moths.

In Hawaii, scattered through the islands along shores, both humid and arid, and planted as an ornamental. Recorded from Niihau. As further evidence of introduction, Degener noted that fossil pollen of this species had not been found in Hawaii and that no peculiar or endemic insects were found on the trees here.

Trees may be seen at Kualoa Beach, Maui Zoological and Botanical Garden, and in many other places. One of the largest kou trees in Hawaii grows in front of the Pauhana Inn at Kaunakakai, Molokai.

Special areas--Waimea Arboretum, Foster, Iolani, City.

Range--Now cultivated and established from eastern Africa and tropical Asia through Malaya to tropical Australia and Pacific Islands. Probably distributed to Hawaii and other islands by early inhabitants as well as by ocean currents.

Kou is often confused with the more recently introduced kou-haole or Geiger-tree, *Cordia sebestena* L., which is smaller in size and has smaller, rough textured leaves, white fruit, and darker orange flowers.

Other common names--koa (Guam); niyoron (N. Marianas); kalau (Palau); galu (Yap); anau (Truk); ikoik (Pohnpei); ikoak (Kosrae); tauanave (Am. Samoa).

139. Tree-heliotrope

Tournefortia argentea L. f.*

This distinctive small umbrella-shaped evergreen tree with very short trunk, low widely forking branches, and very broad rounded spreading crown of gray green foliage, was introduced on sandy shores. To 20 ft (6 m), with trunk to 1 ft (0.3 m) in diameter and crown to 40 ft (12 m) across, often flowering as a low shrub. Bark light brown or gray, rough, very thick, deeply furrowed into narrow oblong plates and ridges. Outer bark streaky blackish brown, inner bark light brown, fibrous, tasteless. Twigs stout, finely hairy, gray green, becoming brown, with raised half-round leaf scars and buds of small overlapping leaves.

Leaves alternate, crowded near ends of twigs, gray green, covered with tiny pressed hairs, with short stout leafstalk about 3/8 in (1 cm) long. Blades narrowly elliptical or obovate, 3--7 in (7.5--18 cm) long, 1--2 1/4 in (2.5--6 cm) wide, thick and slightly succulent, rounded at apex, widest beyond middle, tapering to long-pointed base, not toothed, with few side veins, dull gray green on both surfaces.

Flower clusters (cymes) terminal, 6--8 in (15 cm) long including long stalk, the many branches curved to one side. Flowers many, crowded, stalkless, erect on horizontal curved or coiled branches, bell-shaped, less than 1/4 in (6 mm) long and broad, composed of 5 rounded hairy gray green sepals, white corolla with short hairy tube and 5 spreading rounded lobes, 5 tiny stamens in notches of corolla, and pistil with conical ovary and slightly 2-lobed stigma.

Fruit rounded, flattened, about 1/4 in (6 mm) in diameter, smooth and shiny, green, slightly watery, containing 2 or 4 large half-round brown nutlets 1/8 in (3 mm) long, embedded in a corky or spongy mass.

It is reported that in India the leaves are eaten raw. They have a slightly salty flavor and might serve in salads or cooked as greens. The corky mass of nutlets may be carried by ocean currents.

Ornamental tree planted and hardy along sandy beaches. Scattered on shores through the Hawaiian Islands. Listed by Hillebrand as in cultivation, apparently before his departure in 1871.

Special area--Waimea Arboretum.

Champion--Height 32 ft (9.8 m), c.b.h. 31.5 ft (9.6 m), spread 46 ft (14.0 m). Puako Kawaihae, Hawaii (1968).

Range--Native from India in tropical Asia to Mauritius, Malaya, tropical Australia, western Indian Ocean islands, Polynesia, and Micronesia.

Other common names--velvetleaf; hunig (Guam); huni (N. Marianas); aseri (Palau); chel (Yap); chen yamolehat (Truk); titin (Pohnpei); srusrun (Kosrae); kiden (Marshalls); tausuni (Am. Samoa).

Botanical synonym--*Messerschmidia argentea* (L. f.) I. M. Johnst.

The genus honors Joseph Pitton de Tournefort (1656--1708), French botanist who established the genus concept.

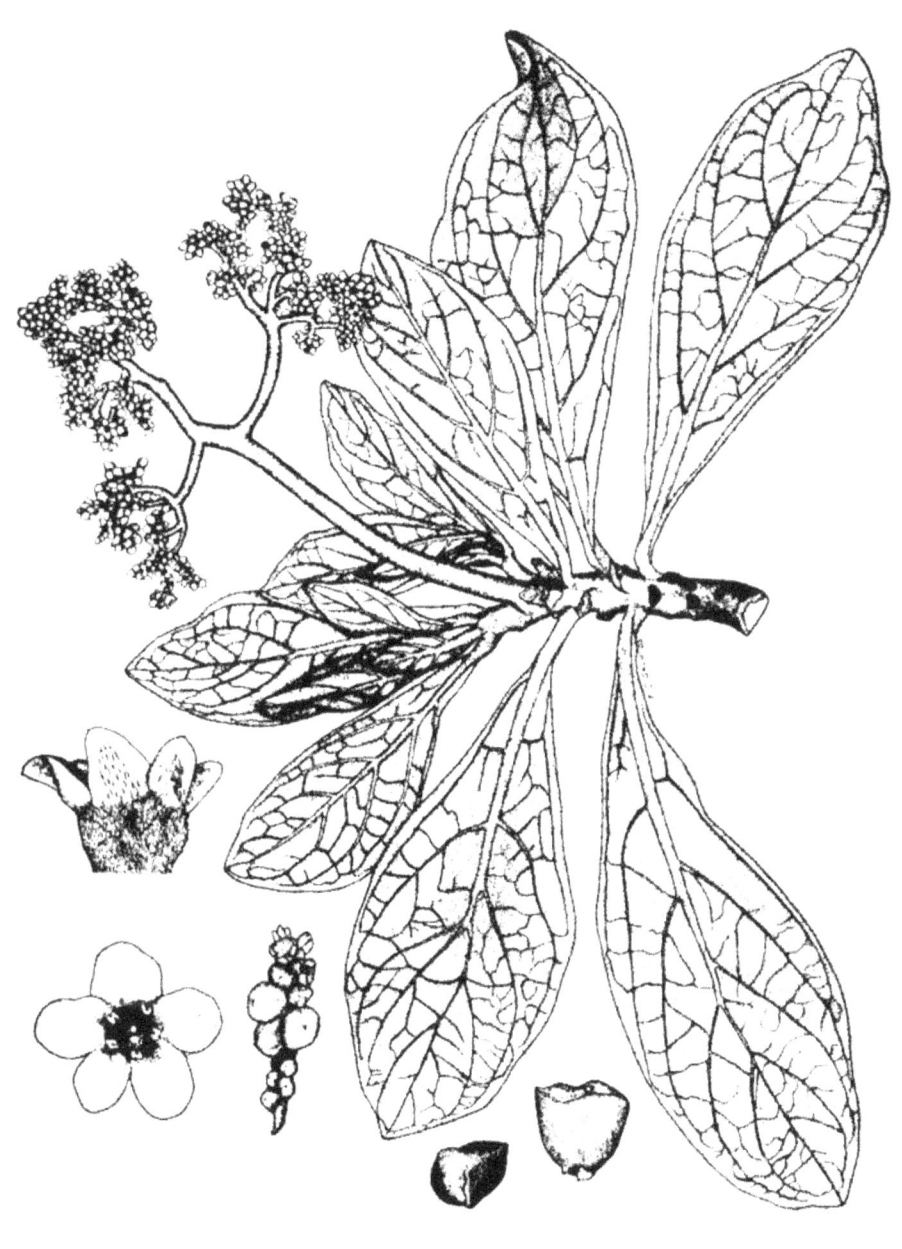

139. Tree-heliotrope *Tournefortia argentea* L. f.*
Flowering twig, 1/2 X; flowers 4 X, and fruits, 1 X (lower left); nutlets (below), 4 X (Degener).

VERBENA FAMILY (VERBENACEAE*)

140. Fiddlewood
Citharexylum caudatum L.*

This shrub or small tree introduced as an ornamental has become naturalized locally. It has paired elliptical leaves and narrow clusters of small slightly fragrant white flowers, and showy shiny pear-shaped or elliptical fruits that turn from green to orange brown or black.

Evergreen shrub or small tree to 50 ft (15 m) high, with trunk 1 ft (0.3 m) in diameter, angled, slightly enlarged at base, and irregular, and with spreading crown. Bark gray, smoothish, becoming finely fissured. Inner bark light brown and whitish streaked, bitter. Twigs long and narrow, light gray brown, hairless, with enlarged nodes and raised leaf-scars.

Leaves opposite, hairless, on leafstalks of 3/8--3/4 in (1--2 cm). Blades elliptical, 2--5 in (5--13 cm) long and 3/4--2 1/4 in (2--6 cm) wide, blunt to short-pointed at apex and short-pointed at base, with edges sometimes slightly turned under, slightly thickened, with few side veins, slightly shiny green above, and beneath dull and paler with many tiny dots visible under a lens. Leafstalks and midveins often orange red.

Flower clusters (racemes) 1 1/2--6 in (4--15 cm) long, terminal and lateral, narrow and unbranched. Flowers many on short stalks of 1/16 in (1.5 mm), about 5/16 in (8 mm) long and broad. Calyx bell-shaped, 1/8 in (3 mm) long and broad, minutely 5-toothed; the white finely hairy corolla with tube nearly 1/4 in (6 mm) long and 5 spreading slightly unequal rounded corolla lobes 1/8 in (3 mm) long; 4 small stamens inserted in corolla tube; and greenish pistil 3/16 in (5 mm) long with 2-celled ovary, slender style, and slightly 2-lobed stigma.

Fruits (drupes) in drooping clusters, nearly 1/2 in (13 mm) long, slightly 2-lobed, fleshy, with calyx remaining at base, containing 2 elliptical shiny brown nutlets 3/8 in (10 mm) long, each 1-seeded.

The light brown hard wood is used elsewhere for posts, not for musical instruments, as reported. This species is a honey plant.

An introduced ornamental in Hawaii, it is a common street tree in Honolulu. When young, highly susceptible to wind damage. Naturalized locally, for example, in wet forests at 2,000 ft (610 m) altitude on Oahu.

Range--Native of West Indies from Bahamas to Jamaica, Puerto Rico, and Dominica. Also from Yucatán, Mexico, and Central America to Colombia. Planted in southern Florida.

Other common names--juniper-berry; péndula de sierra (Puerto Rico).

This species is reported to be one of the many hosts of the black twig-borer, a major insect pest.

The generic name *Citharexylum* is Greek for fiddlewood. The English name fiddlewood and similar ones in French and Spanish for related species apparently were taken from the scientific name without regard to actual use of the wood.

BIGNONIA FAMILY (BIGNONIACEAE*)

141. Primavera, goldtree
Roseodendron donnell-smithii (Rose) Miranda*

Primavera is a large deciduous tree, introduced as an ornamental for its showy masses of large bell-shaped golden yellow flowers when leafless in spring. A poor seeder in Hawaii, it has been tested sparingly as a forest plantation tree. It is recognized also by the large paired, palmately compound leaves with mostly 7 leaflets and the very long, narrow cigarlike seed capsules.

A large tree 60 ft (18 m) or more in height and 3 ft (0.9 m) in trunk diameter, where native to 115 ft (35 m) high and 2--3 ft (0.6--0.9 m), with rounded or spreading crown. Bark light or whitish gray, smoothish or becoming rough, thick, furrowed, and slightly shaggy. Inner bark whitish or pale brown. Twigs stout, with small mealy whitish star-shaped hairs when young and large raised leaf scars.

Leaves opposite, palmately compound (digitate), large, 8--20 in (10-51 cm) long, with long stalk of 5--10 in (13--25 cm) enlarged at ends and mostly 7 (-5), long-stalked leaflets. Blades ovate or elliptical, 2--10 in (5--25 cm) long and 3/4--5 in (2--13 cm) wide, long-pointed at apex, rounded or slightly notched at base, often wavy-toothed on edges, thin, upper surface dull green, lower surface paler, becoming nearly hairless except on veins beneath.

Flower clusters (panicles) to 8 in (20 cm) long at ends of stout leafless twigs, widely branched and spreading, hairy. Flowers, many, crowded, spreading, short-stalked, about 2 in (5 cm) long, composed of deeply 2-lobed thin hairy calyx 5/8 in (15 mm) long; bright yellow bell-shaped finely hairy corolla about 2 in (5 cm) long with broad tube and 5 rounded slightly unequal spreading lobes 1 1/2 in (4 cm) across; 4 stamens in 2 pairs within tube and attached near base,

140. Fiddlewood *Citharexylum caudatum* L.*
Fruiting twig, flowers (upper right), 1 X.

also 1 tiny sterile; and pistil with conical hairy ovary, slender style, and 2 stigmas.

Fruits (seed capsules) very long and narrow, cigar-like, 12--16 in (30--40 cm) long, about 1 in (2.5 cm) in diameter, rough, finely hairy, with 10--12 longitudinal ridges, flattened, 2-celled, splitting lengthwise in 2 parts. Seeds many, flattened, bordered by circular papery wing 3/4 in (2 cm) long.

The wood is light yellow or whitish, lightweight, of medium to rather coarse texture, straight to wavy grain. Resembles satinwood and was sometimes called white mahogany. Is is fairly strong, lightweight (sp. gr. 0.45), easy to work when straight grained, finishes smoothly, but is not resistant to decay. Wood with rippled or roey figure slices poorly and is very difficult to dry flat as veneer.

Known in international commerce as primavera, the wood is used especially for veneering, cabinetwork, furniture, and flooring. Locally, it serves also for general construction. It is exported from Guatemala and Mexico to the United States.

Introduced to Hawaii many years ago as an ornamental street tree, this species has been tried sparingly in the forests. The Division of Forestry has planted only 1,200 trees in the forest reserves, mostly on Oahu and Maui. It has great potential as both a timber tree and a beautiful flowering plant. Goldtree is quite common around Honolulu. It can be found at the former Hawaiian Sugar Planters' Association Experiment Station on Keeaumoku Street, the park at the foot of Makiki Heights Drive, at Queen's Hospital, and in front of the Department of Education Building. The tree will grow to about 3,000 ft (914 m) altitude.

Champion--Height 87 ft (26.5 m), c.b.h. 10.9 ft (3.3 m), spread 79 ft (24.1 m). State Forestry Arboretum, Hilo, Hawaii (1968).

Range--Native of southern Mexico, Guatemala, El Salvador, and Honduras, but introduced in other tropical areas. An uncommon ornamental in Puerto Rico.

Botanical synonyms--*Tabebuia donnell-smithii* Rose, *Cybistax donnell-smithii* (Rose) Seibert.

The drawing by the botanical artist Charles Edward Faxon (1846-1918), is from the original publication (1892) by Joseph Nelson Rose (1862-1928), botanist at the Smithsonian Institution. The current generic name means Rose's tree. The specific name honors the discoverer, John Donnell Smith (1829--1928), botanist and ship captain from Baltimore, MD.

142. African tuliptree

Spathodea campanulata Beauv.*

This handsome ornamental and shade tree is planted for its showy masses of large brilliant orange-red to scarlet tuliplike flowers in erect clusters mostly at top of crown. A prolific seeder, as well as a root sprouter, the species has become naturalized.

Large tree 50--80 ft (15--24 m) high and 1--1 1/2 ft (0.3--0.5 m) in trunk diameter, with dense irregular crown of large spreading branches, evergreen or nearly deciduous. Big trunks develop tall narrow buttresses at base and are slightly broadened and grooved. Bark very light brown, smoothish but becoming slightly fissured; inner bark whitish, bitter. Twigs stout, gray brown, smooth except for corky warts (lenticels).

Leaves opposite or sometimes in 3's, large, pinnate, 1--2 ft (30--61 cm) long. Leaflets usually 11--17 (sometimes 5--19), paired except at end, 3--6 in (7.5--15 cm) long and 1 1/2--3 in (4--7.5 cm) broad, on short stalks of 1/8 in (3 mm), abruptly short-pointed, rounded and slightly oblique at base, a little thickened with edges a little turned under, almost hairless, upper surface green to dark green with sunken veins and slightly shiny, lower surface rusty hairy.

Flower clusters (racemes) terminal, about 4 in (10 cm) high and 8 in (20 cm) across. Flower buds numerous crowded, horn-shaped, 1-2 in (2.5--5 cm) long on stout greenish stalks of the same length, orange brown, curving inward to the center. Those around outside open a few at a time and drop off about 2 days later. The flowers have a most unusual flattened shape, with light brown calyx 2--2 1/2 in (5--6 cm) long, curved and pointed like a horn, splitting open on outer side, minutely hairy and with longitudinal ridges. The tubular orange-red to scarlet corolla about 4 in (10 cm) long and 2 by 3 in (5 by 7.5 cm) broad has an enlarged irregular bell-shaped tube 2 in (5 cm) across and curved downward and 5 broad unequal lobes with crisp wavy edges narrowly bordered with gold. There are 4 pale yellow stamens 2--2 1/2 in (5--6 cm) long, with dark brown anthers, inserted in corolla tube in 2 pairs and projecting barely beyond. The pistil on a disk consists of an oblong 2-celled ovary 1/4 in (6 mm) long, a long slender curved, pale yellow style about 3 in (7.5 cm) long, and 2-lobed red stigma.

Pods (capsules) 1 to several, large lance-shaped or boat-shaped, green to dark brown, 5--10 in (13--25 cm) long, 1 1/2 in (4 cm) wide, and 7/8 in (22 mm) thick, slightly flattened, long-pointed, erect and pointing upward from stout stalks, splitting open on 1 side. Seeds numerous, very thin and papery, with light brown center bordered by transparent wing 1/2--1 in (13--25 mm) across. Flowering and fruiting probably through the year.

Wood is very lightweight, soft, whitish yellow, without distinction between sapwood and heartwood, very brash and coarse-textured. The wood is not utilized in Hawaii.

This ornamental and shade tree is propagated by seeds, cuttings, and root cuttings. It grows very rapidly but requires nearly full light. Measured trees in Puerto Rico increased in trunk diameter as much as 2 in (5 cm)

141. Primavera, goldtree *Roseodendron donnell-smithii* (Rose) Miranda*
Two fruits and seed (left), flower cluster, and leaf, 2/3 X; opened flower (lower right), enlarged (Rose).

a year. As trees are broken by high winds and frequently become hollow and hazardous in age, planting near buildings or along roads is not advised. Also, the superficial root system makes this species undesirable for planting near houses and sidewalks. The trees produce sprouts from the roots, sometimes becoming like weeds. This species has been tried elsewhere for coffee shade but is not recommended for that purpose. Unopened flower buds contain water, ill smelling and tasting, which squirts out when the buds are squeezed, pinched, or pricked with a pin. The name "fountain-tree" has been suggested by this character. Children play with these buds like water pistols. Old dry empty pods when widely open make very realistic toy boats, which, however, close up in water.

Planted and naturalized in lowlands of Hawaii. It was aerially seeded in the Panaewa and lower Waiakea Forest Reserves near Hilo in 1928 and occupies much of the forest in the area seeded. The Division of Forestry has planted more than 30,000 trees in the forest reserves, mostly on Maui and Hawaii.

Special areas--Waimea Arboretum, Foster, Tantalus, Waiakea.

Champion--Height 83 ft (25.3 m), c.b.h. 15.5 ft (4.7 m) spread 56 ft (17.1 m). Kainaliu, North Kona, Hawaii (1968).

Range--Native of tropical West Africa. Widely planted in tropical regions around the world, including southern Florida and Puerto Rico and Virgin Islands.

Other common names--tuliptree, fountain-tree, firebell, spathodea; tulipàn africano (Puerto Rico, Spanish); rarningobche (Yap).

Botanical synonym--*Spathodea nilotica* Seem.

The first trees were introduced into Hawaii apparently by Hillebrand more than a century ago, according to Degener. About 1915, Rock brought seeds from Java.

A horticultural variety with golden yellow flowers is a recent introduction and provides some variation in color. It has been named for the late Lester W. "Bill" Bryan, who served for 40 years with the Territorial and later State forestry staff in Hawaii until his retirement in 1961 as Deputy State Forester.

MYOPORUM FAMILY (MYOPORACEAE)

143. Naio, false-sandalwood

Myoporum sandwicense Gray

One of the common trees of the Island of Hawaii from sea level to timberline, in both dry and wet forests through the islands. Recognized by its dark gray very thick, rough and irregularly furrowed bark, many crowded narrowly elliptical lance-shaped leaves resinous or sticky when young, small white to pinkish bell-shaped flowers at leaf bases, and whitish oblong rounded fruits.

A small tree of 30 ft (9 m) in height and 1 ft (0.3 m) in trunk diameter, with thin rounded crown, or a large tree recorded to 60 ft (18 m) and 3 ft (0.9 m), or a windswept shrub or dwarf shrub of 2 ft (0.6 m) at timberline. Bark on small trunks gray and smoothish, becoming very thick, rough and irregularly furrowed or scaly and divided into small plates, the trunk slightly angled. Inner bark light brown, slightly bitter. Twigs greenish with tiny hairs when young, becoming brown, with raised half-round leaf-scars.

Leaves many, alternate, crowded near end of twig, mostly hairless, with flattened yellowish leafstalk of 3/8 in (1 cm). Blades 2--6 in (5--15 cm) long and 3/8--2 in (1--5 cm) wide, broadest near middle and tapering to long point at both ends, rarely with toothed edges, thin or slightly thickened, side veins not visible, above dull green, paler beneath, with gland-dots visible under lens.

Flowers many fragrant, 1--9 on slender stalks of 1/4--1/2 in (6--13 mm), clustered at leaf base and scattered along twig, bell-shaped, about 1/2 in (13 mm) across, composed of green calyx of 1/8 in (3 mm) deeply 5-lobed; pinkish or whitish corolla about 5/16 in (8 mm) long with short tube and mostly 5--8 elliptical curved spreading lobes; stamens mostly 5 in notches of corolla; and pistil with elliptical ovary mostly 5-celled (4--8) with 1 ovule in each cell and slender style.

Fruits (drupes) many on slender stalks at leaf bases and on twigs back of leaves, oblong or rounded, about 5/16 in (8 mm) in diameter, whitish or tinged with purple, with calyx and style remaining, juicy, bitter, becoming dry, wrinkled, light brown. Stone large, with 5--8 cells and seeds.

Sapwood is pale brown and heartwood dark yellowish brown, moderately heavy (sp. gr. 0.55), hard, and fine textured. The wood has an attractive figure imparted by dark zones in the growth rings. It gives off an odor somewhat like sandalwood during drying, but the odor is short-lived. After the supply of sandalwood was exhausted, this wood was shipped to China as a substitute, but was not accepted. A fire made from the wood is almost unbearably fragrant, according to Degener (1930, 261-271). Considered a good firewood by most upland ranchers.

Timbers of this species were among those preferred for frames of the Hawaiian houses. It was also used for fishing torches because of its good burning characteristics. A number of large trees have been cut in recent

142. African tuliptree *Spathodea campanulata* Beauv.*
Flower, leaf, and fruits (below), 2/3 X (P. R. v. 1).

years, and the lumber produced has been used for flooring, furniture, and craftwood items. Although not too stable, it has performed well in service.

This handsome plant with many pinkish white flowers is suitable for cultivation as an ornamental shrub. Livestock poisoning has been reported in related species westward.

Common in dry upland forest and brushlands and near sea level mostly as a shrub. Becoming uncommon in many areas formerly occupied because of site disturbance. On Hawaii and Maui, large trees occur in wet and dry forests, on the other islands it is mostly a shrub. Trees may be seen in Hawaii Volcanoes National Park along the Mauna Loa Strip Road and at Kipuka Puaulu and near Pohakuloa on the Saddle Road. It also grows as a large tree in the Waiakea Forest Reserve 'ōhi'a-treefern rain forest and in the kōa-'ohi'a forest near Kulani Cone. It grows from sea level to timberline at 10,000 ft (3,048 m). Recorded from Niihau.

Special areas--Waimea Arboretum, Wahiawa, Bishop Museum, Volcanoes, Kipuka Puaulu.

Champion--Height 70 ft (21.3 m), c.b.h. 17.2 ft (5.2 m), spread 56 ft (17.1 m). Keauhou, North Kona, Hawaii (1968).

Range--Including varieties widespread through the Hawaiian Islands including Niihau, but extinct on Kahoolawe. Not known elsewhere.

Other common names--sandalwood myoporum, 'a'aka (wood and dead trees), naieo, naeo, false sandalwood.

MADDER OR COFFEE FAMILY (RUBIACEAE)

144. 'Ahakea
Bobea sandwicensis (Gray) Hillebr.

The genus *Bobea*, common name 'ahakea, is known only from the Hawaiian Islands and has 4 or fewer species of trees distributed through the islands. They have small paired pale green leaves with paired small pointed stipules that shed early, 1--7 small flowers at leaf bases, with tubular greenish corolla, the 4 lobes overlapping in bud, and small round black or purplish fruit (drupe), mostly dry, with 2--6 nutlets. This species, described below, will serve as an example.

Medium-sized evergreen tree to 33 ft (10 m) high and 1 ft (0.3 m) in trunk diameter. Bark gray, smoothish, slightly warty, fissured, and scaly. Inner bark light brown, bitter. Twig light brown, with tiny pressed hairs and with rings at nodes.

Leaves opposite, with pinkish finely hairy leafstalks of 3/8--5/8 in (1--1.5 cm) and paired small pointed hairy stipules 1/8 in (3 mm) long that form bud and shed early. Blades ovate or elliptical, 2--3 1/2 in (5--9 cm) long and 1--2 in (2.5--5 cm) wide, short- to long-pointed at apex and blunt at base, with edges straight or slightly wavy, thin. Upper surface slightly shiny green, hairless, with pinkish midvein and few curved side veins; lower surface light green with pinkish raised midvein and tiny hairs in vein angles.

Flowers mostly 1--3 at leaf bases on slender stalks of 1/4--1 in (6--25 mm), 5/16 in (8 mm) long, finely hairy, composed of greenish base (hypanthium) 1/8 in (3 mm) long; calyx of 4 spreading elliptical green lobes to 3/16 in (5 mm) long; whitish green corolla 1/4 in (6 mm) long with narrow cylindrical tube and 4 spreading narrow lobes overlapping in bud; 4 stalkless stamens in notches of corolla; and pistil with inferior ovary, short style, and 4--6-lobed stigma.

Fruit (drupe) round, about 3/8 in (1 cm) in diameter, purplish black and slightly shiny, with tiny pressed hairs, with 4 enlarged rounded calyx lobes more than 1/4 in (6 mm) long remaining at apex, containing 2--6 stones (nutlets), each 1-seeded.

Wood yellow, hard and heavy, straight-grained. Used by the Hawaiians for the carved end covers and gunwales of canoes for its yellow appearance and for its wearability. Modern canoes are often painted yellow at the gunwales to simulate 'ahakea wood. Also used for poi boards and paddles.

Scattered in wet to dry forests and on open lava flows at 300--4,000 ft (105--1,220 m) altitude.

Range--Maui, Lanai, Molokai, and Oahu.

Botanical synonym--*Bobea hookeri* Hillebr.

This genus was named in 1830 for M. Bobe-Moreau, physician and pharmacist in the French Marine. Three other species are found on the large islands of Hawaii.

145. Alahe'e
Canthium odoratum (G. Forst.) Seem.

This common handsome shrub or small tree of wide distribution in dry lowlands has a round crown or shiny bright green leaves, which are paired, small, and elliptical, the small fragrant white flowers, and small rounded black fruits. To 20 ft (6 m) high and 4 in (10 cm) in trunk diameter, smaller in exposed areas. Trunk very straight, erect, with distinctly horizontal branches. Bark light to dark gray, smoothish; inner bark pale yellow or light brown, slightly bitter. Twigs hairless, green when young, slightly 4-angled with rings at enlarged nodes, becoming gray and slightly scaly.

Leaves opposite, hairless, with short yellowish or light green leafstalks 1/8 in (3 mm) long and paired small pointed scales (stipules) at base. Blades elliptical,

143. Nalo, false sandalwood *Myoporum sandwicense* Gray
Twig with flowers and fruits, about 2/3 X; flower (below), slightly enlarged (Degener).

144. 'Ahakea *Bobea sandwicensis* (Gray) Hillebr.
Flowering twig (above), fruiting twig (below), 1 X.

145. Alahe'e *Canthium odoratum* (G. Forst.) Seem.
Flowering twig, fruits (upper left), 1 X.

1 1/4--2 1/2 in (3--6 cm) long and 5/8--1 1/4 in (1.5--3 cm) wide, slightly thickened and leathery, blunt at both ends and slightly turned under at edges, above shiny green with few fine veins, beneath dull light green.

Flower clusters (cymose corymbs) less than 1 in (2.5 cm) long at leaf bases, flattened. Flowers many, fragrant, on slender stalks 5/16 in (8 mm) long, composed of green base (hypanthium) less than 1/8 in (3 mm) long bearing 5 calyx teeth and tubular white corolla 3/16 in (5 mm) long bearing 5 finely hairy lobes meeting at edges in bud; 5 short stamens attached in notches of corolla; and pistil with inferior 2-celled ovary, slender style, and slightly 2-lobed stigma.

Fruits (drupes) rounded, 5/16 in (8 mm) in diameter, black, with calyx ring at apex, juicy, containing 2 stones or nutlets.

Wood whitish or light brown, very hard, reported to be durable. Hawaiians used it for tools in tilling the soil and for adze blades for cutting softer woods such as kukui and wiliwili.

It is reported that the leaves provided a black dye.

Common and widespread in dry areas from sea level to 3,800 ft (1,158 m) altitude through the islands, even on exposed windswept slopes.

Special areas--Waimea Arboretum, Wahiawa, Bishop Museum, Kamehameha School Hawaiian Garden, Volcanoes.

Champion--Height 39 ft (11.9 m), c.b.h. 3.7 ft (1.1 m), spread 18 ft (5.5 m). Puuwaawaa, Kailua-Kona, Hawaii (1968).

Range--Through Hawaiian Islands, and in Polynesia but not New Zealand, New Hebrides, or Fiji. One of several native tree species not confined or endemic to Hawaii.

Other common names--walahe'e, plectronia, 'ōhe'e.

Botanical synonym--*Plectronia odorata* (G. Forst.) F. Muell.

146. Coffee

Coffea arabica L.*

Coffee, the source of one of the world's most popular beverages, is an introduced shrub or sometimes a small tree scattered on moist lower mountain slopes and grown in plantations, mainly near Kona, Hawaii. Generally a compact shrub 5--10 ft (1.5--3 m) high, but if not pruned it becomes a small tree 12--15 ft (3.7--4.6 m) high and 3 in (7.5 cm) in trunk diameter, evergreen with spreading foliage.

Bark light gray, thin, much fissured, becoming rough; inner bark whitish and tasteless. Twigs many from main axis, long slender, spreading and slightly drooping, green when young but changing to light brown, with paired long-pointed scales (stipules) 3/16 in (5 mm) long at nodes.

Leaves opposite, hairless, with leafstalks of 1/4--1/2 in (6--13 mm). Blades elliptical, 3--7 in (7.5--18 cm) long and 1 1/2--2 3/4 in (4--7 cm) broad, long-pointed at apex and short-pointed at base, upper surface and edges often slightly wavy, a little thickened, shiny dark green with sunken veins on upper surface, paler green beneath.

Flowers many, fragrant, several together on stalks of 1/8 in (3 mm) at leaf bases along twig, about 1 1/4 in (3 cm) across the 5 long white corolla lobes. Calyx consists of 5 minute teeth on green tubular base (hypanthium) less than 1/8 in (3 mm) long; corolla white and showy, with narrow cylindrical tube 3/8--1/2 in (10--13 mm) long and 5 widely spreading narrow pointed lobes 5/8 in (15 mm) long; stamens 5, white, inserted in mouth of corolla tube; and pistil with 2-celled inferior ovary and slender 2-forked white style.

Berries elliptical, 1/2--5/8 in (13--15 mm) long, red, containing thin fleshy pulp and 2 (sometimes 1) elliptical seeds 5/16--1/2 in (8--13 mm) long, flattened on inner surface. There are about 1,000 coffee beans to a pound (2,208 to a kilo). In Hawaii, flowering is mainly in spring and the coffee harvest season from September to January with peak in November.

The wood is whitish, hard, heavy, and tough, seldom used.

Coffee is an important agricultural crop in many tropical regions. The seeds, which contain caffein, are roasted and ground to produce the familiar drink. This species is the most widely grown of several and has many cultivated varieties. Elsewhere, classed as a honey plant, producing white honey with a characteristic flavor.

It is reported that coffee was introduced into the New World first to Surinam by the Dutch in 1714. The same year a tree was presented to King Louis XIV of France as a peace gesture during the signing of the Treaty of Utrecht. From that royal tree, seedlings were smuggled to Brazil in 1727. Nearly a century later, in 1813, coffee was first planted in Hawaii by Don Marin. In 1818, missionary Samuel Ruffles grew ornamental shrubs at Kona. Many plantations were established mainly between the years 1840 to 1856. Losses from insects and fungus disease caused abandonment of most plantations and replacement by sugar cane.

Coffee is still produced commercially in a narrow belt on the west side of the Island of Hawaii. Kona coffee, known for its unique flavor, is grown on many small farms in a narrow subtropical belt with high rainfall at 1,200--2,000 ft (366--610 m). A coffee mill is located near the seaside village of Napoopoo on Kealakekua Bay. The annual Kona Coffee Festival in early November celebrates the coffee harvest season.

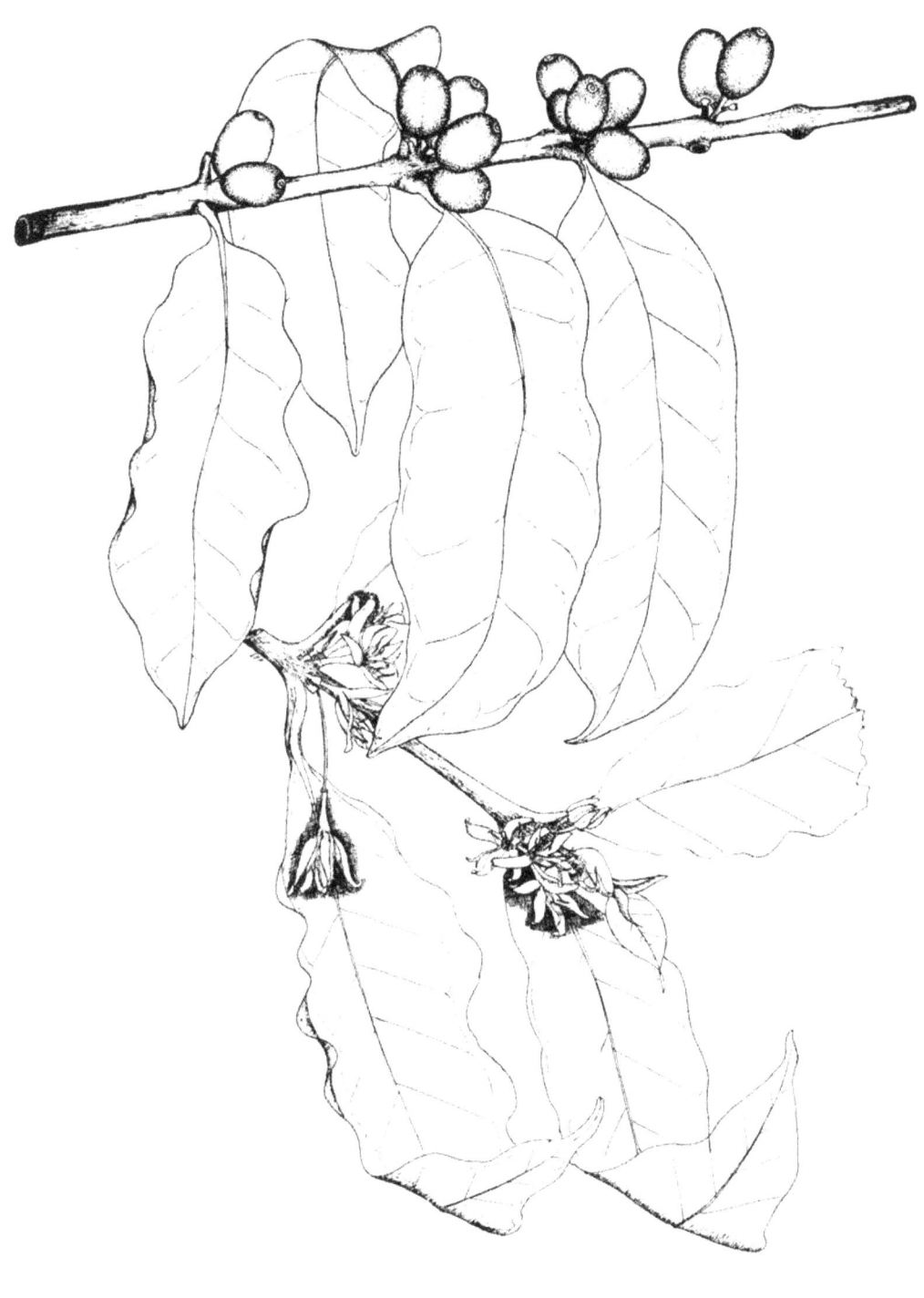

146. Coffee *Coffea arabica* L.*
Fruiting twig (above), flowering twig (below), 2/3 X (P. R. v. 1).

In Hawaii, coffee is naturalized locally on moist lower mountain slopes of the islands of Kauai, Oahu, and Hawaii.

Range--Native of Ethiopia but early introduced into Arabia (14th century) and extensively planted and escaping or naturalized through the tropics. In the New World, coffee is grown commercially from Mexico and Central America south to Brazil and through the West Indies including Puerto Rico and the Virgin Islands. Planted also as a novelty or ornamental shrub in southern Florida and southern California.

Other common names--Kona coffee, Arabian coffee.

147. Pilo

Coprosma montana Hillebr.

The genus *Coprosma*, common name pilo, has about 13 named species of shrubs and small trees through the Hawaiian Islands. These species have paired short-stalked small elliptical leaves with paired scalelike pointed hairy stipules remaining on the slender twigs, small greenish or white flowers male and female on different plants, with tubular corolla 4--9-lobed, 1 to many borne mostly at leaf bases, and small round yellow, orange, or black fruits with 2 nutlets. One example follows.

Evergreen shrub or small tree to 26 ft (8 m) high and 3 in (7.5 cm) in trunk diameter. Bark gray, smooth or slightly fissured. Twigs slender, green to gray, finely hairy.

Leaves opposite, hairless, with short slender leafstalks less than 1/4 in (6 mm) long, and pointed hairy stipules 1/8 in (3 mm) long, remaining at ringed nodes. Blades elliptical or lance-shaped, 3/8--1 in (10--25 mm) long and 1/4--1/2 in (6--13 mm) wide, slightly thickened or thin, blunt or short-pointed at both ends, dull green above, paler beneath.

Flowers male and female on different plants (dioecious), single, stalkless or nearly so on twigs back of leaves, about 1/4 in (6 mm) long. Female flowers have cup-shaped base (hypanthium) with calyx teeth, short tubular corolla with 5--6 lobes curved back, and pistil with inferior 2-celled ovary and 2 styles. Male flowers have as many stamens as corolia lobes, attached near base of tube and extending beyond.

Fruits (drupes) rounded, 1/4 in (6 mm) in diameter, with calyx teeth at apex, shiny yellow to dark orange, turning black, containing 2 half-round nutlets. The fruit has a disagreeable flavor.

Wood light brown, hard. In another species of the genus, whitish yellow and soft.

The genus *Coprosma* is widespread in wet forests and mountain shrublands through the Hawaiian Islands. This species extends in mountains to high altitudes, 6,000--10,000 ft (1,830--3,048 m) on Mauna Kea, Hawaii. Other species reach tree size frequently in forests at 4,000--8,000 ft (1,219--2,438 m).

Special areas--Haleakala, Volcanoes, Kipuka Puaulu.

Champion--Height 35 ft (10.7 m), c.b.h. 3.7 ft (1.1 m), spread 29 ft (8.8 m). Mauna Kea Forest Reserve, Humuula, Hawaii (1968).

Range--Restricted to Maui and Hawaii.

Other common names--pilo kuahiwi (meaning mountain pilo), hupilo.

148. Manono

Gouldia affinis (DC.) Wilbur

The genus *Gouldia*, common name manono, is known only from Hawaii and has numerous variations of shrubs and small trees grouped into 4 species. This species has many varieties and forms. Plants of this genus have paired short-stalked leaves mostly small, oblong, and leathery, with paired blunt stipules that shed early, clusters of many small greenish flowers with 4 lobes meeting at edges in bud, and small round bluish black fruits, 2-celled and many-seeded.

Evergreen shrub, woody vine, or small tree to 26 ft (8 m) tall and 6 in (15 cm) in trunk diameter. Bark gray, smooth to finely fissured; inner bark brownish, slightly bitter. Twigs gray, 4-angled, with enlarged ringed nodes, hairless.

Leaves opposite, with short leafstalks less than 3/8 in (1 cm) long and paired small blunt stipules that shed early. Blades mostly oblong, 2--4 in (5--10 cm) long and 1--1 3/4 in (2.5--4.5 cm) wide, thick and leathery, blunt or short-pointed at both ends, turned under at edges, above shiny green with few inconspicuous curved side veins, beneath dull light green and often slightly hairy.

Flower clusters (panicles) branched, 1--2 in (2.5--5 cm) long, mostly terminal. Flowers many, short-stalked, about 3/8 in (10 mm) long, composed of greenish base (hypanthium) 1/8 in (3 mm) long; 4-toothed greenish calyx; purplish green corolla 1/4--3/8 in (6--10 mm) long with narrow tube and 4 narrow tube and 4 narrow spreading lobes overlapping in bud; 4 short stamens within corolla tube near end; and pistil with inferior 2-celled ovary, many ovules in each cell, and slender style 2-lobed at end.

Fruits (berries) round, 1/4--3/8 in (6--10 mm) in diameter, bluish black, with calyx teeth at apex, 2-celled and many-seeded.

Wood light brown, hard. Uses by Hawaiians were for canoe trim and rigging, none at present.

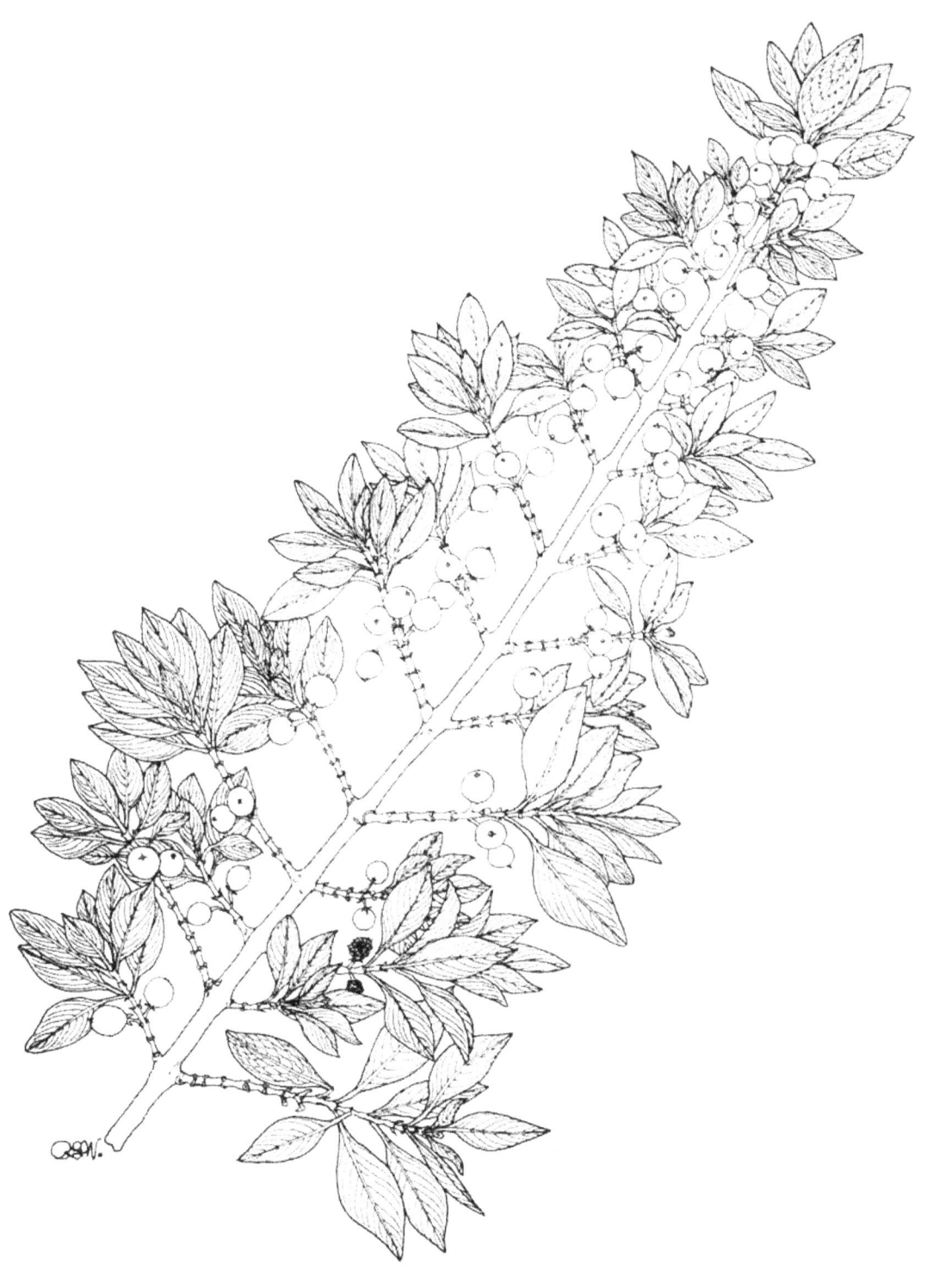

147. Pilo   *Coprosma montana* Hillebr.
Twig with fruits, 1 X.

Widespread in wet forests through the islands, mainly at 900--6,700 ft (274--2,042 m) altitude.
Special areas--Kokee, Volcanoes.
Champion--Height 20 ft (6.1 m), c.b.h. 3 ft (0.9 m), spread 12 ft (3.7 m). Hawaii Volcanoes National Park, Hawaii (1968).

149. Noni, Indian-mulberry (color plate, p. 42)

Morinda citrifolia L.**

This small tree of moist lowlands, especially in gullies, apparently was introduced by the early Hawaiians. It is characterized by paired large elliptical shiny dark green leaves, many small white tubular flowers in balllike clusters, and whitish or yellowish egg-shaped or elliptical multiple fruits slightly resembling small pineapples, fleshy and malodorous.

Small evergreen introduced tree to 20 ft (6 m) high and 5 in (13 cm) in trunk diameter or shrubby. Bark gray or brown, smoothish and slightly warty or scaly, soft. Inner bark light brown and tasteless or slightly irritating. Twigs stout, 4-angled, light green, hairless, with ring scars.

Leaves opposite, hairless, with stout green leafstalks of about 1/2 in (13 mm) and paired rounded scales (stipules) about 1/4 in (6 mm) long at base of each pair, leaving ring scar upon shedding. Blades 5--11 in (13--28 cm) long and 2 1/2--6 1/2 in (6--16.5 cm) broad, not toothed on edges, thin, with sunken curved side veins, upper surface shiny dark green, and lower surface light green with small tufts of hairs in vein angles along midvein.

Flower clusters (heads) elliptical or rounded, about 1 in (25 mm) wide, light green, mostly single on stalks of 1/2 in (13 mm) above leaves. Flowers many, crowded, united at base, more than 1/2 in (13 mm) long. Base (hypanthium) more than 1/8 in (3 mm) long, light green, with very short light green calyx rim; corolla white, nearly 1/2 in (13 mm) long, tubular with 4--6 lobes 3/8--1/2 in (10--13 mm) across; 4--6 stamens 3/16 in (5 mm) inserted near mouth of corolla tube; and pistil composed of inferior 2-celled ovary with slender light green style and 2-lobed stigma.

Multiple fruit (syncarp) from flower head is a compact soft juicy mass 3--4 in (7.5--10 cm) long and 2 1/2 in (6 cm) broad, with irregular warty surface. Individual fruits 1/2 in (13 mm) across, 4--6-sided, each 2-celled and 2-seeded. Seeds more than 1/8 in (3 mm) long. Flowering and fruiting nearly through the year.

Wood bright yellow or yellow brown, soft, finetextured and straight-grained. Not used in Hawaii.

This species was grown by the Polynesians and Hawaiians for dyes. Red for coloring tapa or bark cloth was obtained from the bark and yellow from the trunk and roots. The fruits with fetid cheeselike odor and insipid or unpleasant taste were eaten raw or cooked and were fed to hogs. Fruit juice as an insecticide was an ingredient in a hair shampoo. Leaves, bark, and fruits were used in folk remedies.

The English name *painkiller* refers to use of the leaves in the Virgin Islands, Trinidad, Guyana, and probably elsewhere in alleviating pain. According to different directions, a hot leaf (heated over a fire) or wilted leaf is pressed against the body on painful swellings, a poultice of the leaves is applied to wounds or to the head for headaches, or crushed leaves in lard or camphor oil are put on the face for treatment of neuralgia or head colds.

This small tree is sometimes used as an ornamental. A variety with variegated leaves (*M. citriodora* var. *potteri* Deg.) is occasionally cultivated. It was introduced from Fiji by Degener in 1941.

Planted and naturalized in lowlands to 1,500 ft (457 m) through Hawaii, especially in moist small valleys and pockets of soil in lava rocklands, also frequent in sandy coastal areas. Perhaps persistent on old home sites.

Special areas--Waimea Arboretum, Bishop Museum, Haleakala, Volcanoes.

Range--Native of India, Malaya including East Indies, and tropical Australia and introduced into other tropical areas. Cultivated and in part naturalized through West Indies including Puerto Rico and Virgin Islands and in Guianas. Rarely planted at Key West, Florida.

Other common names--painkiller; morinda, gardenia hedionda, noni (Puerto Rico); lada (Guam, N. Marianas); ngel (Palau); lol (Yap); nen (Truk); weypul (Pohnpei); ee (Kosrae); nen (Marshalls); nonu (Am. Samoa).

The lightweight seeds have an air-sac and may have been disseminated along beaches by floating.

Hawaii has 2 native species of *Morinda*, called noni kuahiwi (meaning noni of the mountains): *Morinda sandwicensis* Deg., Hawaiian morinda, of Oahu, and *Morinda trimera* Hillebr. on Oahu, Maui, and Lanai.

150. Kōpiko

Psychotria hawaiiensis (Gray) Fosberg

The genus *Psychotria* (including *Straussia*) is represented in Hawaii by about 11 species of evergreen shrubs and small trees distributed in wet forests through the islands. They have paired elliptical leaves with

Range--Hawaiian Islands only.
Botanical synonym--*Gouldia terminalis* (Hook. & Arn.) Hillebr.

This genus was dedicated by Asa Gray to his colleague Augustus Addison Gould (1805--66), zoology professor at Harvard University.

148. Manono *Gouldia affinis* (DC.) Wilbur
Fruiting twig (above), flowering twig (below), 1 X.

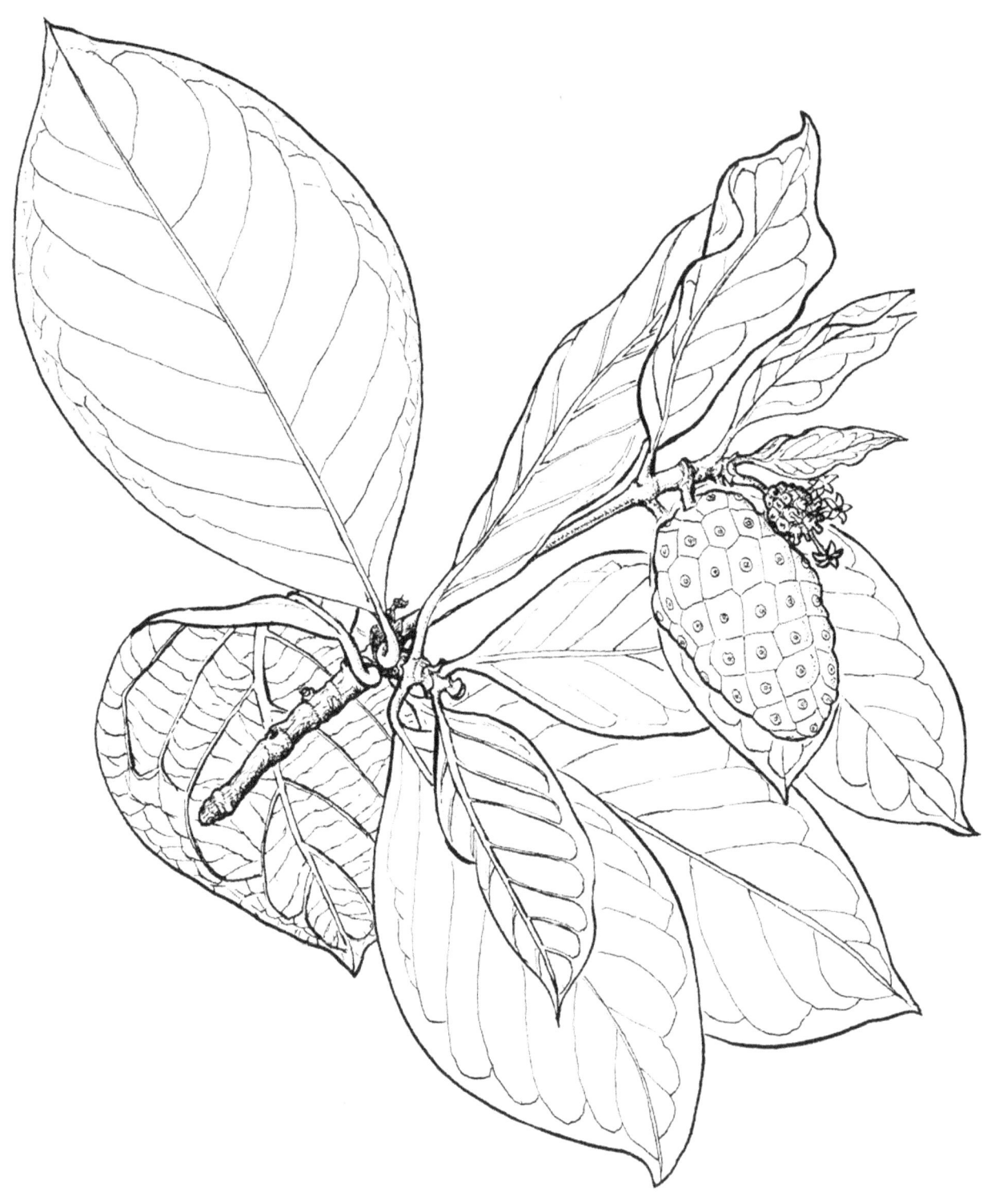

149. Noni, Indian-mulberry *Morinda citrifolia* L.**
Twig with flowers and fruit, 1 X (P. R. v. 1).

150. Kōpiko *Psychotria hawaiiensis* (Gray) Fosberg
Flowering twig, 2/3 X.

paired elliptical broad blunt stipules, terminal erect long-stalked branched flower clusters with many small white flowers, and many small rounded orange fruits. This species is distinguished by short reddish brown hairs on leafstalks, lower leaf surface, and flower clusters.

Shrub or small tree reported to reach 33 ft (10 m) high and 1 ft (0.3 m) in trunk diameter. Bark smooth, blackish. Twigs stout, light green to gray, hairless except at ringed nodes, with large half-round leaf-scars.

Leaves opposite, with large paired light green hairless stipules or scales about 3/8--5/8 in (10--15 mm) long, rounded at apex, falling early and leaving ring scar, and with long stout light green leafstalks 3/4--1 1/2 in (2--4 cm) long. Blades elliptical or obovate, 4--7 in (10--18 cm) long and 2--3 1/2 in (4--9 cm) wide, widest beyond middle, slightly thickened or leathery, blunt or rounded at apex, short-pointed at base, with edges straight, upper surface slightly shiny green and hairless, with midvein and many curved side veins light yellow, lower surface slightly shiny light green with scattered pressed rusty brown hairs especially on the raised veins.

Flower clusters (panicles) terminal, erect, 3--6 in (7.5--15 cm) long, branches with rusty brown hairs. Flowers many, stalked or nearly stalkless, 1/4 in (6 mm) long, composed of green hairy basal cup (hypanthium) with calyx forming wavy border; white corolla with short tube and 4--6 spreading lobes longer than tube; 4--6 short stamens in notches of corolla; and pistil with inferior 2-celled ovary, short style, and 2 stigmas.

Fruits (drupes) elliptical, 1/4--5/16 in (6--8 mm) long, with calyx at apex, 2-seeded.

The hard whitish wood of this and related species was used by the Hawaiians as the anvil or *kua kukukapa* for beating bark into tapa or bark cloth, and also for fuel.

Scattered in understory of moist to dry forests to about 500--5,000 ft (152--1,524 m) altitude.

Special areas--Volcanoes, Kipuka Puaulu.
Champion--Height 45 ft (13.7 m), c.b.h. 4.9 ft (1.5 m), spread 30 ft (9.1 m). Hoomau Ranch, Honomolino, Hawaii (1968).
Range--Hawaii, Molokai, and Maui.
Other common names--opiko, kōpiko 'ula.
Botanical synonym--*Straussia hawaiiensis* Gray.

The Hawaiian small genus *Straussia* with about 8 species has been united with the large widespread tropical genus *Psychotria*, which has about 1,500 species.

BELLFLOWER FAMILY (CAMPANULACEAE)

151. 'Ohawai, haha, tree clermontia   *Clermontia arborescens* (Mann) Hillebr.

Clermontia is a distinctive genus of about 21 species of shrubs and small trees confined to Hawaii. Several are classed as rare. Only a few species reach tree size, and one example will serve for recognition of the group. These plants have few branches, like candelabras, and milky sap. They bear many long narrow leaves at the ends of stout branches. The few large flowers borne on stalks at leaf bases have an irregular curved tubular corolla deeply split to base, varying from green to white or yellowish to purplish. The fruits are large yellow or orange berries 1 in (2.5 cm) or more in diameter, edible but insipid.

Tree clermontia, one of the largest in size and in flower size, is a tree recorded at 15--25 ft (4.6--7.6 m) tall with a trunk to 8 in (20 cm) in diameter, or rarely a shrub, hairless or nearly so throughout. Bark gray, smooth to slightly scaly. Twigs stout, light gray, with many large rounded leaf-scars diagonally arranged below the leaves. White sap or latex flowing abundantly from cuts, almost tasteless.

Leaves many, alternate but crowded together near ends of branches, large, spreading on slender leafstalks of 1 1/4--2 1/2 in (3--6 cm). Blades oblong, 5--7 in (12--18 cm) long and 2 1/4--2 in (3--5 cm) wide, often slightly long-pointed at both ends, wavy toothed on edges, thick and leathery, shiny dark green above, pale beneath.

Flowers 1--2 at leaf base on stalks of about 1 in (2.5 cm) from fleshy stalk of about 3/8 in (1 cm), 2 1/2--3 1/4 in (6--8 cm) long, large and fleshy, hairless, consisting of bell-shaped base (hypanthium) 3/4 in (2 cm) long ending in fleshy calyx of 5 short blunt lobes; fleshy corolla 2--2 3/4 in (5--7 cm) long, greenish or whitish with reddish purple tinge, irregular and much curved, deeply 5-lobed and deeply split to base; stamens 5, separate from corolla, purplish or violet, in column 2 1/2 in (6 cm) long, united by the long anthers; and pistil with inferior 2-celled ovary, numerous ovules, and long style.

Fruit (berry) rounded, 1--1 1/2 in (2.5--4 cm) in diameter, yellow, with 5 long ridges and grooves, and with calyx at apex, 2-celled. Seeds many, about 1/32 in (1 mm) long, rounded but flattened, shiny yellow or brown.

The yellow berries are sometimes eaten by birds and humans.

One of the most common species of the genus, in understory of wet forests at 1,700--6,000 ft (518--1,829 m) altitude.

Range--Molokai, Lanai, and Maui.
Other common name--'oha.

151. 'Ohawai, haha, tree clermontia   *Clermontia arborescens* (Mann) Hillebr.
  Flowering twig, 2/3 X (Degener).

GOODENIA OR NAUPAKA FAMILY (GOODENIACEAE)

152. Naupaka kuahiwi, mountain naupaka
Scaevola gaudichaudiana Cham.

The genus *Scaevola* with about 9 species in Hawaii, mostly shrubs, is easily recognized by the corolla as in a half flower. This example, mountain naupaka, is a much branched evergreen native shrub of wet mountain areas, has the distinctive fragrant white flowers with narrow tubular corolla split to base and 5 narrow lobes spreading on one side. Sometimes a small tree to about 16 ft (5 m) tall and 3 in (7.5 cm) in diameter. Bark light brown. Twigs slender, greenish, hairless except for tufts at base of young leaves.

Leaves alternate, hairless or nearly so, with slender leafstalk about 1/2 in (13 mm) long. Blades elliptical to lance-shaped, 2--5 in (5--13 cm) long and 3/4--2 in (2--5 cm) wide, long-pointed at both ends, with several short inconspicuous yellowish teeth in upper half, thin, shiny green, the sides often slightly unequal, with long fine nearly parallel side veins, with thick yellowish midvein raised beneath.

Flower clusters (cymes) at leaf bases, becoming slightly longer than leaves, with slender forking branches often zigzag, bearing 10--17 flowers, the end flower opening first. Flowers almost 3/4 in (2 cm) long, composed of greenish base (hypanthium) about 1/16 in (1.5 mm) long with tiny rim of 5-toothed calyx; irregular (bilaterally symmetrical), narrow tubular corolla white but turning yellowish, about 5/8 in (1.5 cm) long, split to base and 5 narrow 3-pointed lobes about 3/8 in (1 cm) as in a half flower; 5 slender stamens almost as long as tube, withering early; and pistil with inferior 2-celled ovary, slender curved greenish style, and enlarged stigma. Flowering nearly through the year.

Fruit (drupe) elliptical, about 1/4 in (6 mm) long with calyx rim at apex, hairless, shiny purplish black (whitish in a form), with purple juice, containing large elliptical pointed 2-celled stone. Seeds 1 or 2.

Common in understory of open wet forests and in open areas at 600--2,600 ft (183--792 m) altitude.

Special area--Waimea Arboretum.

Range--Kauai and Oahu only.

Other common names--naupaka, Gaudichaud scaevola, half-flower.

This species honors Charles Gaudichaud-Beaupré (1789--1854), French botanist on the world scientific cruise of the *Uranie*. He collected in Hawaii in 1819 and published the first report on the Hawaiian flora. It has been known also by the name of a closely related species, *Scaevola chamissonana* Gaud. That species is a slightly smaller shrub of Molokai, Lanai, Maui, and Hawaii. It has leaves with fewer branching side veins, tufts of hairs in vein angles, and several prominent sharp teeth; large flowers 1--1 1/4 in (2.5--3 cm) long, with longer pointed lobes, white often with violet lines; and larger fruits more than 1/2 in (13 mm) long.

Botanical synonyms--*Scaevola taccada* (Gaertn.) Roxb., *S. frutescens* Krause.

Naupaka kahakai or beach naupaka, *Scaevola sericea* Vahl, is a common native much branched shrub on beaches through the Hawaiian Islands and other Pacific Islands to the western Indian Ocean. It has fleshy silky hairy stout twigs, elliptical blunt-pointed, thick brittle, light green leaves finely hairy beneath, yellowish white flowers, and white fruits. Other common names--naupaka kai, huakekili, aupaka.

The odd half-flower of this genus has been the source of several Hawaiian legends. The following is from Degener (1933--1986): "...two lovers were forced to be separated by the man's departure for war on another island. On taking leave, he tore a *naupaka* blossom in half, saying: 'When the flower is whole again I shall return.' To this day the *naupaka* flower is seemingly only half a flower as a sad reminder that the man never returned from battle."

Neal (1965, p. 820-821) had two other versions, quoted in full:

> According to an old Hawaiian chant, the ocean naupaka was born of heaven and earth. The story of the naupaka flower, though of recent origin, is one of the best known of Hawaii. It has several variants. Two lovers quarreled and the maiden tore a naupaka flower in two and declared she would not love her old sweetheart again until he should bring to her a perfect flower. He searched in vain all over the islands, for these flowers, whether they grew on the seashore, on the plains, or in the mountains, had become but half flowers. And it is said that he died of a broken heart.

> Another story tells of a beautiful stranger who fell in love with a village youth. When he turned from her and went back to his sweetheart, the beautiful woman followed him and tore him from her embrace. Anger blazed about the woman, and they knew that she was Pélé, goddess of volcanoes. She pursued the youth into the mountains hurling lava after him. Then the gods took pity on him and transformed him into a half flower, the naupaka. Pélé shrieked with rage and fled on a river of lava to the ocean. She overtook the maid, whom the gods turned into a beach naupaka. The lovers are forever separated, for the half flowers of the youth still bloom alone in the mountains, and the half flowers of the maiden blossom alone on the beach.

152. Naupaka kuahiwi, mountain naupaka *Scaevola gaudichaudiana* Cham.
Twig with flowers and fruits, 1/2 X; flower, 1 X, seeds and fruit
(lower left), 3 X; leaf (lower right), 1 X (Degener).

SELECTED REFERENCES

Besides titles cited in the text, this list includes references with additional information about the native and introduced forest trees of Hawaii.

Anonymous. 1938. Trees: reforestation, reserves, continue good work. Sales Builder (Honolulu) 11(11): 2-22. [Source of monkey-pod introduction data.]

Alston, A. S. 1973. Coconut palm timber. Fiji Dep. For. Fiji timbers and their uses, No. 60: 1-10.

Amerson, A. Binion, Jr., W. Arthur Whistler, and Terry D. Schwaner. Richard C. Banks, ed. 1982. Wildlife and wildlife habitat of American Samoa. II. Accounts of flora and fauna. U. S. Dep. Int. Fish Wild. Serv. 151 p.

American Forestry Association. 1979. Champion trees of Hawaii. Am. For. 80(5): 26-31, 34-35.

Arnold, Harry L. 1944. Poisonous plants of Hawaii. Honolulu, HI: Tongg Publishing Co. 71 p. [Reprinted 1968. Rutland, VT; Tokyo: Charles E. Tuttle Co.]

Baldwin, Roger E. 1979. Hawaii's poisonous plants. Hilo, HI: Petroglyph Press. 112 p.

Ballard, Robert D. 1983. Exploring our living planet. Washington, DC: National Geographic Society. 366 p.

Barrett, Wilfredo H. G. 1958. Las plantas cultivadas en la República Argentina. Araucariáceas. vol. 1, fasc. 19. Argent. Inst. Bot. Agric. 26 p.

Beccari, Odoardo, and Joseph F. Rock. 1921. A monographic study of the genus Pritchardia. Bernice P. Bishop Mus. Mem. 8(1): 1-77.

Becker, Richard E. 1976. The phytosociological position of tree ferns (Cibotium spp.) in the montane rain forests on the island of Hawaii. Ph. D. dissertation, Univ. of Hawaii, 368 p.

Becker, Richard E. 1984. The identification of Hawaiian tree ferns of genus Cibotium. Am. Fern J. 74: 97-100.

Beckwith, Martha. 1940. Hawaiian mythology. New Haven, CT: Yale Univ. Press. 575 p. [Reprinted 1970. Honolulu: Univ. Press of Hawaii.]

Bendtsen, B. A. 1964. Some strength and related properties of yagrumo hembra (*Cecropia peltata*) from Puerto Rico. U.S. Dep. Agric. For. Serv. For. Prod. Lab. Res. Note FPL-53: 1-7.

Blakely, W. F. 1965. A key to the eucalypts. 3d ed. Australia Forestry and Timber Bureau. 359 + 24 p.

Boas, I. H. 1947. The commercial timbers of Australia: their properties and uses. Australia: Council for Scientific and Industrial Research. 364 p.

Bolza, E., and N. H. Kloot. 1963. Mechanical properties of 174 Australian timbers. Aust. CSIRO Div. For. Prod. Technol. Pap. 25: 1-112.

Brewbaker, J. L., D. L. Pluckness, and V. Gomez. 1972. Varietal variation and yield trials of Leucaena leucocephala (koa haole) in Hawaii. Hawaii Agric. Exp. Stn. Res. Bull. 166: 1-29.

Brown, Forest Bullen Harkness. 1922. The secondary xylem of Hawaiian trees. Bernice P. Bishop Mus. Occas. Pap. 8(6): 215-371.

Bryan, L. W. 1947. Twenty-five years of forestry work on the Island of Hawaii. Hawaii Plant. Rec. 51(1): 1-80.

Bryan, L. W., and Clyde M. Walker. 1966. A provisional check list of some common native and introduced forest plants in Hawaii. U.S. Dep. Agric. For. Serv. Pac. Southwest For. Range Exp. Stn. Misc. Pap. 69: 1-34. [1962 publication revised 1966.]

Bryan, William Alanson. 1915. Natural history of Hawaii. Honolulu: Hawaiian Gazette Co. 596 p.

Burgan, Robert E., and H. C. Wesley, Jr. 1971. Species trails at the Waiakea Arboretum . . . tree measurements in 1970. U.S. Dep. Agric. For. Serv. Pac. Southwest For. and Range Exp. Stn. Res. Note PSW-240: 1-6.

Carlquist, Sherwin. 1965. Island life: a natural history of the islands of the world. Garden City, NY: Natural History Press, 451 p.

Carlquist, Sherwin J. 1970. Hawaii, a natural history: geology, climate, native flora and fauna above the shore line. Garden City, NY: Am. Mus. Nat. Hist., Natural History Press, 463 p.

Carlquist, Sherwin. 1974. Island biology. New York: Columbia Univ. Press. 660 p.

Carlquist, Sherwin. 1980. Hawaii, a natural history: climate, native flora and fauna above the shore line. 2d. ed. Honolulu: Pacific Tropical Botanical Garden. 468 p.

Carlson, Norman I., and L. W. Bryan. 1959. Hawaiian timber for the coming generations. Honolulu: Trustees of the Bishop Estate. 112 p.

Chippendale, G. M., ed. 1968. Eucalyptus buds and fruits; illustrations of the buds and fruits of the genus. Austr. For. Timber Bur. Leafl. 63.

Chippendale, G. M. 1976. Eucalyptus nomenclature. Austr. For. Res. 7: 69-107.

Chock, Alvin K. 1968. Hawaiian ethnobotanical studies. I. Native food and beverage plants. Econ. Bot. 22: 221-238.

Clay, Horace F., and James C. Hubbard. 1962. Trees for Hawaiian gardens. Univ. Hawaii Coop. Ext. Serv. Bull. 67: 1-101.

Clay, Horace F., and James C. Hubbard. 1977a. The Hawaii garden: tropical exotics. Honolulu: Univ. Press of Hawaii. 267 p.

Clay, Horace F., and James C. Hubbard. 1977b. The Hawaii garden: tropical shrubs. Honolulu: Univ. Press of Hawaii. 295 p.

Corn, Carolyn A., and William H. Hiesey. 1973. Altitudinal variation in Hawaiian Metrosideros. Am. J. Bot. 60: 991-1002.

Dallimore, W., and A. Bruce Jackson. 1967. A handbook of the Coniferae and Ginkgoaceae. 4th ed, revised by S. G. Harrison. New York: St. Martin's Press. 729 p.

Dawson, J. W. 1970. Pacific capsular Myrtaceae. 2. The Metrosideros complex: M. collina group. Blumea 18: 441-445.

Degener, Otto. 1930. Illustrated guide to the more common or noteworthy ferns and flowering plants of Hawaii National Park, with descriptions of ancient Hawaiian customs and an introduction to the geologic history of the islands. Honolulu: Honolulu Star-Bulletin. 312 p.

Degener, Otto. 1933-1973. Flora Hawaiiensis or the new illustrated flora of the Hawaiian Islands. 7 vol. (vol. 6 and 7 by Otto Degener and Isa Degener.) Honolulu, HI.

Degener, Otto. 1973. Plants of Hawaii National Park illustrative of plants and customs of the South Seas. Ann Arbor, MI; Braun-Brumfield, 314 p. [Reprint, first published 1930.]

Degener, Otto, and Isa Degener. 1983. Plants of Hawaii National Parks. Ed. 3, 481 p. (page proof)

Degener, Otto, and Isa Degener. 1986. Flora Hawaiiensis. 2d ed. 4 vol. Honolulu: 1192 p.

Degener, Otto, Isa Degener, and Noah Pekelo, Jr. 1975. Hawaiian plant names, their botanical and English equivalents. Flora Hawaiiensis p. X O-- X 28.

Desch, H. E. 1957. Manual of Malayan timbers. Malay. For. Rec. 15, vol. 1 and 2. [Reprinted. Singapore: Tien Woh Co. 762 p.]

Edmondson, C. H. 1955. Resistance of woods to marine borers in Hawaiian waters. Bernice P. Bishop Mus. Bull. 217. 91 p.

Emerson, Nathaniel B. 1909. Unwritten literature of Hawaii: the sacred songs of the Hula, collected and translated with notes and an account of the hula. U.S. Bur. Amer. Ethnol. Smithson. Inst. Bull. 38, 288 p. [Reprinted 1965. Rutland, VT; Tokyo: Charles E. Tuttle Co.]

Ewan, Joseph. 1974. The botany of Cook's voyages (or around the world on six shillings a day). Bull. Pac. Trop. Bot. Gard. 4(4): 65-75.

Fagerlund, Gunnar O., and Arthur L. Mitchell. 1944. A checklist of the plants, Hawaii National Park, Kilauea--Mauna Loa Section with a discussion of the vegetation. Hawaii Natl. Park, Nat. Hist. Bull. 9, 76 p.

Falanruw, Marjorie C., Jean E. Maka, Thomas G. Cole, and Craig D. Whitesell. 1989. Names of trees and shrubs of the Caroline, Marshall, and Mariana Islands. U.S. Dep. Agric. For. Serv. Pac. Southwest For. Range Exp. Stn. Resource Bull. [in press]

Food and Agriculture Organization of the United Nations. 1979. Eucalypts for planning. FAO For. Ser. No. 11: 1-677.

Fosberg, F. R. 1937. The genus Gouldia (Rubiaceae). Bernice P. Bishop Mus. Bull. 147: 1-82.

Fosberg, F. R. 1939. Diospyros ferrea (Ebenaceae) in Hawaii. Bernice P. Bishop Mus. Occas. Pap. 15(10): 119-131.

Fosberg, F. R. 1948. Derivation of the flora of the Hawaiian Islands. In: Zimmerman, Elwood C., Insects of Hawaii, vol. 1, Introduction. Honolulu: Univ. of Hawaii Press. p. 107-119.

Fosberg, F. R. 1961. Guide to excursion III, Tenth Pacific Science Congress. Tenth Pacific Science Congress and Univ. of Hawaii, 207 p.

Fosberg, F. R. 1964. Studies in Pacific Rubiaceae. V. The Hawaiian species of Psychotria L. Brittonia 16: 255-271.

Fosberg, F. R. 1972. Guide to excursion III, Tenth Pacific Science Congress. Rev. ed. Honolulu: Univ. of Hawaii with assistance from Hawaiian Botanical Gardens Foundations, Inc., 249 p.

Fosberg, F. R. 1973. The name of the octopus tree. Baileya 19: 45-46.

Fosberg, F. R., and Derral Herbst. 1975. Rare and endangered species of Hawaiian vascular plants. Allertonia 1(1): 1-72.

Frear, Mary Dillingham. 1929. Our familiar island trees: Sponsored by the Outdoor Circle of Honolulu. Boston: R. G. Badger, 161 p.

Fujii, David M. 1980. The Nuuanu eucalyptus planting: growth, survival, stand development after 64 years. U.S. Dep. Agric. For. Serv. Pac. Southwest For. Range Exp. Stn. Res. Note PSW-318: 1-5.

Fullaway, David T. 1972. Norfolk Island pine culture. Univ. Hawaii Coop Ext. Serv. Circ. 453: 1-16.

Gerhards, Charles C. 1964. Limited evaluation of physical and mechanical properties of Nepal alder grown in Hawaii. U. S. Dep. Agric. For. Serv. For. Prod. Lab Res. Note FPL-036: 1-9.

Gerhards, Charles C. 1965. Physical and mechanical properties of saligna eucalyptus grown in Hawaii. U.S. Dep. Agric. For Serv. For. Prod. Lab. Res. Pap. FPL-23: 1-12.

Gerhards, Charles C. 1966a. Physical and mechanical properties of Molucca albizzia grown in Hawaii. U.S. Dep. Agric. For. Serv. For. Prod. Lab. Res. Pap. FPL-55: 1-8.

Gerhards, Charles C. 1966b. Physical and mechanical properties of blackbutt eucalyptus grown in Hawaii. U.S. Dep. Agric. For. Serv. For. Prod. Lab. Res. Pap. FPL-65: 1-8.

Gerhards, Charles C. 1967. Physical and mechanical properties of "Norfolk-Island-pine" grown in Hawaii. U.S. Dep. Agric. For. Serv. For. Prod. Lab. Res. Pap. FPL-73: 1-8.

Gillett, G. W. 1966. Hybridization and its taxonomic implications in the Scaevola gaudichaudiana complex of the Hawaiian Islands. Evolution 20: 506-516.

Haass, Judith E. 1977. The Pacific species of Pittosporum Banks ex Gaertn. (Pittosporaceae). Allertonia 1: 73-167.

Hall, Norman, R. D. Johnson, and G. M. Chippendale. 1975. Forest trees of Australia. Canberra: Australian Dep. Agric., For. Timber Bur., Australian Govt. Printing Service. 334 p.

Hall, William L. 1904. The forests of the Hawaiian Islands. U.S. Dep. Agric. Bur. For. Bull. 48: 1-29.

Hamilton, Richard A., and W. Yee. 1962. Mango varieties in Hawaii. Hawaii Farm Sci. 11(3): 3-5.

Handy, E. S., and E. G. Handy. 1972. Native plants in old Hawaii. Bernice P. Bishop Mus. Bull. 233: 1-641.

Hargreaves, Dorothy, and Bob Hargreaves. 1964. Tropical trees of Hawaii. Portland, OR: Hargreaves Indus. 65 p.

Hargreaves, Dorothy, and Bob Hargreaves. 1970. Tropical trees of the Pacific. Hailua, HI: Hargreaves Co. 64 p.

Hasselwood, E. L., and G. E. Motter, eds. 1966. Handbook of Hawaiian weeds. Honolulu: HI: Hawaiian Sugar Planters Assoc. Exp. Stn., 479 p.

Hawaii, University, Department of Geography. 1983. Atlas of Hawaii. 2d ed. Honolulu: Univ. of Hawaii Press, 238 p.

Hawaii Natural History Association. 1965. Kipuka Puaula self-guiding nature trail: Hawaii Volcanoes National Park.

Hawaii Weed Conference. 1957. Common weeds of Hawaii. Published by Univ. Hawaii Agric. Ext. Serv., Board Comm. Agric. Forestry, Exp. Stn., Hawaii Sugar Planters Assoc., Pineapple Res. Inst. Hawaii, 16 p.

Herbst, Derral. n.d. Field guide to the Awaawapuhi Trail, Kokee Hawaii. Pac. Trop. Bot. Gard. Guide 1: 1-13.

Hillebrand, William. 1888. Flora of the Hawaiian Islands: a description of their phanerogams and vascular cryptogams. Heidelberg: Carl Winter: 673 p. [Reprinted 1965. New York: Hafner Publishing Co.]

Hirano, Robert T., and Kenneth M. Nagata. 1972. A checklist of indigenous and endemic plants of Hawaii in cultivation at the Harold L. Lyon Arboretum. Honolulu: Univ. of Hawaii, Harold L. Lyon Arboretum.

Honda, N., W. H. C. Wond, and R. E. Nelson. 1967. Plantation timber on the island of Kauai--1965. U.S. Dep. Agric. For. Serv. Pac. Southwest For. Range Exp. Stn. Res. Bull. PSW-6: 1-34.

Hosaka, Edward F. 1940. A revision of the Hawaiian species of Myrsine (Suttonia, Rapanea), (Myrsinaceae). Bernice P. Bishop Mus. Occas. Pap. 16(2): 25-76.

Hubbard, Douglass H. 1952. Ferns of Hawaii National Park. Hawaii Nat. Hist. Assoc., Hawaii Nature Notes 5(1): 1-40.

Hubbard, Douglass H., and Vernon R. Bender. 1960. Trailside plants of Hawaii National Park. rev. ed. Hawaii Nat. Hist. Assoc., Hawaii Nature Notes 4(1): 1-28. [Ed. 1, 1950]

Hunt, Frances A. 1986. National register of big trees. Am. Forest 92(4): 21-52.

Johnston, R. D., and R. Marryatt. 1962. Taxonomy and nomenclature of eucalypts. Aust. For. Timber Bur. Leafl. 92: 1-24.

Judd, Charles S. 1916. The first algaroba and royal palm in Hawaii. Hawaiian Forester and Agriculturist 13(9): 330-333.

Judd, Charles S. 1920. The Australian red cedar. Hawaii. For. Agric. 17(3): 57-59.

Judd, Charles S. 1938. Trees and other native plants used by early Hawaiians. Mimeogr.; on file, Hawaii Div. Forestry and Wildlife, Honolulu, HI.

Kauai Outdoor Circle. 1976. Ornamental trees recommended for the Garden Island by the Kauai Outdoor Circle. Lihue, HI.

Kelly, Stan. 1959. Eucalypts. Text by G. M. Chippendale and R. D. Johnston. Melbourne, Sydney: Thomas Nelson (Australia). 82 p. [250 color plates] [reprinted 1969].

Kingston, R. S. T., and C. J. E. Risdon. 1961. Shrinkage and density of Australian and other Southwest Pacific woods. Austr. CSIRO Div. For. Prod. Tech. Pap. 13: 1-65.

Knapp, Rudiger. 1975. Vegetation of the Hawaiian Islands. [Trans. by Alvin Y. Yoshinaga and Hugh H. Iltis.] Hawaiian Bot. Soc. Newsl. 14: 95-121.

Korte, Karl H. 1967? The history of forestry in Hawaii: from World War II to the present. 3 p. [Reprinted from Aloha Aina.]

Krauss, Beatrice H. 1974. Ethnobotany of Hawaii--Compilation of Botany 105 notes. Univ. Hawaii, Dep. Bot. 248 p.

Küchler, A. W. 1970. Potential natural vegetation of Hawaii. *In*: The national atlas of the United States of America. U.S. Dep. Interior, Geol. Surv. 92 p., map (folio).

Kuck, Loraine, and Richard C. Tongg. 1958. Hawaiian flowers and flowering trees: a guide to tropical and semi-tropical flora. Rutland, VT; Tokyo: Charles E. Tuttle Co. 158 p.

Lamb, Samuel H. 1936. The trees of the Kilauea--Mauna Loa Section, Hawaii National Park. U.S. Dep. In. Natl. Park Serv. Hawaii Natl. Park Nat. Hist. Bull. 2, 32 p.

Lamb, Samuel H. 1981. Native trees and shrubs of the Hawaiian Islands. Santa Fe, NM: Sunstone Press. 159 p.

Lamoureaux, Charles H. 1976. Trailside plants of Hawaii's National Parks: Hawaii Natural History Association. 80 p.

LeBarron, Russell K. 1970. The history of forestry in Hawaii: from the beginning through World War II. 3 p. [Reprinted April 1970. Aloha Aina.]

Little, Elbert L., Jr. 1969. Native trees of Hawaii. Am. Forest 75(2): 16-17, 44-45.

Little, Elbert L., Jr., and Frank H. Wadsworth. 1964. Common trees of Puerto Rico and the Virgin Islands. U.S. Dep. Agric. For. Serv. Agric. Handb. 249. 1-548.

Little, Elbert L., Jr., Frank H. Wadsworth, and José Marrero. 1967. Arboles comunes de Puerto Rico y las Islas Vírgenes. Rio Piedras, PR: Editorial, Universidad de Puerto Rico, 827 p.

Little, Elbert L., Jr., Roy O. Woodbury, and Frank H. Wadsworth. 1974. Trees of Puerto Rico and the Virgin Islands, vol. 2, U.S. Dep. Agric. For. Serv. Agric. Handb. 449: 1-1,024.

Littlecott, Lorna C. 1969. Hawaii first. Am. Forests 75(2): 12-15, 59-63.

Longwood, Franklin R. 1961. Puerto Rican woods: their machining, seasoning, and related characteristics. U.S. Dep. Agric. For. Serv. Agric. Handb. 105: 1-98.

Lutz, John F., and C. G. Roessler. 1964. Veneer and plywood characteristics of Nepal alder. Unpublished report. U.S. Dep. Agric. For. Serv. For. Prod. Lab.

MacCaughey, Vaughan. 1917. An annotated list of the forest trees of the Hawaiian Archipelago. Bull. Torrey Bot. Club 44: 145-157.

Maiden, J. H. 1902-1924. The forest flora of New South Wales. Sydney: Govt. of New South Wales. 8 vol.

Maiden, J. H. 1903-1933. A critical revision of the genus Eucalyptus. Sydney: Govt. of New South Wales, 8 vol.

Merlin, Mark David. 1976. Hawaiian forest plants. Honolulu: Oriental Publishing Co. 68 p.

Merlin, Mark David. 1978. Hawaiian coastal plants and scenic shorelines. Honolulu: Oriental Publishing Co. 68 p.

Metcalf, Mervin E., Robert E. Nelson, Edwin Q. P. Petteys, and John M. Berger. 1978. Hawaii's timber resources 1970. U.S. Dep. Agric. For. Serv. Pac. Southwest For. Range Exp. Stn. Resour. Bull. PSW-15. 1-20.

Mill, Susan W., Warren L. Wagner, and Derral R. Herbst, comp. 1985. Bibliography of Otto and Isa Degener's Hawaiian floras. Taxon 34(2): 229-259.

Miller, Carey D., Katherine Bazore, and Mary Bartow. 1955. Fruits of Hawaii: description, nutritive value, and use. 2d ed. Honolulu: Univ. of Hawaii Press, 197 p.

Mueller, Baron Ferd. von. 1879-1884. Eucalyptographia. A descriptive atlas of the eucalypts of Australia and the adjoining islands. Melbourne and London: 10 vol.

Nagata, Kenneth M. 1971. Hawaiian medicinal plants. Econ. Bot. 25: 245-254.

Neal, Marie C. 1965. In gardens of Hawaii. Rev. ed. Bernice P. Bishop Mus. Spec. Publ. 50: 1-924.

Nelson, Robert E. 1965. A record of forest plantings in Hawaii. U.S. Dep. Agric. For. Serv. Pac. Southwest For. Range Exp. Stn. Resour. Bull. 1: 1-18.

Nelson, Robert E. 1967. Records and maps of forest types in Hawaii. U.S. Dep. Agric. For. Serv. Pac. Southwest For. Range Exp. Stn. Resour. Bull. PSW-8: 1-22.

Nelson, Robert E., and Nobuo Honda. 1966. Plantation timber on the island of Hawaii--1965. U.S. Dep. Agric. For. Serv. Pac. Southwest For. Range Exp. Stn. Resour. Bull. PSW-3: 1-52.

Nelson, Robert E., and E. M. Hornibrook. 1962. Commercial uses and volume of Hawaiian tree fern. U.S. Dep. Agric. For. Serv. Pac. Southwest For. Range Exp. Stn. Tech. Pap. 73: 1-10.

Nelson, Robert E., and Thomas H. Schubert. 1976. Adaptability of selected tree species planted in Hawaii forests. U.S. Dep. Agric. For. Serv. Pac. Southwest For. Range Exp. Stn. Resour. Bull. PSW-14: 1-22.

Nelson, Robert E., and Phillip R. Wheeler. 1963. Forest resources of Hawaii--1961. Berkeley, CA: Hawaii Dep. Land Nat. Resour. in coop. with U.S. Dep. Agric. For. Serv. Pac. Southwest For. Range Exp. Stn. 48 p.

Nelson, Robert E., W. H. C. Wong, Jr., and H. L. Wick. 1968. Plantation timber on the island of Oahu--1966. U.S. Dep. Agric. For. Serv. Pac. Southwest For. Range Exp. Stn. Resour. Bull. PSW-10: 1-52.

Oliver, W. R. B. 1935. The genus Coprosma. Bernice P. Bishop Mus. Bull 132: 1-207.

The Outdoor Circle. 1972. Keep Hawaii green with The Outdoor Circle's planting charts for the home garden. Honolulu.

Pardo, Richard. 1978. The AFA national register of big trees. Am. For. 84(4): 17-46.

Penfold, A. R., and J. L. Willis. 1961. The eucalypts: botany, cultivation, chemistry and utilization. London: Leonard Hill, 551 p.

Peters, C. C., and J. F. Lutz. 1966. Some machining properties of two wood species grown in Hawaii, Molucca albizzia and Nepal alder. U.S. Dep. Agric. For. Serv. For. Prod. Lab. Res. Note FPL-0117: 1-18.

Pickford, Gerald D. 1962. Opportunities for timber production in Hawaii. U.S. Dep. Agric. For. Serv. Pac. Southwest For. Range Exp. Stn. Misc. Pap. 67: 1-11.

Pope, Willis T. 1929. Manual of wayside plants of Hawaii. Honolulu, HI: Advertiser Publishing Co. 289 p. [Reprinted 1968. Rutland, VT; Tokyo: Charles E. Tuttle Co.]

Porter, John R. 1972. Hawaiian names for vascular plants. Univ. Hawaii Coll. Trop. Agric. Exp. Stn. Pap. 1: 1-64.

Pryor, Lindsay D. 1976. The biology of eucalypts. Institute of Biology, Studies in Biology No. 61, 82 p.

Pryor, L. D., and L. A. S. Johnson. 1971. A classification of the eucalypts. Canberra: Australian National Univ. 102 p.

Pukui, Mary Kawena, Samuel H. Elbert, and Esther T. Mookini. 1974. Place names of Hawaii. Rev. ed. Honolulu: Univ. Press of Hawaii. 298 p.

Pukui, Mary Kawena, and Samuel H. Elbert. 1986. Hawaiian dictionary; Hawaiian-English; English-Hawaiian. Rev. ed. Honolulu: Univ. Press of Hawaii, 582 p.

Rao, Aragula Sathyanarayana. 1956. A revision of Rauvolfia with particular reference to the American species. Ann. Mo. Bot. Gard. 43: 253-354.

Record, Samuel J., and Robert W. Hess. 1943. Timbers of the New World. New Haven: Yale Univ. Press. 640 p.

Reyes, L. J. 1938. Philippine woods. Philippine Dep. Agric. Comm. Tech. Bull. 7: 1-536.

Richmond, G. B. 1963. Species trials at the Waiakea Arboretum, Hilo Hawaii. U.S. Dep. Agric. For. Serv. Pac. Southwest For. Range Exp. Stn. Res. Pap. PSW-4: 1-21.

Ripperton, J. C., and E. Y. Hosaka. 1942. Vegetation zones of Hawaii. Hawaii Agric. Exp. Stn. Bull. 89: 1-60.

Rock, Joseph F. 1913. The indigenous trees of the Hawaiian Islands. Honolulu: Published under patronage. 518 p.

Rock, Joseph F. 1916. The sandalwoods of Hawaii: a revision of the Hawaiian species of the genus Santalum. Hawaii Bd. Agric. For. Div. For. Bot. Bull. 3: 1-43.

Rock, Joseph F. 1917a. The ornamental trees of Hawaii. Honolulu: Published under patronage. 210 p.

Rock, Joseph R. 1917b. The ohia lehua trees of Hawaii: a revision of the Hawaiian species of the genus Metrosideros Banks. Hawaii Bd. Agric. For. Div. For. Bot. Bull. 4: 1-76.

Rock, Joseph F. 1917c. Hawaiian trees--a criticism. Bull. Torrey Bot. Club 44: 545-546.

Rock, Joseph F. 1919a. A monographic study of the Hawaiian species of the tribe Lobelioideae, family Campanulaceae. Bernice P. Bishop Mus. Mem. 7(2): 1-394.

Rock, Joseph F. 1919b. The arborescent indigenous legumes of Hawaii. Hawaii Bd. Agric. For. Div. For. Bot. Bull. 5: 1-53.

Rock, Joseph F. 1920. The leguminous plants of Hawaii: being an account of the native, introduced and naturalized trees, shrubs, vines, and herbs, belonging to the family Leguminosae. Honolulu: Hawaiian Sugar Planters Assoc. Exp. Stn. 234 p.

Rock, Joseph F. 1974. The indigenous trees of the Hawaiian Islands. 2d ed. Pacific Tropical Botanical Garden, Iawai, Kauai, HI. Rutland, VT; Tokyo: Charles E. Tuttle Co. 548 p. [Reprint of Rock 1913, with updated nomenclature by Derral Herbst and an introduction by Sherwin Carlquist.]

Rose, J. N. 1892. A new Tabebuia from Mexico and Central America: Tabebuia Donnell-Smithii n. sp. Bot. Gaz. 17: 418-419.

Scott, M. H. 1953. Utilization notes on South African timbers. Union S. Afr. Dep. For. Bull. 36: 1-95.

Sherff, Earl Edward. 1939. Additional studies of the Hawaiian Euphorbiaceae. Field Mus. Nat. Hist. Bot. Ser. 22: 467-576. [Notes on Antidesma L., p. 568-576.]

Sherff, Earl Edward. 1942. Revision of the Hawaiian members of the genus Pittosporum Banks. Field Mus. Nat. Hist. Bot. Ser. 22: 467-566.

Sherff, Earl Edward. 1947. A preliminary study of Hawaiian species of the genus Rauvolfia. Field Mus. Nat. Hist. Bot. Ser. 23: 321-331.

Sherff, Earl Edward. 1951. New entities in the genus Cheirodendron Nutt. ex Seem. (Fam. Araliaceae) from the Hawaiian Islands. Bot. Leafl. 5: 2-14.

Sherff, Earl Edward. 1952a. Further studies of Hawaiian Araliaceae: Additions to Cheirodendron Helleri Sherff and a preliminary treatment of the endemic species of Reynoldsia A. Gray. Bot. Leafl. 6: 6-19.

Sherff, Earl Edward. 1952b. Additions to our knowledge of the genus Tetraplasandra A. Gray (fam. Araliaceae). Bot. Leafl. 6: 19-41.

Sherff, Earl Edward. 1952c. Contributions to our knowledge of the genus Tetraplasandra A. Gray and Reynoldsia A. Gray (fam. Araliaceae) in the Hawaiian Islands. Bot. Leafl. 7: 7-17.

Sherff, Earl Edward. 1952d. Munroidendron, a new genus of Araliaceous trees from the Island of Kauai. Bot. Leafl. 7: 21-24.

Sherff, Earl Edward. 1953. Further notes on the genus Tetraplasandra A. Gray (fam. Araliaceae) in the Hawaiian Islands. Bot. Leafl. 8: 2-13.

Sherff, Earl Edward. 1954. Revision of the genus Cheirodendron Nutt. ex Seem. for the Hawaiian Islands. Fieldiana: Botany 29: 1-45.

Sherff, Earl Edward. 1955. Revision of the Hawaiian members of the genus Tetraplasandra A. Gray. Fieldiana Bot. 29: 49-142.

Sherff, Earl Edward. 1965. Some additions to the genus Dodonaea L. (fam. Sapindaceae). Am. J. Bot. 32: 202-214.

Sinclair, Mrs. Francis, Jr. [Isabella]. 1885. Indigenous flowers of the Hawaiian Islands. London: Sampson Low, Marston, Searle, and Rivington. 44 p. [44 color plates.].

Skolmen, Roger G. 1963a. Robusta eucalyptus wood: its properties and uses. U.S. Dep. Agric. For. Serv. Pac. Southwest For. Range Exp. Stn. Res. Pap. PSW-9: 1-12.

Skolmen, Roger G. 1963b. Wood density and growth of some conifers introduced to Hawaii. U.S. Dep. Agric. For. Serv. Pac. Southwest For. Range Exp. Stn. Res. Pap. PSW-12: 1-20.

Skolmen, Robert G. 1967a. Heating logs to relieve growth stresses. For. Prod. J. 17(7): 41-42.

Skolmen, Roger G. 1967b. Specific gravity and shinkage of Elaeocarpus joga wood from Guam. U.S. Dep. Agric. For. Serv. Pac. Southwest For. Range Exp. Stn. Res. Note PSW-163: 1-2.

Skolmen, Roger G. 1968. Wood of koa and black walnut similar in most properties. U.S. Dep. Agric. For. Serv. Pac. Southwest For. Range Exp. Stn. Res. Note PSW-174: 1-4.

Skolmen, Roger G. 1973. Characteristics and amount of brittleheart in Hawaii-grown robusta eucalyptus. Wood Sci. 6(1): 22-29.

Skolmen, Roger G. 1974a. Lumber potential of 12-year-old saligna eucalyptus trees in Hawaii. U.S. Dep. Agric. For. Serv. Pac. Southwest For. Range Exp. Stn. Res. Note PSW-288: 1-7.

Skolmen, Roger G. 1974b. Some woods of Hawaii . . . properties and use of 16 commerical species. U.S. Dep. Agric. For. Serv. Pac. Southwest For. Range Exp. Stn. Gen. Tech. Rep. PSW-8: 1-30.

Skolmen, Roger G. 1978. Vegetative propagation of Acacia koa Gray. In: Proceedings of the Second Conference in Natural Sciences, Hawaii Volcanoes National Park, June 1-3, 1978. Honolulu: Univ. of Hawaii. p. 260-273.

Skolmen, Roger G. 1980a. Plantings on the forest reserves of Hawaii 1910-1960: reference distribution. Honolulu, HI: U.S. Dep. Agric. For. Serv. Pac. Southwest For. Range Exp. Stn. 441 p. (On file, Honolulu, HI)

Skolmen, Roger G. 1980b. Growth of four unthinned Eucalyptus coppice stands on the island of Hawaii. In: Proceedings IUFRO/MAB/Forest Service Symposium, Sept. 8-12, 1980. Rio Piedras, PR: U.S. Dep. Agric. For. Serv. Inst. Trop. For. p. 87-95.

Skolmen, Roger G. 1983. Growth and yield of some eucalypts of interest to California. In: Proceedings of the Workshop on eucalypts in California, June 14-16, 1983. Sacramento, CA. U.S. Dep. Agric. For. Serv. Pac. Southwest For. Range Exp. Stn. Gen. Tech. Rep. PSW-69: 49-57.

Skolmen, Roger G. 1986a. Acacia (Acacia koa Gray). In: Bajaj, Y. P. S., ed. Biotechnology in agriculture and forestry. Berlin, Heidelberg: Springer-Verlag, p. 375-384.

Skolmen, Roger G. 1986b. Performance of Australian provenances of Eucalyptus grandis and Eucalyptus saligna in Hawaii. U.S. Dep. Agric. For. Serv. Pac. Southwest For. Range Exp. Stn. Res. Pap. PSW-181: 1-8.

Skolmen, Roger G., and Charles C. Gerhards. 1964. Brittleheart in Eucalyptus robusta grown in Hawaii. For. Prod. J. 14(12): 549-554.

Skolmen, Roger G., and Marion O. Mapes. 1976. Acacia koa plantlets from somatic callus tissue. J. Hered. 67: 114-115.

Skottsberg, C. 1927. Artemisia, Scaevola, Santalum, and Vaccinium of Hawaii. Bernice P. Bishop Mus. Bull. 43: 1-89.

Skottsberg, C. 1938. Astelia and Pipturus of Hawaii. Bernice P. Bishop Mus. Bull. 117: 1-77.

Skottsberg, C. 1944. Hawaiian vascular plans. IV. Medd. Goteborgs Bot. Tradg. 15: 275-531. [Metrosideros, p. 402-409.]

Sohmer, S. H. 1972. A revision of the genus Charpentiera (Amaranthaceae). Brittonia 24: 283-312.

Sohmer, S. H. 1977. Psychotria L. (Rubiaceae) in the Hawaiian Islands. Lyonia Occas. Pap. Harold L. Lyon Arbor. 1: 103-186.

Sohmer, S. H., and R. Gustafson. 1987. Plants and flowers of Hawai'i. Honolulu: Univ. of Hawaii Press. 160 p.

St. John, Harold. 1954. Review of Mrs. Sinclair's "Indigenous flowers of the Hawaiian Islands:" Hawaiian plant studies 23. Pacific Sci. April 1954: 140-146.

St. John, Harold. 1973. List and summary of the flowering plants in the Hawaiian Islands. Pac. Trop. Bot. Gard. Mem. 1: 1-519.

St. John, Harold. 1977a. The variations of Alphitonia ponderosa (Rhamnaceae): Hawaiian Plant Studies 59. Phytologia 35: 177-182.

St. John, Harold. 1977b. Observations on Hawaiian Panicum and Sapindus: Hawaiian plant studies 61. Phytologia 36: 465-467.

St. John, Harold. 1977c. Revision of the genus Pittosporum in Hawaii: Hawaiian plant studies 64. Phytologia 38: 75-98.

St. John, Harold. 1979. Metrosideros polymorpha (Myrtaceae) and its variations. Hawaiian plant studies 88. Phytologia 42: 215-218.

St. John, Harold. 1982. Vernacular names used on Ni'ihau Island: Hawaiian plant studies 69. Bernice P. Bishop Mus. Occas. Pap. 25(3): 1-10.

St. John, Harold. 1984. Revision of the Hawaiian species of Santalum (Santalaceae): Hawaiian plant studies 109. Phytologia 55: 217-226.

St. John, Harold. 1985. Monograph of the Hawaiian species of Pleomele (Liliaceae): Hawaiian plant studies 103. Pac. Sci. 39(2): 171-190.

St. John, Harold. 1986. Revision of the Hawaiian Diospyros (Ebenaceae): Hawaiian plant studies 120. Phytologia 59: 389-405.

Stemmermann, Lani. 1981. Observation on the genus Santalum (Santalaceae) in Hawai'i. Pac. Sci. 34: 41-54.

Stone, Benjamin C. 1962. A monograph of the genus Platydesma (Rutaceae). J. Arnold Arbor. 43: 410-427.

Stone, Benjamin C. 1967. A review of the endemic genera of Hawaiian plants. Bot. Rev. 33: 216-257.

Stone, Benjamin C. 1969. The genus Pelea A. Gray (Rutaceae: Evodiinae): a taxonomic monograph. Phanerogamarum Monographiae Tomus III. Lehre: J. Cramer, 180 p.

Stone, Benjamin C. 1970. The flora of Guam. Micronesia 6: 1-659.

Streets, R. J. 1962. Exotic trees in the British Commonwealth. Oxford: Claredon Press, 765 p.

Swain, E. H. F. 1928. The timbers and forest products of Queensland. Brisbane: Queensland For. Serv. Govt. Printer. 500 p.

Thrum, T. G. 1890. List of indigenous Hawaiian woods, trees and large shrubs. Hawaiian Almanac and Annual for 1891, p. 87-91.

U.S. Department of the Interior, Fish and Wildlife Service. 1976. Endangered and threatened wildlife and plants. Proposed endangered status for some 1700 U.S. vascular plant taxa. Federal Register 41(117): 24524-24572, June 16, 1976.

U.S. Department of Agriculture, Forest Service, Forest Products Laboratory. 1987. Wood handbook: wood as an engineering material. U.S. Dep. Agric. For. Serv. Agric. Handbk. 72.

U.S. Postal Service. 1988. 1988 national five-digit zip code & post office directory. U.S. Postal Serv. Publ. 65: 1-2,460.

Wagner, Warren L., Derral R. Herbst, and Seymour H. Sohmer. 1989. Manual of the flowering plants of Hawaii. Honolulu: Bernice P. Bishop Museum. [In press.]

Whitesell, Craig D. 1964. Silvical characteristics of koa (Acacia koa Gray). U.S. Dep. Agric. For. Serv. Pac. Southwest For. Range Exp. Stn. Res. Pap. PSW-16: 1-12.

Whitesell, Craig D. 1970. Early effects of spacing on loblolly pine in Hawaii. U.S. Dep. Agric. For. Serv. Pac. Southwest For. Range Exp. Stn. Res. Note PSW-223: 1-3.

Whitesell, Craig D., and Gerald A. Walters. 1976. Species adaptability trails for man-made forests in Hawaii. U.S. Dep. Agric. For. Serv. Pac. Southwest For. Range Exp. Stn. Res. Pap. PSW-118: 1-30.

Wilder, Gerrit Parmile. 1911. Fruits of the Hawaiian Islands. Rev. ed. Honolulu: Hawaiian Gazette Co., 247 p.

Wilson, Peter G., and John T. Waterhouse. 1982. A review of the genus Tristania R. Br. (Myrtaceae): a heterogeneous assemblage of five genera. Austr. J. Bot. 30: 416-446.

Wong, W. H. C., Jr., Nobuo Honda, and Robert E. Nelson. 1967. Plantation timber on the island of Lanai--1966. U.S. Dep. Agric. For. Serv. Pac. Southwest For. Range Exp. Stn. Res. Bull. PSW-7: 1-18.

Wong, W. H. C., Jr., Herbert L. Wick, Nobuo Honda, and Robert E. Nelson. 1967. Plantation timber on the island of Molokai--1967. U.S. Dep. Agric. For. Serv. Pac. Southwest For. Range Exp. Stn. Res. Bull. PSW-9: 1-25.

Wong, W. H. C., Jr., Herbert L. Wick, Nobuo Honda, and Robert E. Nelson. 1969. Plantation timber on the island of Maui--1967. U.S. Dep. Agric. For. Serv. Pac. Southwest For. Range Exp. Stn. Res. Bull. PSW-11: 1-42.

Youngs, Robert L. 1960. Physical, mechanical, and other properties of five Hawaiian woods. U.S. Dep. Agric. For. Serv. For. Prod. Lab. Rep. 2191: 1-34.

Youngs, Robert L. 1964. Hardness, density, and shrinkage characteristics of silk-oak from Hawaii. U.S. Dep. Agric. For. Serv. For. Prod. Lab. Res. Note FPL-74: 1-14.

INDEX OF COMMON AND SCIENTIFIC NAMES

Common and scientific names adopted in headings for the 152 numbered species and the page numbers where descriptions begin are in heavy (boldface) type, the scientific names in heavy (boldface) italics. Other common names appear in ordinary (roman) type. Common names in the English language are indexed under the last word. Other scientific names, including synonyms, are in italics. Family names, common and scientific, are shown in capitals, the scientific also in italics.

'a'aka, 294
'a'ali'i, 180, 182
'a'ali'i-ku ma kua, 182
'a'ali'i ku makani, 182
'a'ali'i-mahu, 266
abas, 250
acacia, black, 134
acacia, black-wattle, 132
acacia, blackwood, 134
Acacia confusa, 128
Acacia decurrens, 134
acacia, green-wattle, 134
***Acacia koa, 128*, 134**
Acacia mearnsii, 132
Acacia melanoxylon, 134
acacia, small Philippine, 128
a'e, 162, 186
AGAVACEAE, 84
AGAVE FAMILY, 84
aguacate, 124
'ahakea, 294
a'ia'i, 104
'aiea, 178
'ākia, 202
akiahala, 126
ākōlea, 106
'ala'a, 270
alageta, 124
alahe'e, 294
alani, 158
Albizia falcata, 136
Albizia falcataria, 134
albizia, Molucca, 134
Albizia moluccana, 136
Albizia saman, 150
alcanfor, 122
alder, Nepal, 96
alelaila, 164
Aleurites moluccana, 166
Alexandrian-laurel, 200
algarroba, 150
alligator-pear, 124
almendro, 208
almond, 208
Alnus nepalensis, 96
aloalo, 190
Alphitonia excelsa, 188
Alphitonia ponderosa, 186
AMARANTH FAMILY, 116
AMARANTHACEAE, 116
'ama'u, 48
'ama'uma'u, 44, 48
ANACARDIACEAE, 170

anau, 284
Angophora costata, 208
Angophora lanceolata, 210
Antidesma platyphyllum, 168
apas, 250
APOCYNACEAE, 280
apricot, 124
AQUIFOLIACEAE, 178
ARALIA FAMILY, 256
ARALIACEAE, 256
Araucaria angustifolia, 50
Araucaria brasiliana, 50
araucaria, columnar, 50
Araucaria columnaris, 50
araucaria, Cook, 54
Araucaria cookii, 54
araucaria, Cunningham, 56
Araucaria cunninghamii, 54
Araucaria excelsa, 54
ARAUCARIA FAMILY, 50
Araucaria heterophylla, 54
araucaria, parana, 50
ARAUCARIACEAE, 50
árbol de pan, 100
Ardisia elliptica, 266
ardisia, elliptical-leaf, 268
ardisia, shoebutton, 266
Ardisia solanacea, 268
Artocarpus altilis, 98
Artocarpus communis, 100
Artocarpus incisus, 100
arudo, 100
as, 208
asasa, 202
aseri, 284
ash, Hawaiian, 278
ash, Shamel, 278
ash, tropical, 276
āulu, 118, 182, 274
Australian-pine, 88, 90
avocado, 124

badrirt, 196
bamboo, common, 76
bamboo, feathery, 78
bambú, 78
Bambusa vulgaris, 76
banahl, 98
banalo, 196
bang-beng, 196
bangalay, 214
banyan, Chinese, 102
banyan, Malayan, 104

bata, 124
bayahonda, 152
BAYBERRY (WAXMYRTLE) FAMILY, 92
beadtree, 164
beautyleaf, 200
beefwood, 90
BELLFLOWER FAMILY, 304
BETULACEAE, 96
BIGNONIA FAMILY, 286
BIGNONIACEAE, 286
BIRCH FAMILY, 96
BITTERSWEET FAMILY, 180
biyuch, 200
blackbutt, 226
black-wattle, 134
blackwood, 134
blackwood, Australia, 134
BLECHNACEAE, 48
BLECHNUM FERN FAMILY, 48
bluegum, 221, 234
bluegum, southern, 221
bluegum, Sydney, 234
bluegum, Tasmanian, 221
Bobea hookeri, 294
Bobea sandwicensis, 294
bop, 76
BORAGE FAMILY, 280
BORAGINACEAE, 280
bottle-brush, 244
brassaia, 262
Brassaia actinophylla, 262
breadfruit, 98
Brisbane-box, 254
Broussaisia arguta, 124
Broussiasia arguta pellucida, 126
Broussiasia pellucida, 126
Brunoniana sandwicensis, 104
brushbox, 250
btaches, 200
BUCKTHORN FAMILY, 186
BUCKWHEAT FAMILY, 114

cajeput-tree, 240
***Calophyllum inophyllum, 196*, 206**
CAMPANULACEAE, 304
camphor-tree, 122
camphor-tree, Japanese, 122
candelabra-tree, 50
candlenut, 168
candlenut-tree, 166
Canthium odoratum, 294

315

CASHEW FAMILY, 170
casia de Siam, 138
Cassia glauca, 138
cassia, Siamese, 136
Cassia siamea, 136
Cassia surattensis, 138
casuarina, Cunningham, 88
Casuarina cunninghamiana, 86
Casuarina equisetifolia, 88, 90
CASUARINA FAMILY, 86
Casuarina glauca, 88, *90*
casuarina, horsetail, 88, *90*
casuarina, longleaf, 88, *90*
casuarina, river-oak, 86
CASUARINACEAE, 86
cayeput, 244
cayeputi, 244
Cecropia mexicana, 102
Cecropia obtusifolia, 100
Cedela toona, 166
cedar, Bermuda, 74
CELASTRACEAE, 180
charcoal tree, 98
Charpentiera obovata, 116
Cheirodendron kauaiense, 256
Cheirodendron gaudichaudii, 258
Cheirodendron platyphyllum, 256
Cheirodendron trigynum, 256, *258*
chel, 284
chen yamolehat, 284
chinaberry, 162
chinaberry, umbrella, 164
chinatree, 164
CHOCOLATE FAMILY, 196
Christmas-berry, 172, 176
Chrysophyllum polynesicum, 270
Cibotium chamissoi, 44, 48
Cibotium glaucum, 46
Cibotium hawaiiense, 46
Cibotium menziesii, 44
Cinnamomum camphora, 122
Citharexylum caudatum, 286
CITRUS FAMILY, 156
Clermontia arborescens, 304
clermontia, tree, 304
Coccoloba uvifera, 114
coco, 82
coco-palm, 82
coconut, 78
coconut-palm, 82
Cocos nucifera, 78
cocotero, 82
Coffea arabica, 296
coffee, 296
coffee, Arabian, 298
COFFEE FAMILY, 294
coffee, Kona, 298
COMBRETACEAE, 206
COMBRETUM FAMILY, 206
Cook-pine, 54

Coprosma montana, 298
coralbean, India, 142
coralbean, variegated, 144
coraltree, Hawaiian, 142
coraltree, Indian, 144
coraltree, variegated, 144
Cordia subcordata, 280
Cordyline fruticosa, 84
Cordyline terminalis, 84
CORYNOCARPACEAE, 176
Corynocarpus laevigatus, 176
cryptomeria, 64
Cryptomeria japonica, 64
Cuban-bast, 190
CUPRESSACEAE, 66
Cupressus arizonica, 66
Cupressus benthamii, 68
Cupressus lindleyi, 68
Cupressus lusitanica, 68
Cupressus macrocarpa, 68, 70
Cupressus sempervirens, 70
Cybistax donnell-smithii, 288
cypress, Arizona, 66
cypress, Arizona rough, 68
cypress, Arizona smooth, 68
CYPRESS FAMILY, 66
cypress, Italian, 70
cypress, Mexican, 68
cypress, Monterey, 70

daog, 200
devilpepper, 280
DICKSONIACEAE, 44
Diospyros ferrea sandwicensis, 276
Diospyros hillebrandii, 274
Diospyros sandwicensis, 276
Dodonaea viscosa, 180, 182
DOGBANE FAMILY, 280
dogdog, 100
Dodonaea eriocarpa, 182
Dodonaea sandwicensis, 182
Dodonaea stenocarpa, 182
dracaena, 84
Dracaena aurea, 86
dracaena, common, 84
dracaena, golden, 86
Drypetes forbesii, 202

earpod-tree, 150
EBENACEAE, 274
EBONY FAMILY, 274
ee, 300
eet, 200
ekoa, 144
ELAEOCARPACEAE, 188
Elaeocarpus bifidus, 188
ELAEOCARPUS FAMILY, 188
ēlama, 274
ELM FAMILY, 96
elodechoel, 98
emajagua, 194

emajagua excelsa, 192
emajagüilla, 196
Enterolobium cyclocarpum, 150
EPACRIDACEAE, 266
EPACRIS FAMILY, 266
ermall, 194
erythrina, Hawaiian, 142
Erythrina indica, 144
Erythrina monosperma, 142
Erythrina sandwicensis, 138
Erythrina variegata, 142
eucalypt, 210
eucalyptus, 210
Eucalyptus, 208, *210,* 250
eucalyptus, amammanit, 218
eucalyptus, bagras, 218
eucalyptus, bangalay, 214
eucalyptus, beakpod, 232
eucalyptus, blackbutt, 226
eucalyptus, bluegum, 221
Eucalyptus botryoides, 214
Eucalyptus camaldulensis, 214
Eucalyptus citriodora, 216
Eucalyptus deglupta, 218
Eucalyptus globulus, 221
Eucalyptus grandis, 222, 232
eucalyptus, gray-ironbark, 224
eucalyptus, kinogum, 226
eucalyptus, lemon-gum, 216
Eucalyptus maculata citriodora, 218
Eucalyptus microcorys, 222, 226
Eucalyptus multiflora, 232
Eucalyptus naudiniana, 218
Eucalyptus paniculata, 224
Eucalyptus pilularis, 226
eucalyptus, red-ironbark, 234
eucalyptus, red-mahogany, 230
Eucalyptus resinifera, 226
eucalyptus, river-redgum, 214
eucalyptus, robusta, 112, 230
Eucalyptus robusta, 112, 220, 222, 226, ***230,*** 232
eucalyptus, rosegum, 222, 232
Eucalyptus rostrata, 216
eucalyptus, saligna, 222, 230, 232
Eucalyptus saligna, 220, 222, 224, 226, 230, ***232,*** 254
Eucalyptus sideroxylon, 234
eucalyptus, swamp-mahogany, 232
eucalyptus, Sydney bluegum, 324
eucalyptus, tallowwood, 222
eucalyptus, Tasmanian blue, 221
Eugenia cumini, 234
Eugenia jambolana, 236
Eugenia jambos, 236
Eugenia malaccensis, 238

Eugenia sandwicensis, 240
EUPHORBIACEAE, 166

fach, 76
Fagara oahuensis, 162
fala, 76
faliap, 238
false-mulberry, Hawaiian, 104
false-sandalwood, 290
faniap, 238
fau, 194
fetau, 200
Ficus microcarpa, 102
Ficus nitida, 104
Ficus retusa, 104
fiddlewood, 286
fig, 104
fig, India-laurel, 104
firebell, 290
firetree, 92
FLACOURTIA FAMILY, 200
FLACOURTIACEAE, 200
Flindersia brayleyana, 156
flooded-gum, 222
fountain-tree, 290
FOUR O'CLOCK FAMILY, 118
Fraxinus uhdei, 276
fresno, 278

gaal, 194
gabgab, 144
galu, 284
ganitnityuwan tangantan, 144
gardenia hedionda, 300
gaugau, 144
gatae, 144
GINSENG FAMILY, 256
glorybush, 254
glorybush, Hawaiian, 256
glorybush, Urville, 256
goldtree, 286
GOODENIA FAMILY, 306
GOODENIACEAE, 306
Gouldia affinis, 298
Gouldia terminalis, 300
GRAMINEAE, 76
grape, 116
GRASS FAMILY, 76
green-wattle, 134
grevilea, 112
grevillea, Banks, 110
Grevillea banksii, 106
Grevillea robusta, 110
guabang, 250
guahva, 250
guamá americano, 146
guamuche, 146
gaumuchil, 146
guarumo, 102
guava, 236, **248**
guava, Cattley, 248
guava, common, 250
guava, purple strawberry, 248

guava, strawberry, 248
guayaba, 250
gum-myrtle, lanceleaf, 208
gumorni spanis, 148
gunpowder-tree, 96
GUTTIFERAE, 196

hā, 240
ha'a, 168
ha'āmaile, 168
hāawa, 128
haha, 304
ha'ikū, 110
ha'iku ke'oke'o, 112
haiti-haiti, 196
hala, 74, 86
hala lihilihi'ula, 76
halapepe, 84, **86**
halapia, 76
hala'ula, 76
half-flower, 306
hame, 168
hamehame, 168
hao, 280
hāpu'u, 48
hāpu'u-'i'i, 44
hāpu'u-pulu, 46
hau, 180, 190, **192,** 194
Hawaiian-heather, 266
Hawaiian-olive, 278
Hawaiian-star-pine, 56
hawane, 84
hea'e, 162
hei'i, 44
Heimerliodendron brunoianum, 118
Hibiscus arnottianus, **188**
Hibiscus elatus, **190**
hibiscus, linden, 194
hibiscus, native white, 188
hibiscus, punaluu, 190
hibiscus, sea, 192
Hibiscus tiliaceus, 180, 190, **192,** 194
hō'awa, 126
hollock, 208
HOLLY FAMILY, 178
holly, Hawaiian, 178
hona, 266
hoop-pine, 54
hōpue, 106
huni, 284
hunig, 284
hupilo, 298

iedel, 172
ikoak, 284
ikoik, 284
Ilex anomala, **178**
Ilex hawaiensis, 178
Ilex sandwicensis, 178
'iliahi, 114
'iliahi-a-lo'e, 112
'ilinia, 164
īnia, 164

India-almond, 206
Indian-lilac, 164
Indian-mulberry, 300
ipilipil, 144
ironbark, 234
ironbark, grey, 224
ironbark, red, 234
ironbark, white, 224
iron-gum, lemon-scented, 218
ironwood, 88
ironwood, common, 90
ironwood, longleaf, 92
ironwood, salt-marsh, 92
ironwood, shortleaf, 90
iru, 82
isou, 200

jaboncillo, 186
jagüey, 104
jambolan, 236
jambolan-plum, 236
Jambosa jambos, 238
Jambosa malaccensis, 238
Japanese-cedar, 64
Java-plum, 234
jhalna, 208
juniper, Bermuda, 74
juniper-berry, 286
Juniperus bermudiana, **74**

kāawa'u, 178
kafu, 76
kahili-flower, 106
kalamona, 138
kalau, 194, 284
kalia, 188
kamachili, 146
kamalindo, 154
kamani, 196, 206
kamani, false, 200, **206**
kamani'ula, 206
kamani haole, 206
kamanu, 200
kamarere, 218
kamatsiri, 146
kanawao, 124
kanawau, 126
kānehoa, 266
kangit, 172
karaka, 176
KARAKA FAMILY, 176
karaka-tree, 178
karakanut, 178
karakanut, New Zealand, 176
kassod-tree, 138
kauila, 186
kauila māhu, 258
kaulu, 182, 274
kawai, 266
k āwa'u, 178, 266
keahi, 270
kel, 208
ki, 84
kiawe, 142, **150**

317

kidel, 238
kiden, 284
kilife, 194
kilulo, 196
kipar, 76
koa, 128, 134, 168, 284
koa, false, 144
koa, Formosa, 128
koa haole, 144
koki'o ke'oke'o, 188
kōlea, 268, 270
kōlea lau-li'i, 268
kōlea lau nui, 268
kolomona, 138
kōpiko, 300
kōpiko 'ula, 304
kotal, 208
kou, 280
kuawa, 248
kukui, 166
kupuwao, 126

lada, 300
la-i, 84
lama, 168, 274, 276
lampuaye, 182
lapalapa, 256
lasiandra, 256
Lasiandra urvilleana, 256
lauhala, 76
LAURACEAE, 122
laurel de la India, 104
LAUREL FAMILY, 122
LEGUME FAMILY, 128
LEGUMINOSAE, 128
lehua 'ōhi'a, 246
lelah, 164
lemai, 100
lemon-scented-gum, 218
Leptospermum scoparium,
 240
leucaena, 144
Leucaena glauca, 144
Leucaena leucocephala, 144
licorice, 148
lilaila, 164
lilac, 164
lilac, Persian, 164
lili-koa, 144
lo, 194
lol, 300
lopā-samoa, 144
Lophostemon confertus, 254
loulu, 82
lu, 82
lueg, 200
lulk, 104
lumbang, 168

ma, 100
maa, 100
Maba hillebrandii, 274
Maba sandwicensis, 276
macupa, 238

MADDER FAMILY, 294
Madras-thorn, 146
mahoe, 194
mahoe, blue, 190, 192
mahoe, seaside, 196
MAHOGANY FAMILY, 162
māhu, 258
mai, 100
maiele, 266
majagua, 192
majagua azul, 192
majó, 192
Malay-apple, 238
MALLOW FAMILY, 188
MALVACEAE, 188
mamaki, 104
mamane, 152
mamani, 154
mamano, 154
manako, 170
mānele, 182
manga, 172
Mangifera indica, 170
mangle, 204
mangle colorado, 204
mango, 170
MANGOSTEEN FAMILY, 196
mangrove, 204
mangrove, American, 204
mangrove, common, 204
MANGROVE FAMILY, 204
mangrove, red, 204
Manila-tamarind, 146
manono, 298
manuka, 240
manzana malaya, 238
ma'u, 48
maua, 200
ma'uma'u, 48
mehame, 168
mehamehame, 168
Melaleuca quinquenervia,
 240
MELASTOMATACEAE, 254
MELASTOME FAMILY, 254
Melia azedarach, 162
MELIACEAE, 162
melochia, 196
Melochia indica, 196
Melochia umbellata, 196
mesechelangel, 182
mesegerak, 236
Messerschmidia argentea, 284
messmate, red, 230
messmate, swamp, 232
mesquite, 152
Metrosideros collina, 246
Metrosideros polymorpha, 44,
 48, *244*
meu, 46
MEZEREUM FAMILY, 202
miiche, 208
milo, 194
Mindanao-gum, 218

moen, 76
mohs, 100
mokihana, 156, 158
monkeypod, 146
MORACEAE, 98
Moreton-Bay-pine, 56
morinda, 300
Morinda citrifolia, 300
mountain-apple, 238
mugga, 234
MULBERRY FAMILY, 98
MYOPORACEAE, 290
MYOPORUM FAMILY, 290
Myoporum sandwicense, 290
Myrica faya, 92
MYRICACEAE, 92
MYRSINACEAE, 266
MYRSINE FAMILY, 266
Myrsine lessertiana, 268, 270
Myrsine meziana, 268
Myrsine sandwicensis, 268
MYRTACEAE, 208
MYRTLE FAMILY, 208

naeo, 294
naieo, 294
naio, 290
nani-o-Hilo, 176
naupaka, 306
NAUPAKA FAMILY, 306
naupaka kuahiwi, 306
naupaka, mountain, 306
nawao, 126
neleau, 172
nen, 300
neneleau, 172
nesoluma, Hawaiian, 270
Nesoluma polynesicum, 270
Nestegis sandwicensis, 278
NETTLE FAMILY, 104
New-Guinea-gum, 218
New-Zealand-tea, 240
ngel, 300
ni, 82
niyoron, 284
noni, 300
nonu, 300
Norfolk-Island-pine, 54
nu, 82
nuez, 168
nuez de India, 168
niu, 78, 126
nunu, 104
niyog, 82
nizok, 82
NYCTAGINACEAE, 118

o'a, 188
octopus-tree, 260
'oha, 304
'ōhai, 146
'ohawai, 304
'ohe, 76, 260, 262
'ōhe'e, 296

'ohe-kukuluae'o, 260
'ohelo, false, 202
'ohemakai, 258
'ohe'ohe, 262
'ōhi'a ai, 238
'ōhi'a hā, 240
'ōhi'a lehua, 44, 48, **244**
'ōhi'a loke, 236
'oka kilika, 112
'okapua, 110
'ōlapa, 256, **258**
'ōlapalapa, 258
OLEACEAE, 276
OLIVE FAMILY, 276
olomea, 180, *192*
olopua, 278
ongor, 76
opiko, 304
'opiuma, 146, *150*
ōpuhe, 106
Osmanthus sandwicensis, 278
otaheita, 196

pago, 194
pahr, 144
pā'ihi, 240
painkiller, 300
palama, 236
PALM FAMILY, 78
PALMAE, 78
palo de pan, 100
pā-makini, 190
pan, 100
pana, 100
panapéh, 100
PANDANACEAE, 74
pandanus, 76
Pandanus odoratissimus, 76
Pandanus tectorius, 74, 86
paniap, 238
panu, 196
pāpala, 116
pāpala képau, *118*, 122
paperbark, 240
paperbark-tree, 244
par, 144
paraiso, 164
parana-pine, 50
Paraserianthes falcataria, 136
Pariti tiliaceum, 194
Pariti tiliaceum elatum, 192
Paritium elatum, 192
Paritium tiliaceum, 194
pasilla, 164
pear, 124
Pelea anisata, 156, 158
pelea, Clusia-leaf, **158**
Pelea clusiifolia, 158
péndula de sierra, 286
peppertree, Brazil, 176
Perrottetia sandwicensis, 180, *192*
Persea americana, 124

persimmon, Hillebrand, **274**
pi'ao palao'an, 78
pilo, 298
pilo kea, 158
pilo kuahiwi, 298
pimienta de Brazil, 176
pi'ohi'a, 126
PINACEAE, 56
pine, cluster, 58
PINE FAMILY, 56
pine, insignis, 60
pine, jelecote, 58
pine, loblolly, 60
pine, maritime, 60
pine, Mexican weeping, 58
pine, Monterey, 60
pine, oldfield, 64
pine, seaside, 60
pine, shortleaf, 64
pine, slash, 56
pino australiano, 88, 90
pino de Australia, 88, 90
Pinus elliottii, 56, 64
Pinus insignis, 60
Pinus patula, 58
Pinus pinaster, 58
Pinus radiata, 60
Pinus taeda, 60
Pipturus albidus, 104
Pipturus brighamii, 106
Pipturus gaudichaudianus, 106
Pipturus hawaiiensis, 106
Pipturus helleri, 106
Pipturus oahuensis, 106
Pipturus pachyphyllus, 106
Pipturus pterocarpus, 106
Pipturus rockii, 106
Pipturus skottsbergii, 106
Pisonia brunoniana, 118
Pisonia inermis, 118
Pisonia sandwicensis, 118
Pithecellobium dulce, 146, 150
Pithecellobium saman, 146
PITTOSPORACEAE, 126
Pittosporum cauliflorum, 128
Pittosporum cladanthum, 128
Pittosporum confertiflorum, 126, 128
PITTOSPORUM FAMILY, 126
Pittosporum halophiloides, 128
Pittosporum halophilum, 128
Pittosporum lanaiense, 128
Planchonella puulupensis, 274
Planchonella sandwicensis, 274
Platydesma spathulatum, 158
platydesma, spatula-leaf, **158**
plectronia, 296
Plectronia odorata, 196
Pleomele aurea, 84, ***86***
polo, 196
POLYGONACEAE, 114

pomarrosa, 238
pomarrosa malaya, 238
pone, 196
portiatree, 194
Pouteria auahiense, 274
Pouteria ceresolii, 274
Pouteria rhynchosperma, 274
Pouteria sandwicensis, 270
Pouteria spathulata, 274
pride-of-India, 162
primavera, 286
pritchardia, 82
Pritchardia, 82
Prosopis pallida, 142, *150*
PROTEA FAMILY, 106
PROTEACEAE, 106
Pseudomorus sandwicensis, 104
Psidium cattleianum, 248
Psidium guajava, 236, ***248***
Psidium littorale, 248
Psilorhegma glauca, 138
Psychotria hawaiiensis, 300
pua, 278
pua aloalo, 190
pua'a olomea, 180
pu'aha, 126
puakiawe, 266
puhala, 76
pūkiawe, 266
punk-tree, 244
púpūkiawe, 266

Queensland-maple, 156

raintree, 148
rakich, 200
rapanea, Hawaiian, 270
Rapanea lessertiana, 268
Rapanea sandwicensis, 270
rarningobche, 290
Rauvolfia degeneri, 280
Rauvolfia forbesii, 280
rauvolfia, Hawaiian, **280**
Rauvolfia helleri, 280
Rauvolfia mauiensis, 280
Rauvolfia molokaiensis, 280
Rauvolfia remotiflora, 280
Rauvolfia sandwicensis, 280
redcedar, Australian, 166
redcedar, Bermuda, 74
redgum, 216
redgum, Murray, 216
redgum, river, 216
red-mahogany, 230
redwood, 66, **230**
redwood, California, 66
redwood, coast, 66
REDWOOD FAMILY, 64
Reynoldsia degeneri, 260
reynoldsia, Hawaiian, **258**
Reynoldsia hillebrandii, 260
Reynoldsia hosakana, 260
Reynoldsia huehuensis, 260
Reynoldsia mauiensis, 260

Reynoldsia oblonga, 260
**Reynoldsia sandwicensis,
258**
Reynoldsia venusta, 260
RHAMNACEAE, 186
Rhizophora mangle, 204
Rhizophora mangle samoensis, 204
Rhizophora samoensis, 204
RHIZOPHORACEAE, 204
Rhus chinensis sandwicensis, 172
Rhus sandwicensis, 172
Rhus semialata sandwicensis, 172
river-gum, 216
roble de sada, 112
roble australiano, 112
Rockia sandwicensis, 122
roro, 144
rose-apple, 236
rose-gum, 222
**Roseodendron donnell-
smithii, 286**
RUBIACEAE, 294
RUE FAMILY, 156
RUTACEAE, 156

sadleria, 48
Sadleria cyatheoides, 44, **48**
sakan, 168
Samanea saman, 150
samāu, 148
sandalwood, coast, 112
sandalwood, false, 294
SANDALWOOD FAMILY, 112
sandalwood, Freycinet, 114
sandalwood myoporum, 294
SANTALACEAE, 112
Santalum ellipticum, 112
***Santalum freycinetianum,
114***
SAPINDACEAE, 180
***Sapindus oahuensis,* 182**
***Sapindus saponaria,* 182**
Sapindus thurstonii, 186
SAPODILLA FAMILY, 270
SAPOTACEAE, 270
sau, 136
SAXIFRAGACEAE, 124
SAXIFRAGE FAMILY, 124
Scaevola frutescens, 306
scaevola, Gaudichaud, 306
***Scaevola gaudichaudiana,
306***
Scaevola taccada, 306
schefflera, 262
***Schefflera actinophylla,* 260**
***Schinus terebinthifolia,* 172**
Sciacassia siamea, 138
scrambled-eggs, 138
screwpine, 74, 86
SCREWPINE FAMILY, 74
seagrape, 114
senna, glossy-shower, 138
Senna surattensis, 138
Sequoia sempervirens, 66

she-oak, 90, 92
she-oak, river, 88
Sideroxylon sandwicense, 274
silk-oak, 110
silkwood, 156
silver-oak, 112
soapberry, 186
SOAPBERRY FAMILY, 180
soapberry, Oahu, 182
soapberry, wingleaf, 182
***Sophora chrysophylla,* 152**
spathodea, 290
***Spathodea campanulata,
288***
Spathodea nilotica, 290
spotted-gum, 218
SPURGE FAMILY, 166
srofaf, 208
srusrun, 284
southern-mahogany, 214
STERCULIACEAE, 196
Straussia hawaiiensis, 304
***Streblus pendulinus,* 104**
Streblus sandwicensis, 104
Styphelia douglasii, 266
styphelia, Kamehameha, 266
***Styphelia tameiameiae,* 266**
Stylurus banksii, 110
Stylurus robusta, 112
sugi, 64
sumac, Hawaiian, 172
Suttonia lessertiana, 268
Suttonia sandwicensis, 270
swamp-mahogany, 232
swamp-oak, 92
***Syncarpia glomulifera,* 250**
Syncarpia laurifolia, 250
Syzygium cumini, 236
syzygium, Hawaiian, 240
Syzygium jambos, 238
Syzygium malaccense, 238
Syzygium sandwicense, 240

Tabebuia donnell-smithii, 288
talie, 208
talisai, 208
tallowwood, 224
taln tangan, 144
tamarind, 154
tamarind, wild, 144
tamarindo, 154
***Tamarindus indica,* 154**
tangan-tangan, 144
tantan, 144
tauanave, 284
tausuni, 284
TAXODIACEAE, 64
teatree, 240
teatree, broom, 240
telentund, 144
Terminalia catappa, 200, **206**
***Terminalia myriocarpa,* 208**
***Tetraplasandra hawaiiensis,
262***

Thespesia populnea, 194
thipwopu, 208
THYMELAEACEAE, 202
ti, 84
tibet, giant, 148
tibuchina, 256
Tibouchina semidecandra, 256
***Tibouchina urvilleana,* 254**
tigers-claw, 144
titin, 284
toa, 90
toon, Australian, 164
toon, Burma, 166
***Toona ciliata,* 164**
***Tournefortia argentea,* 284**
TREEFERN FAMILY, 44
treefern, Hawaiian, 44, **46**
tree-heliotrope, 284
***Trema orientalis,* 96**
***Tristania conferta,* 250**
tropical-almond, 200 **206**
trumpet, 102
trumpet-tree, 100
tulipán africano, 290
tuliptree, 290
tuliptree, African, 288
tun, 166
turpentine-myrtle, 250
turpentine-tree, 250
tutui, 168

'ula'ula, 110
ULMACEAE, 96
'ulu, 98
ulupua, 278
umbrella-tree, 164, 206, 262
***Urera glabra,* 106**
URTICACEAE, 104
uva de mer, 116
uva de playa, 116
uvero, 116
Urea sandwicensis, 106

velvetleaf, 284
VERBENA FAMILY, 286
VERBENACEAE, 286
vinegar-tree, 254

waiawi, 248
waimea, 106, 180
walahe'e, 296
West-Indian-almond, 208
weypul, 300
wi 'awa'awa, 154
Wikstroemia basicordata, 202
Wikstroemia degeneri, 202
Wikstroemia elongata, 202
Wikstroemia eugenioides, 202
Wikstroemia haleakalensis, 202
Wikstroemia isae, 202
Wikstroemia lanaiensis, 202
Wikstroemia leptantha, 202
Wikstoremia macrosiphon, 202
***Wikstroemia oahuensis,* 202**

Wikstroemia palustris, 202
Wikstroemia recurva, 202
Wikstroemia sellingii, 202
Wikstroemia vaccinifolia, 202
wilelaiki, 176
wiliwili, 138
wiliwili, Indian, 144

xylosma, 200
Xylosma hawaiiense, 200

yanangi, 128
youenwai, 238

Zanthoxylum oahuense, 162
zarcilla, 144

www.ingramcontent.com/pod-product-compliance
Lightning Source LLC
Chambersburg PA
CBHW081154020426
42333CB00020B/2498